# Food Preservation and Safety

Iowa State University Press / Ames

# Food Preservation and Safety

## PRINCIPLES AND PRACTICE

Shirley J. VanGarde ■ Margy Woodburn

**Shirley VanGarde** has worked for the Continental Can Company, taught in community colleges, done extensive volunteer work, and preserved the food her family raises on their Oregon acreage. She is an Oregon State University Cooperative Extension agent with the Metro Food Preservation Hotline.

**Margy Woodburn** is professor and head of Oregon State University's Nutrition and Food Management Department.



♾ Printed on acid-free paper in the United States of America

First edition, 1994

Library of Congress Cataloging-in-Publication Data

VanGarde, Shirley J.
Food preservation and safety: principles and practice / Shirley J. VanGarde, Margy Woodburn.—1st ed.
p. cm.
Includes index.
ISBN 0-8138-2133-9
1. Food—Preservation. 2. Food—Storage. I. Woodburn, Margy J. II. Title.
TX601.V36 1994
641.4—dc20 94-18050

# CONTENTS

# PREFACE

The goal of all consumers is top quality and safe food whether they prepare it or eat out. These are complex issues, however, requiring knowledge of the sciences and recent research findings. We have taken this information from scientific journals and scholarly texts and written it in easy-to-understand language with practical examples. We also provide consumers and food professionals with answers to both the "why" and the "how to" questions of storing and preserving food.

This book gives the basic criteria for keeping food palatable and safe from time of harvest to consumption through a variety of storage methods. Quality and microbial changes that can be controlled in drying, pickling, fermentation, curing, freezing, refrigeration, canning, and sugaring are covered in depth. Commercially preserved foods are also discussed to reduce confusion over functions of additives listed on package labels and to describe the variety of choices in products. This text is set up so food professionals already working in the field and consumers needing an immediate answer to storage questions can easily use it as a reference. Book I, Principles, provides the scientific background that is a base for decision making. Book II, Practice, is designed for teachers or demonstrators who need additional applications and examples.

# BOOK I

# Principles

# 1 Quality Changes in Foods during Storage

Consumers, food processors, food chemists, and farmers and ranchers all evaluate food quality. An alert, questioning mind is all that is required. Whether to a gourmet, who considers food to be a creative experience, or to a person hurriedly eating lunch between appointments, food quality is an important factor. If the quality is poor, the food is not eaten and the nutrients it contains are wasted. Preserving food in such a manner that it retains good quality is second in importance only to preserving food safety. Since nutrient loss often parallels quality loss, both goals can be achieved at the same time.

Most consumers agree the physical and chemical changes that occur during storage are undesirable, but the level at which food is considered unacceptably spoiled is an individual decision. Standards for top quality food vary from country to country, state to state, and among individuals. In cultures where refrigeration and rapid food distribution systems are unavailable, rancid meats and oils are common and nonrancid food is regarded with suspicion. In coastline areas where fresh crab is commonly caught, freshness is considered a product standard. In states where seafood arrives through a complex transportation system, freshness is not easily affordable and off-odors that occur with time in storage are considered the norm. Additionally, individuals within any region have different food preferences—often stemming from childhood experiences. Common standards for food acceptability in the United States have been set by the food industry, but home food preservers usually have realistic acceptance of some quality defects that occur without commercial additives.

## Evaluating Preservation Methods

The goal of food preservation is to increase the time for keeping food safe while retaining quality and nutrients. Each year, many recipe books containing food preservation information are published. Some of their recipes are excellent, some result in poor quality products, and some methods produce food that is potentially unsafe to consume. Householders must be able to sort through the information available and select good methods for each food they wish to store. A good food preservation method must

**1.** KEEP THE FOOD ACCEPTABLE. Food with unacceptable appearance, odors, or taste is discarded, which represents a waste of time and money. Householders must estimate the length of storage they will need in selecting the best preservation method. Low temperatures are effective, but refrigeration and freezing are appropriate for different storage periods. Root cellaring preserves fruits for shorter periods than drying. The use of the product will often determine if it is to be dried, canned, or frozen.

**2.** KEEP THE FOOD SAFE. Early food preservationists did not understand the relationship between microorganisms and illness, so they used trial and error in establishing preservation processes. Safe processes are now widely available, but unreliable ones are also passed along. Following sound methods, using recommended equipment, monitoring the food during storage, and evaluating it before consumption are necessary for safety.

**3.** RETAIN NUTRIENTS. Preserved foods should contribute more than protein, carbohydrates, and fiber. Some vitamin losses occur during preservation processes but many of these can be minimized by following sound directions. Often vitamin losses that occur during long storage parallel quality losses. Estimating the amount to preserve for one season's use and storing it properly addresses this problem.

**4.** BE AN ASSET TO THE FOOD DISTRIBUTION CHAIN. This is important in commercial operations. Canning in glass jars makes some foods more marketable, but there are increased shipping losses due to breakage. Drying grain before storage decreases losses due to mold, but it increases milling losses. In the 1990s, reducing nonreusable packaging and environmentally sound manufacturing practices are also an issue.

**5.** BE TECHNICALLY FEASIBLE. Commercial aseptic packaging—canning in presterilized containers—is not available to householders, nor are many of the additives that give commercial foods a long shelf life. Fermentation which requires low technology can be safely practiced in nonindustrialized countries, but freezing and pressure canning often cannot.

**6.** CONSERVE AVAILABLE ENERGY RESOURCES. Drying is an excellent preservation method in hot, dry climates but becomes more difficult in humid areas. Freezing requires electricity (or gas) throughout the storage period. Canned foods do not require energy during storage, but their initial processing uses more water and heat energy than root cellaring, drying, or fermentation.

## ■ Quality Spoilage

Food spoils in one of two ways. It can lose acceptability (quality spoilage) or it can produce illness (safety spoilage). The majority of household food discards are due to quality spoilage during storage. These discards represent significant economic losses to families, as 25% of the food purchased is often discarded. The United States Department of Agriculture (USDA) has established a thrifty food plan on which food allowances for low-income people are based. The thrifty plan allows for only 5% of the purchased food to be discarded. Decreasing quality discards in these households is essential, and decreasing them for all households is desirable.

Quality changes in a food do not cause foodborne illnesses. Discard of quality-spoiled food is a matter of individual preference. Determining which changes are related to safety and which are solely quality is the most important decision where human health is involved. Evaluating the type of quality change and the reasons for its occurrence can decrease discards. Quality changes include physical and chemical changes. Some changes in foods are not clearly categorized as solely physical or chemical, but for simplicity we have placed them with the predominant process in preserved foods. At the household level some of these changes are preventable, some can be decreased, and some will occur with certain preservation methods.

### Physical Changes

Physical changes in food do not result in the formation of new compounds. Physical changes include the effect of gravity (heavy particles settle), molecules going into or out of solution (dissolving or crystallizing), water evaporating, and molecules moving to different parts of a food (osmosis and leaching). Some physical changes are easily reversed such as in redispersing heavy particles that have settled to the bottom of Italian salad dressing. The following physical changes are common in food preservation:

**1.** CHANGE OF STATE. Liquid water becomes a solid when food is frozen and causes ice crystal damage to the food structure. Water can also become a vapor and evaporate, which is a problem in refrigerated food that is not well covered. Drying is a problem with root cellaring too. Initial moisture loss from fruits and vegetables is usually accompanied by vitamin loss, and further loss results in death of plant cells and discard. Moisture uptake (vapor to liquid) is a spoilage problem in dried foods. Sugar crystals can form in jelly, ruining the texture.

**2.** SEPARATIONS. Separations are generally physical changes but chemical interactions may also be involved. The most common is the separation of water and oil in emulsions. These include milk, cheese, salad dressings, mayonnaise, and batters. When dairy emulsions such as cheese and milk are frozen, the water that was involved in the emulsion becomes crystallized (ice) and the emulsion is broken; when the ice melts, the fat separates from the product.

In gels, separation results in a watery layer and toughening of the solid portion. The number of bonds between large molecules such as starches and proteins increases. When a starch-starch bond forms, it replaces a starch-water bond. The bonded water gave the food a more fluid texture, so the net result of this change is a firmer food and free water. The formation of these bonds is enhanced when storage temperature is lowered—until the freezing point is reached. Starch gels, such as in gravies and cornstarch or tapioca desserts, become firmer and a water layer may form on top as water is squeezed out during refrigerated storage. These starch-starch bonds are part of the process when bread stales; thus staling of baked goods is accelerated at refrigeration temperatures.

Pectin gels such as those in jams and jellies also become firmer with storage. Usually jams and fruit syrups remain fluid enough to contain the displaced water, but jellies stored for long periods may have a liquid layer on top. During several years storage, a fruit syrup may acquire the firm texture of a jam.

Meat proteins are in a gel-like structure. During freezing, more protein-protein bonds form. When the meat is thawed, the displaced water is called drip. Freezing and thawing a meat several times results in increased drip. A thin layer of liquid will form on cottage cheese, yogurt, and sour cream during long refrigerated storage times due to increased protein-protein bonds. This liquid can be temporarily dispersed by stirring. Large amounts of sour cream are discarded because this change is confused with a microbiological one.

**3.** DISSOLVING. Water soluble compounds such as vitamins C and the B complex, plant pigments, and flavors may leach from the food to water during preservation processes.

### Chemical Changes

Chemical reactions in foods are not usually reversible since they involve the formation of new compounds. The Maillard and oxidative reactions are responsible for the majority of household quality discards so their effects and prevention are stressed throughout this book. The following are chemical reactions that spoil the quality of preserved foods:

**1.** MAILLARD BROWNING is a color, flavor, odor, and sometimes texture change which results from a chemical reaction between proteins and carbohydrates. It is named after the Frenchman Maillard (pronounced MAY-YARD) who discovered it. Maillard browning is responsible for the desirable browning of most heated foods such as bread crusts, roasted meats, and roasted coffee beans. This browning is accelerated by heat and occurs very quickly in ovens, slowly at room temperature, and very slowly at refrigerator temperatures. The browning that occurs during room temperature storage is a major cause of quality changes in preserved food. Very small amounts of the protein or carbohydrate substrates are needed for Maillard browning to occur. The proteins of enzymes and those in the thin layer of cell membranes in fruits and vegetables is enough. In light-colored foods such as dried apples the effects are particularly noticeable.

Maillard browning is also accelerated by low moisture content. Commercially dried milk and home dried foods which are both low moisture and stored at room temperature are frequently discarded due to this reaction. Maillard browning becomes more pronounced with increased storage time.

Light does not accelerate Maillard browning but does increase other color changes. This problem is addressed commercially by packaging foods in opaque materials. Householders need to store their glass jars and plastic bags in dark places. Avoid a sunny window in the pantry or leaving a light on to decrease humidity or discourage pests.

**2.** OXIDATIVE RANCIDITY is a chemical change in an unsaturated bond of a fat or oil. Polyunsaturated fats are particularly susceptible because they contain more of these reaction sites. Off-odors and -flavors of oxidative rancidity make it a major quality spoiler of stored foods.

Some commercial foods, such as potato chips and dried vegetables, are packaged in nitrogen instead of air or in vacuum to decrease rancidity. Very small amounts of oxidized lipid, such as in the cells of green peas, can render food unacceptable. Rancidity reactions in vegetables are enhanced (catalyzed) by enzymes. Blanching treatments for foods to be frozen are designed to destroy these enzymes with heat.

Rancidity is accelerated by moderate heat. Temperature differences between room temperature storage, refrigerated storage, and frozen storage have a significant effect on rancidity reactions. Home canned beef is usually stored at room temperatures so the rancid flavor it acquires limits its shelf life. Commercially, antioxidants (butylated hydroxyanisole [BHA], butylated hydroxytoluene [BHT], and tertiary butylhydroquinone [TBHQ]) are added to canned meats and other high-fat foods to significantly retard rancidity, but these additives are not available to householders. Monitoring storage conditions and storage time is the best householder defense against rancidity in meats. Storing shortening in the refrigerator instead of in the cupboard greatly increases its shelf life due to slowing rancidity reactions, but refrigeration is not effective for all foods. Refrigerated foods, such as cooked beef, pork, and chicken, exhibit rancid flavors and odors within a day; but, if frozen, the process is delayed several months. Ice cream slowly becomes rancid at 0F (−18C) in home freezers, but the reaction is a much lesser problem at commercial frozen storage temperatures of −35F (−37C). Light also accelerates rancidity so dark storage delays this chemical reaction too.

**3.** OTHER OXIDATION REACTIONS. Vitamins C and E, which are antioxidants themselves, can be destroyed by other oxidation reactions, as can some plant pigments and flavor compounds. Vitamin C is used as a reducing agent in household pretreatment dips to prevent oxidative color changes in fruits and vegetables. Browning of apples and other light fruits is an enzyme-catalyzed reaction that occurs when the fruit is cut. The pigments of flour will bleach during storage through an oxidative process. The net result is a whiter flour that also produces baked goods with higher volumes. Bleached flour, which is available commercially, has an additive to provide the same result.

**4.** HYDROLYSIS is the splitting of molecules in a chemical reaction that involves water. For example, sucrose splits to form two different sugars—glucose and fructose—during the cooking of conventional jelly. The amount of these sugars that is formed greatly affects the end product. When vegetables are blanched and canned fruits and vegetables are heat-processed, certain components of their cell walls, such as hemicelluloses, are softened by hydrolysis and a softer food results. During the extraction step of jelly making, pectin is formed by hydrolysis of another plant compound (protopectin). This pectin formation is critical for gelling.

### Changes Effected by Temperature

Enzymes in foods are destroyed by heat to stop quality deterioration during storage, but some heat-labile compounds, such as vitamins (thiamin) and artificial sweeteners, are altered adversely by heat.

Heat breaks down barriers in plants, and substances, such as the pigment chlorophyll, are then free to combine with plant acids in other parts of the cell. Chlorophyll forms a new compound, pheophytin, which is an olive-green color instead of the original bright green. The pigment in canned green beans and spinach is pheophytin.

Low storage temperatures can alter the metabolism of some fruits and vegetables, such as potatoes, which are stored best at root cellar temperatures, and cause them to accumulate sugars instead of starch. Low temperatures, however, slow metabolism which greatly increases the storage time of most fruits and vegetables. Commercially, the optimum temperature is determined for each cultivar (variety) of apple and other products.

## Summary

Differentiating between safety and quality changes that occur in stored foods is critical for human health. Knowing which quality changes in preserved foods are unacceptable to household members and how to minimize them is a great asset to budgets. The safety principles upon which each preservation method is based and the effects on quality of steps in popular recipes are identified in the chapters that follow. Armed with this knowledge, householders can preserve top quality foods that compete well with commercial products.

## References

Bennion, M. 1990. Introductory foods. 9th ed. New York: Macmillan.

Bowers, J., ed. 1992. Food theory and applications. 2d ed. New York: Macmillan.

Expert Panel on Food Safety and Nutrition. 1993. Scientific status summary—browning of foods: Control by sulfates, antioxidants, and other means. Food Technol. 47(10):75.

Fennema, O., ed. 1985. Food chemistry. New York: Marcel Dekker.

Freeland-Graves, J.H. and Peckham, G.C. 1987. Foundations of food preparation. 5th ed. New York: Macmillan.

McGee, H. 1990. The curious cook. San Francisco: North Point Press.

McGee, H. 1984. On food and cooking. New York: C. Scribner's Sons.

McWilliams, M. 1993. Foods: Experimental perspectives. 2d ed. New York: Macmillan.

Trager, J.G., Jr. 1970. The foodbook. New York: Grossman Publishing.

## CONSUMER QUESTIONS

**Q.** *Why does nonfat dry milk sometimes become tan or brownish?*
**A.** This is due to a chemical reaction between protein and carbohydrates called the Maillard reaction. The milk is safe to consume, but will have off-flavors and off-odors, and may be difficult to rehydrate. Maillard browned foods do not cause foodborne illness; microorganisms are not involved. Storage for a shorter time period prevents this reaction. It also happens more slowly if food is stored in a cooler place.

**Q.** *Why does shredded, dried coconut become tan?*
**A.** Maillard reaction, see above.

**Q.** *What if a newly purchased package of dried fruits contains brown peaches and apricots?*
A. Maillard and related reactions, see above. In addition, early in the drying process oxidative browning may have occurred. These reactions cause losses of commercial foods too. Sulfite is sometimes used to prevent these changes.

**Q.** *What has caused butter stored in the refrigerator to have an off-odor?*
**A.** Rancidity occurs slowly during storage even at low temperatures. Foods which contain large amounts of lipid (fats) show these effects readily. Refrigeration slows rancidity, but it will continue to occur slowly. Although safe to eat, many consumers do not consider the off-flavor to be acceptable. For dairy products, lipase (an enzyme) activity may split the fat molecules and release volatile fatty acids, which have off-

odors. Another possibility is that the butter simply absorbed other components from the air in the refrigerator, for example, onion.

**Q.** *Should oil in a deep fry kettle that has a strong odor and smokes be saved to use again or discarded?*
**A.** Discard because the quality of the products that would be fried would be poor. There are also health concerns with continued long-term consumption of such abused oil. After several heatings, oil will smoke more easily—lower smoke point—and also catch on fire more easily—lower flash point. So that you will have longer use from your frying oil, strain it after each use and add some fresh oil each time.

**Q.** *Prepared foods often separate when refrigerated. Is this a sign of spoilage?*
**A.** As long as the refrigerator temperature was at least as low as 40F (4C), this is a quality change. Gels shrink, forcing out water, or emulsions separate. Sometimes stirring or whipping will restore the original texture.

**Q.** *How can I determine if food that shows changes is safe to eat?*
**A.** What is its temperature history? If it is a perishable food and has been between 40F (4C) and 140F (60C) for more than 3 hours, it should be discarded.

Is it moldy? Since for many molds, the risk of toxicity is not known, moldy food should be trimmed (if solid) or discarded. In food stored at room temperature such as bread or nuts, the appearance of mold on any part generally indicates that more widespread molding may be occurring and all should be discarded. For fresh fruits and vegetables, small areas may be trimmed, but generally moldy products should be discarded.

Most changes in quality are not evidence of unsafe food but should be carefully considered.

# 2 Foodborne Illness

Quality may be defined differently by each person, but safe food is a universal desire. There are three main causes of foodborne illness.

**1.** Chemicals added to foods intentionally or as an incidental result of their use in production, processing, or distribution.

**2.** Poisonous plants and animals.

**3.** Microorganisms, including bacteria, molds, viruses, and parasites.

Surveys have found that the general public is most concerned about the first of these but public health professionals have identified the third as the most serious problem.

## Additives

Additives are used in foods to provide color and flavor, to preserve natural qualities of the food such as in prevention of rancidity, to increase the nutritive value by adding vitamins and minerals in fortified and enriched foods, to improve functional properties such as salt which does not cake, and to improve texture. The commercial food distribution system also uses some additives to maintain quality during the time required for food to reach the household level. The fastest growing approved additive category is for low- or noncaloric sweeteners.

Chemicals added to foods as additives must be approved for use by the Food and Drug Administration (FDA) and used within the guidelines for approval, which generally include a maximum level. Approved additives must serve a functional purpose which does not deceive the consumer, be harmless—with a margin of safety added in permitted use levels—and there must be analytical methods for detecting the additive and its metabolic products. By 1958 additives had become frequently used ingredients and new ones were being introduced regularly. To address this new technology, Congress required the FDA to establish a prior approval policy. At that time the additives which had a history of safe use, called GRAS (generally recognized as safe) did not have to go through the approval process but since then have been reviewed. Requests for approval of new additives must now be accompanied by extensive research data, a costly process for industry.

Most additives must be listed on the label as an ingredient, with the exception of colors and flavors, to protect the product formula.

Under the labeling act of 1990 foods that have an FDA approved standard of identity are no longer exempt from the requirements of ingredient listing.

Unintentional additives may occur in any part of the pathway of food from production to consumption. Residues of pesticides and other chemicals used in agricultural production are regulated by the FDA. Use in the field is controlled by the Environmental Protection Agency under the authority of the Federal Insecticide, Fungicide, and Rodenticide Act (FIFRA). A tolerance of a specified limited quantity of the chemical in specific food(s) may be approved. Examples of these are pesticides, herbicides, fungicides, and growth promoters (including antibiotics) used in agriculture during the growing of the plant or animal; lubricants used in machinery; and package components such as those in plastics that come into contact with food or equipment surfaces. Sources for more information are listed in the reference section.

Unintentional food contamination may occur at several steps in the food pathway. Environmental pollutants can enter water and food. The deliberate placement of a poisonous substance in food as a criminal act is a potential risk but rarely occurs. More likely, but still rare, is the use of food containers that are unsuitable. At the household level, the most frequent is the use of galvanized cans or pails for food preparation or fruit beverages; the zinc in the metal coating is readily dissolved and causes illness in those who drink the punch or lemonade. Lead was a concern from the solder used in the seams of tin cans but this use has been rapidly phased out and ended in 1991. Of more concern is lead dissolved into food or beverages from pottery containers, lead crystal decanters, or other wares. Crystal items should be used only to serve beverages and not for storage; in addition, the FDA recommends that crystal not be used for foods or beverages for infants and young children or pregnant women since the effect of lead on the brain and nervous system of children is especially serious.

Consumers also need to avoid plastic containers or wraps that were not intended for food use. These, such as plastic drop cloths or some brands of garbage cans, contain plasticizers to improve the durability of the plastic but which are not FDA approved. A similar misuse is microwave heating in plastic containers which were not intended for use at high temperatures.

## ■ New Varieties

Agricultural biotechnology has been suspected by consumers of altering the healthfulness of plants and animals. Genes can be transferred from the same or other species to create plants or animals that have more desirable characteristics (MacDonald, 1990). Traditional cross-breeding of plants and animals achieves this transfer but not selectively and generally only within closely-related species. The Food and Drug Administration has announced plans to regulate these as it does any new varieties. However, more specific guidelines are expected. The first genetically-engineered food, a tomato that softens more slowly, was approved in May of 1994. Consumers can learn about these new products from newspaper and other sources.

## ■ Poisonous Plants and Animals

The second category, poisonous plants and animals, may be avoided by using only foods which are generally known to be edible. With the increasing popularity of herbs, care must be taken to use only those which are safe (Tyler, 1993). Wild mushrooms are the most frequent cause of problems, especially for immigrants not familiar with the native mushrooms. Fish in tropical and subtropical waters may ingest smaller fish that have eaten poisonous algae and themselves become unsafe to eat; the resulting ciguatera poisoning may be fatal. Related to this is paralytic shellfish poisoning (PSP) from shellfish which have fed in waters with toxic plankton. Public health monitoring of waters and shellfish, so that problem areas are identified and seafood not used until shown to again be safe, is the only prevention.

## Microbial Foodborne Illnesses

Foodborne illnesses occur when food containing large enough numbers of pathogenic microorganisms (foodborne infections) or a toxin produced by microorganisms (foodborne intoxication) is consumed. Foodborne infections can be caused by bacteria that either produce toxins as they multiply inside the intestinal tract (toxicoinfections) or invade the intestinal mucosa where they multiply or pass to the blood and the system.

In order to decide whether or not a food is safe to eat or if a food handling practice is wise, it is necessary to have some basic information about microorganisms. Several factors must be in place for an illness to occur:

**1.** The pathogenic microorganisms or the preformed toxin must be present in the food as eaten.

**2.** The pathogenic bacteria that are present must be in sufficient numbers to cause illness or to have produced toxin.

**3.** A sufficient quantity of this contaminated food, which is capable of overwhelming the person's resistance, must be eaten.

There are three sources of the microorganisms: the raw food itself and ingredients, contamination from food handlers, and contamination from surfaces, utensils, and other equipment. If quality defects accompany the microbial spoilage, it is unlikely that the food will be consumed by most people in the United States. However, large numbers of pathogenic organisms and toxins can be present in food that appears to be top quality. This is the area of most concern and the reason for the recommendation to discard food held at unsafe temperatures (50–125F, 10–52C) for over 3 hours. The recommended temperatures of 40F (4C) or below for storing cold food and 140F (60C) or above for holding hot foods provides a desirable margin of safety. Holding temperatures may not be adequately controlled or measured.

No one knows the exact number of foodborne illness cases since many of the less serious are treated at home, go unreported, and often are diagnosed by householders as "the flu," but estimates are made based on epidemiological data available. An economist in the United States Department of Agriculture (Roberts, 1989) estimates that foodborne illnesses cost the U.S. economy 4.8 billion in 1987, taking into account medical costs and productivity losses. Many feel this is an underestimate. *Campylobacter*, *Salmonella*, and *Staphylococcus aureus* are the major organisms in terms of both frequency of occurrence and cost to the economy (Table 2.1). Together they accounted for $3.7 billion in losses.

To understand microbial food hazards, one must begin with the natural bacteria present in foods. The microbial flora will vary greatly with the type of food and with treatments it has received. In raw foods, organisms are present from soil, water, animal feeds, and fertilizers, and those introduced from humans and by cross contamination with handling. When foods undergo preparation and preservation treatments, only the organisms capable of surviving the treatment are found live in the food.

Microorganisms have specific requirements for growth. The nutrients and inhibitory substances present, oxygen, carbon dioxide, temperature, moisture, pH, and the stage of growth of the organism interact to determine growth

**Table 2.1. Annual number of bacterial foodborne illness cases in the United States**

| Pathogenic Organism | Cases[a] |
|---|---|
| *Campylobacter jejuni, C. coli* | 2,100,000 |
| *Salmonella* (non-typhi) | 1,920,000 |
| *Staphylococcus aureus* | 1,513,000 |
| *Shigella* | 90,000 |
| *Escherichia coli* | 50,000 |
| *Clostridium perfringens* | 10,000 |
| *Vibrio* (non-cholerae) | 10,000 |
| *Bacillus cereus* | 5,000 |
| *Yersinia enterocolitica* | 3,250 |
| *Listeria monocytogenes* | 1,581 |
| *Salmonella typhi* | 480 |
| *Clostridium botulinum* | 180 |
| *Vibrio cholerae* | 25 |

[a]Based on CDC 1987 statistics.

rates and possible toxin production. A food that is highly perishable has a pH above 4.6, is nutritious, and has adequate moisture. Since the water must be available to the bacteria, moisture is determined as water activity expressed as $a_w$. A potentially hazardous food has an $a_w$ above 0.85.

Bacteria compete with each other so that the growth of pathogenic bacteria is favored if the number of competing bacteria are low. Thus raw foods generally spoil with resulting loss of quality but do not become unsafe. Cooked foods may become unsafe but show no quality loss. One exception is seafood. If fish, especially those of the scromboid type (tuna, mackerel, and bonito) and mahi mahi, are not kept cold, bacteria normally present on the fish multiply and produce histamine. The symptoms are varied and occur usually shortly after the fish is eaten. Fortunately they are generally mild and soon pass.

The objective of food preservation is to kill the organisms or to control the environment so spoilage and pathogenic microorganisms do not increase and toxins cannot be produced. The food then will have a longer shelf life and be safe. Canning removes oxygen and kills the pathogenic organisms present that could grow in that particular canned food. Drying reduces the $a_w$ to a level that inhibits growth. Refrigeration is an unfavorable temperature that slows growth of all organisms and prevents growth of most pathogens. Freezing goes further toward creating unfavorable temperatures and stops growth. Pickling increases the acidity and sometimes the ionic concentration to one which microorganisms cannot tolerate for multiplication. Curing involves the use of inhibitory substances such as nitrite and an unfavorable ionic concentration such as high salt levels. Since a mild cure is preferred, it must be combined with another preservation method, usually refrigeration. Root cellaring—common storage—maintains an environment where the plant has slowed metabolic processes but remains alive so its natural defenses against microorganisms protect it from invasion.

## Microbial Growth Curve

Microorganisms that are present or are introduced into a favorable environment begin

**Table 2.2. Summary of foodborne illnesses**

| Organism | Type | Incubation Period | Mode of Transmission |
|---|---|---|---|
| Intoxications | | | |
| *B. cereus* | bact | 1–7 hr | food |
| *C. botulinum* | bact | 12–48 hr | food |
| *S. aureus* | bact | 1–7 hr | food |
| Infections | | | |
| *C. jejuni* | bact | 3–5 days | food, water, pets, f-o[a] |
| *C. perfringens* | bact | 6–24 hr | food |
| *E. coli* | | | |
| enterotoxigenic | bact | 12–72 hr | food, water, pets, f-o |
| enteropathogenic | bact | 2–6 days | (above) |
| enteroinvasive | bact | 2–3 days | (above) |
| enterohemorrhagic | bact | 3–5 days | (above) |
| Hepatitis A | virus | 30 days | food, water, f-o |
| *L. monocytogenes* | bact | 3–5 days | food, water |
| *Salmonella* | bact | 4–48 hr | food, water, f-o |
| *Shigella* | bact | 1–7 days | food, water, people, f-o |
| *T. spiralis* | roundworm | 7 days | food |
| *Y. enterocolitica* | bact | 2–7 days | food, water, people, pets, f-o |
| *V. cholerae* | bact | 8–72 hr | f-o |
| *V. parahaemolyticus* | bact | 4–96 hr | water, food |

[a]Fecal-oral route abbreviated f-o.

Adapted from Centers for Disease Control Morbidity and Mortality Weekly Reports 39 (RR-14):7. 1990.

to grow and multiply. Their numbers increase through phases that result in a typical growth curve shown in Figure 2.1. The initial lag phase is characterized by little change in numbers. This is followed by a phase of acceleration in which the growth rate increases slowly. The logarithmic (exponential) phase has the most rapid increase in numbers. It is from this phase that foods most often become unsafe to consume. By this point, the food may be rejected as unpalatable, depending upon the organisms. The multiplication rate slows as the nutrients are used and/or products from metabolism accumulate and inhibit growth. The stationary phase is one of constant numbers. It may be followed by the death (decline) phase in which numbers decrease. The population size may go completely down to zero or to less than the number initially present but food quality is usually so poor that the decreased numbers are of no importance. Figure 2.1 only represents the growth of one species. Typically growth curves of other bacteria would be occurring simultaneously or in sequence.

It is important that the food preparation and preservation method chosen lengthen the lag phase as much as possible to increase the shelf life and safety of the food. To do this

**A.** START WITH LOW NUMBERS. Produce with broken skin, bruises, or past optimum maturity has higher numbers of bacteria. Washing and blanching foods can reduce the microorganism load. Droplets from sneezing and coughing carry many microorganisms.

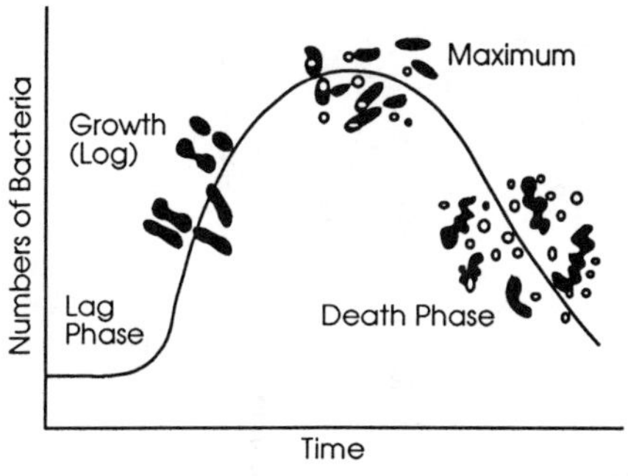


**2.1.** Growth curve for one species of bacteria.

Utensils instead of hands should be used to handle ready-to-eat foods, and contamination from dirty surfaces avoided. Heating the food reduces the numbers of microorganisms and may be rigorous enough to kill all vegetative cells and even spores.

**B.** DO NOT ADD ORGANISMS WHICH ARE IN THE LOGARITHMIC STAGE. These organisms are rapidly growing. When introduced into a new environment, they go through a lag phase, but a very short one. They may have been growing on unclean containers, equipment, utensils, or humans. Householders should make a habit of washing can openers, cutting boards, and their hands as part of food preparation. Mixing fresh food with stored food also may increase the rate of spoilage.

**C.** PROVIDE A LESS THAN IDEAL GROWTH ENVIRONMENT. The more unfavorable growth conditions are for a given organism, the longer the lag phase (Fig. 2.1). Most bacteria that cause foodborne illness grow best at temperature, pH, ionic concentration, and moisture levels that are similar to those of the human body. Therefore, food preservation methods are based on changing these factors: either raising or lowering the storage temperature sufficiently, decreasing the pH (increasing is not practical for foods as human taste buds are sensitive to alkaline compounds and they are objectionable), or decreasing the availability of water by drying or binding the water with other compounds (salt or sugar). Changing the oxidation-reduction potential (as in canning and other vacuum packaging) eliminates growth possibilities for many organisms but favors the multiplication of clostridia.

The control of bacteria, molds, and viruses that can cause foodborne illness is even more important than preserving the quality of the food. Although over 200 diseases can be transmitted via foods, only those shown in Table 2.3 are major problems. The prevention of foodborne illness depends upon the application of our research-based knowledge of the survival and growth requirements of these organisms.

**Table 2.3. Comparison of foodborne illnesses in the United States**

| Disease | Cause | Incubation Time (hr) | Symptoms | Duration | Mortality | Foods Commonly Involved |
|---|---|---|---|---|---|---|
| Botulism | Toxin—*C. botulinum* | 18–36 (few hr to 8 days) | May be early gastrointestinal symptoms, dizziness, headache. Dry skin, mouth, throat; constipation, no fever, weakness or paralysis of muscles, double vision, respiratory failure. | 10–21 days or longer | 23% | Low- or medium-acid foods (meats and vegetables that are cooked or canned, preserved fish products.) (Infant botulism—spores in food or other sources.) |
| Staphylococcus food poisoning and *Bacillus cereus* | Enterotoxins—*Staphylococcus aureus* and *Bacillus cereus* | 1–7 | Nausea, vomiting, abdominal cramping, diarrhea. Headache, sweating, prostration. Low temperature. | 1–2 days | Very low | Meats, especially ham, fowl, and their gravy; meat sandwiches; salads containing protein; cream sauces; custard-filled goods and cakes; dairy products. |
| Perfringens | Toxin, produced in intestine. *Clostridium perfringens* | 8–22 | Mild abdominal pain and diarrhea. | 1–2 days | Very low | Cooked meats, gravies, stuffings, prepared or semiprepared mixtures. |
| Salmonellosis | Infection—Salmonellae species | 6–48 | Nausea, vomiting, abdominal pain, diarrhea—sudden onset. Usually fever. Chills, headache, prostration. May be a generalized infection. | 1–14 days | 1% | Meat and poultry products, eggs, cross-contaminated foods. Raw milk. |
| Campylobacteriosis | Infection—*Campylobacter jejuni* | 1–11 days | Diarrhea, abdominal pain, fever. May be a generalized infection. | 7–21 days | 1% | Undercooked poultry or meat, eggs or cross-contaminated foods. Raw milk. |
| *Vibrio parahaemolyticus* | Infection—*V. parahaemolyticus* | 12–24 | Violent epigastric pains, nausea, vomiting, and diarrhea. Fever, headache, and chills may occur. | 1–2 days | Very low | Marine fish, shellfish, and crustaceans; raw or undercooked, or foods cross-contaminated from these sources. |

**Table 2.3. *continued***

| Disease | Cause | Incubation Time (hr) | Symptoms | Duration | Mortality | Foods Commonly Involved |
|---|---|---|---|---|---|---|
| Listeriosis | *Listeria monocytogenes* | 1 day to several wk | Flulike, abortion or illness of the newborn, sepsis, meningitis. | 3–15 days depending on form | 0–70% depending on form, age, immune status | Uncooked foods, especially raw milk and unpasteurized soft cheeses. Cross-contaminated ready-to-eat foods. |
| *Escherichia coli* infections | Infection—pathogenic strains of *E. coli.* | 12 hr–6 days | Range from mild to severe diarrhea, may include vomiting, fever. Serotype 0157:H7 can lead to bloody diarrhea, brain damage, and kidney failure. | 1 day to 1 mo | Low to 30% | Undercooked ground beef. Raw milk. Other animal products. Cross-contamination from human or animal product. |

Others: Brucellosis, *Shigella,* Group A *Streptococcus,* A & E hepatitis, *Vibrio cholerae, Yersinia enterocolitica,* viruses such as Rotaviruses and Norwalk virus, *Giardia* and amoebae, *Trichinella spiralis.*

## Food Poisoning

### Botulism

Botulism is one of the best known food intoxications because it can be fatal or leave the victim with extensive neurological damage. It occurs rarely now, but the severity of this disease makes studying botulism important. The mortality rate peaked in the United States in 1915 at 75%. It currently is down to approximately 20% due to improved diagnosis and treatment. Botulism can result from any of the following:

**1.** Eating inadequately processed, low-acid, canned foods.

**2.** Consuming a cooked low-acid food (for example, casseroles, meat pies) that had been held at a temperature that permitted growth—that is, neither cold nor hot.

**3.** Infected wounds.

**4.** From *C. botulinum* growing in the intestinal tract of an infant or in an adult who has had extensive gastrointestinal surgery.

Of 355 cases of botulism occurring in the United States between 1976 and 1984, 4% of the outbreaks occurred from restaurant food; but these accounted for 42% of the number of people afflicted. In an Illinois restaurant outbreak, onions sauted in margarine (the fat made anaerobic conditions) and held warm were the culprit when served with sandwiches. In New Mexico and Colorado potato salad made from leftover baked potatoes, which were foil-wrapped and stored unrefrigerated before salad preparation, was the probable cause of poisoning.

Once ingested, most of the botulinal toxin in the food is absorbed in the upper part of the small intestine. It travels through the circulatory system to the cholinergic nerves and causes paralysis of the muscles innervated by this system, including the diaphragm necessary for breathing. Symptoms of botulism poisoning can develop from 2 hours to 14 days after the toxin is consumed but 12 to 36 hours is the usual range. Gastrointestinal upset may occur before the neurological symptoms, especially for type E, but often does not. The most common neurological symptoms include blurred vision and other vision disorders, difficulty swallowing and speaking, generalized weakness, and dizziness. Treatment includes respiratory assistance, sometimes removal of unabsorbed toxin from the GI tract, and injection of antitoxin to neutralize the toxin in the circulatory system.

*Clostridium botulinum* organisms are classified by the type of toxin produced. Eight different toxins are known, but the great majority of human poisonings are from A, B, and E types. All types grow only in the absence of oxygen (anaerobic), but differ by A and proteolytic type B being more resistant to adverse conditions.

*Clostridium botulinum* spores are commonly present in soil and dust in all parts of the world and so may be on any food. In the United States, type A spores are found most frequently west of the Rocky Mountains; type B spores in the Mississippi Valley and Great Lakes areas. Type E spores are present in natural waters such as the Great Lakes and the Pacific coast and thus in fish.

Growth requirements vary for these types. The minimum pH for growth of types A and B is 4.8; 5.3 is the minimum for type E. Interactions with other food components may allow the organism to multiply at a lower pH; however, this has been demonstrated in laboratory experiments and not yet found to occur in practice. On the basis of pH, foods are divided into two groups for perishability and processing (canning) requirements (Table 2.4). If the pH is less than or equal to 4.6 the food is classified as acid. These do not permit the germination and multiplication of *C. botulinum*. If the pH is above 4.6, it is potentially hazardous and, if canned, requires the use of a pressure canner or commercial retort.

The minimum water activity ($a_w$) for growth of types A and B is 0.95 (10% NaCl); 0.97 (5% NaCl) for type E and nonproteolytic strains of type B. Thus, salted foods must be quite salty to be preserved. Sugar is used for some preserves, such as pumpkin butter. An English firm which produced a hazelnut filling for yogurt discovered the importance of sugar as a preservative when their substitution of a

**Table 2.4. Common pH ranges of foods**

| Food | pH |
|---|---|
| Low acid | |
| Meat and poultry | 5.6–5.8 |
| Fish (most species) | 6.0–6.8 |
| Milk | 6.3–6.5 |
| Cheese (cheddar) | 5.9 |
| Beans (green) | 5.3 |
| Beans (lima) | 5.7 |
| Carrots | 5.0–6.0 |
| Corn | 6.3 |
| Potatoes (wt and swt) | 5.3–5.6 |
| Spinach | 5.5–6.0 |
| Squash | 5.0–5.4 |
| Watermelon | 6.0 |
| Acid | |
| Apples | 2.9–3.3 |
| Oranges | 3.1–4.3 |
| Peaches | 3.5–3.9 |
| Pears | 3.6–4.4 |
| Plums | 2.8–4.6 |
| Tomatoes[a] | 4.0–4.6 |

[a]Firm, disease-free, mature-ripe tomatoes are in the acid range. However, tomatoes may have higher pH values under some growing conditions and if mold occurs.

low-calorie sweetener resulted in botulinum toxin production.

The temperature range for growth of types A and B is 50 to 122F (10– 50C), with 99F (37C) being optimum. The minimum temperature for growth of type E and nonproteolytic strains of type B is 39F (4C) and 86F (30C) is optimum. Twelve to 48 hours at 68–113F (20–45C) is the usual incubation time for the *C. botulinum* bacteria to increase in numbers and produce toxin (Bryan, 1979). Refrigerating low-acid foods promptly prevents type A and most type B multiplication.

Spoilage may often be detected by off-odor and gas production in the food. Type A and some type B botulinum organisms are proteolytic. Proteolysis in canned meats usually results in foul odors as the proteins are broken down to yield putrefactive compounds. Gas may also be produced. Examining the jar seal and contents before serving is a good way to detect this spoilage. However, spoiled products may appear normal, especially foods with a strong natural aroma and those which contain little protein.

Type E and some type B botulinum organisms are nonproteolytic, thus it is possible for these bacteria to produce toxin in foods without the spoilage being evident. These types produce the maximum amount of toxin at 79F (26C), but in laboratory tests, multiplication and toxin production in food with high initial numbers of the organism occurred at 38F (3C) after 3 weeks storage. This research should not cause undue consumer alarm, as the scenario of low-temperature type E or B toxin production is uncommon. Raw foods would show other types of spoilage and be discarded early; for example, storage of fresh fish in a household refrigerator for 3 weeks is unlikely; other microorganisms would produce quality changes by the end of the first week causing the consumer to discard the fish. The possible problem is with lightly smoked or prepared foods.

Botulinum toxin is readily inactivated by heat; 185F (85C) or above makes it nontoxic to humans and animals. Lower temperatures for longer times are also effective. As a margin of safety, cooked or canned foods may be heated at boiling for 10 minutes or to 185F (85C) in the oven (casseroles will bubble in the center). Many cases of botulism from underprocessed home canned foods in the past were prevented because the contents were boiled rather than served directly from the jar.

*C. botulinum* spores are heat stable. Boiling temperatures do not inactivate them, but, in contrast, stimulate germination. However, the other bacteria, which in a raw food would compete successfully, are killed so that typical spoilage will not occur. Destruction of any spores present by heating at temperatures above boiling and holding for a period of time—which depends on the physical characteristics of each specific food—is required for canning these low-acid foods. Canning at home is safe when directions from reliable sources are followed.

Low-acid, sealed canned foods (meat, fish, poultry, vegetables, legumes) are an excellent growth medium for surviving spores to germinate, grow, and produce toxin. Cooked foods have anaerobic interiors, so if there is botulism contamination, growth is possible. These foods include those tightly wrapped or with a coating of breading, and all portions of foods below a shallow surface layer. The advice "Keep cold foods cold and hot foods hot" prevents the problem.

Infant botulism, in contrast to the previously discussed adult illness from botulinum toxin, is caused not by the presence of the toxin in the baby's food but by toxin production in the intestinal tract after the infant has ingested *C. botulinum* spores. Seldom does this illness result in death but recovery may be prolonged. Early symptoms include constipation and weakness, which may result in an inability to nurse and "floppiness." Most ill infants are younger than 6 months but it has occurred in those up to 1 year old. It is not understood why some infants become ill while most do not; breast- and bottle-fed infants are equally susceptible. Honey has been identified as having been fed to some of these infants and spores have also been isolated from the honey. Therefore it is recommended that honey should only be given to children over 12 months of age. This is advised from a nutritional standpoint also.

In summary, botulism is a rare but potentially fatal or permanently damaging disease. Therefore preventive measures should be consistently practiced. Adult botulism intoxications can be controlled by: (1) Processing home canned, low-acid foods adequately by using a pressure canner and correct times to kill the heat-resistant botulism spores; (2) Refrigerating perishable foods after cooking; (3) Adding acids to low-acid foods or nitrites to cured meats to prevent germination of spores and vegetative cell growth; and (4) Drying food or adding sugar or salt to reduce the available water as a preservation method.

### Staphylococcal Food Poisoning

Staphylococcal food poisoning is caused by eating food which contains one of the heat stable staphylococcal enterotoxins—heat-resistant proteins produced in a food by staphylococcal growth. It is common throughout the world and perhaps the most frequent foodborne disease in the United States. The growth of *Staphylococcus aureus* is selectively favored in cooked foods and slightly salty foods such as ham. Many (30%) of the reported outbreaks result from unsafe home food preparation, particularly salads with cooked vegetables or meat and combination dishes. Although easily prevented by keeping food either cold or hot, *S. aureus* is likely to remain a major cause of foodborne illness since absence of contamination from humans cannot be guaranteed.

Presence of the staphylococci bacteria does not create a problem; enterotoxin is present only after their multiplication in the food. The more organisms present, the more toxin can be produced in a short period of time. With optimum growth conditions, including room temperature, staphylococci double their numbers every 20 minutes after about a 1 hour lag phase. Usually minimum counts of $5 \times 10^5$ (500,000) are present in foods containing *S. aureus* toxin. Very small amounts of enterotoxin (1 $\mu$g to 100 ng) can cause illness. There are at least 7 enterotoxins, which differ in their chemical structure. The toxins are very heat resistant. They are not destroyed by normal cooking, baking, boiling. Therefore, once staphylococcal growth has occurred, the food must be discarded. Food preparers can decrease the likelihood of contaminating the food by washing hands after nose and mouth contact and using utensils instead of hands. The most important control is preventing bacterial growth by chilling food promptly or holding it hot.

Human nose and skin are the main sources of *S. aureus*. A common scenario (Fig. 2.2) for the transmission involves a low-acid food being cooked, thereby eliminating the possibility of competing organisms. The food is then contaminated by a human and kept at optimum temperature for *S. aureus* growth, either by being left at room temperature or by the slow cooling of a large amount of food. The bacteria then multiply and produce toxin. Outbreaks have occurred in food held at room temperature only 4 hours. Some public health laboratories are equipped to determine enterotoxin levels in a suspect food; however, this is not done routinely.

Symptoms of staphylococcal food poisoning usually appear 2 to 4 hours after ingestion of contaminated food, but the range is 30

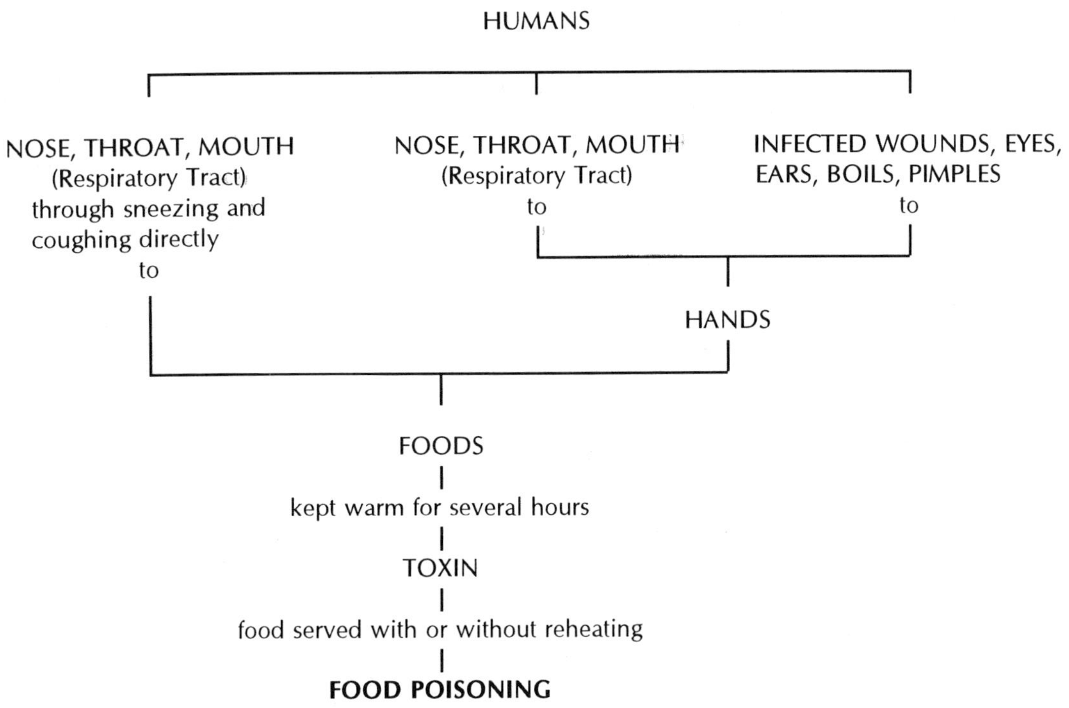


**2.2.** Sources and routes for staphylococcal food poisoning.

minutes to 8 hours. Greater amounts of toxin consumed result in shorter incubation periods. Vomiting and retching are usually the first symptoms and persist for 1 to 4 hours. Diarrhea usually follows the initial vomiting by 15 minutes and lasts 2 to 7 hours. Though people poisoned by *S. aureus* may believe death is imminent, mortality is low. Young children and the elderly are more likely to be severely affected by the large fluid losses and the accompanying electrolyte imbalances.

*Staphylococcus aureus* does not produce spores so it is readily destroyed in cooking. Most of the problems are a result of contamination of cooked foods by humans. This can occur in the boning of cooked chicken, mixing pasta salad, the slicing of ham, and any other handling of food that is to be consumed without further heat treatment. *S. aureus* can actively grow in foods with $a_w$'s as low as 0.86. This feature is important because most of the other bacteria which compete with *S. aureus* for nutrients or which could grow and produce acid or off-flavors are unable to grow in such an environment (Table 2.3). The common food vehicles for *S. aureus* poisoning are ham and other cured and lightly cured meats; cooked beef, pork, and poultry; egg and other protein or starch-containing salads; and dairy products (Notermans and Heuvelman, 1983).

The optimum temperature for growth and enterotoxin production is 99F (37C) and growth is rapid in the temperature range of 68 to 104F (20–40C). However, enterotoxin production has been observed in laboratory media at temperatures ranging from 59 to 110F (15 to 43C). The standard recommendation of holding perishable foods below 40F (4C) or above 140F (60C) gives householders a good margin of safety. The time-temperature history of a specific perishable food is the information needed to estimate if it is safe to eat.

### *Bacillus cereus* Foodborne Illness

*Bacillus cereus* as a cause of illness is seldom reported, but it is believed to be a fairly common problem throughout the world. *B. cereus* is found in the environment and produces spores so it may be in food ingredients and dust. It is an aerobic organism with the capability of also growing anaerobically. Some strains produce one or both of two toxins: diarrhea-producing enterotoxin which is not heat stable and an emetic (vomiting) toxin which can withstand 259F (126C) for 90 minutes (Bryan, 1979). Clearly, prevention of toxin production is more practical in foods than toxin inactivation through heat treatment.

The vomiting form of the illness may begin from 15 minutes to 11 hours after eating since the toxin has already been produced in the food. The diarrheal form is characterized by an incubation period in the body of 8 to 16 hours since this toxin is produced in the digestive tract. The duration is usually less than 12 hours. In approximately one-quarter of the cases, both vomiting and diarrhea are experienced by the unfortunate victim.

The pH range for growth is 4.9–9.3; optimum growth temperature is 86–95F (30–35C); and the maximum salt concentration tolerated by these bacteria is 9%. *B. cereus* is common to soil and dust, so grain foods—rice, legumes, wheat, bakery products—are likely to contain the spores, as are mixtures of ingredients such as pudding, sauce, and soup mixes. The most common vehicles for the vomiting form of this illness are boiled and fried rice (Bryan, 1979). Large numbers of this organism, $10^6$ to $10^9$ per gram, are usually found in implicated rice. This organism has also created problems in refried beans.

Since keeping spores of a soil organism out of food is impractical, *B. cereus* is controlled by holding cooked, perishable food at temperatures at or below 40F (4C) or above 140F (60C) to prevent multiplication of the bacteria and toxin production.

### Toxins from Molds

Molds are widely distributed in the environment and may occur as part of the normal flora of food, on inadequately sanitized food processing equipment, or as airborne contaminants. Since molds are poor competitors and grow slowly, their growth is often a problem under conditions that are unfavorable to bacterial growth. Foods with low-pH, low-moisture, high-salt or -sugar concentrations, and those stored at refrigeration temperatures are susceptible to molds.

Molds themselves can be responsible for food spoilage. They are able to utilize carbohydrates, including pectins, organic acids, proteins, and lipids as substrates causing off-odors, off-flavors, and surface discoloration. *Aspergillus* and *Penicillium*, the most common molds isolated from foods in home refrigerators (Torrey and Marth, 1977a and 1977b), are strongly lipolytic, converting large amounts of neutral fats into fatty acids. Off-flavors and -odors are associated with these changes in fats.

There are also health hazards due to consumption of moldy foods that contain mycotoxins. Mycotoxicoses—illness from eating toxin that was produced by a mold—have been documented since 1826 (ICMSF, 1978). Over 150 species of fungi (molds are a type of fungi) are capable of producing toxic substances. These substances may be present in the spores or released into the food. There are accounts of mycotoxicoses occurring throughout recorded history. St. Anthony's Fire (now known as ergotism) was a problem in the Middle Ages. The last major outbreak was in 1951 in France. This illness is rare now due to grain harvesting practices. Alimentary toxic aleukia (ATA) is caused by toxicogenic fungi, primarily *Fusarium*, releasing toxin into grain. ATA afflicted a large number of Russians in World War II when manpower was not available at wheat harvest time. The grain lay in the fields over winter before it was harvested in the spring and then consumed, poisoning many. Rice may also be contaminated with several species of *Penicillium* and their toxins. These fungi have been responsible for outbreaks of an illness called yellow rice disease that has been a problem in the Philippines. These three illnesses are rare today because past outbreaks have been learning experiences for the world. Grains are harvested promptly and grain storage facilities are monitored closely so that mold growth does not occur. However, grain may be damaged during growth or as it ripens and then mold before harvest, the most recent example being aflatoxins in corn harvested after a drought. *Aspergillus flavus* produces mycotoxins called aflatoxins. The problem was recognized in the 1960s when over 100,000 turkeys died from ingesting peanut meal contaminated with aflatoxins.

Ingesting mold toxins can cause acute or chronic symptoms depending upon the dosage and/or frequency. Incidents of acute mycotoxicoses are rare; but cause and effect relationships are fairly easily determined. There is now concern that delayed cancer or organ damage may result from repeated ingestion of subacute levels of mycotoxins (ICMSF, 1978). The exact effect and mechanism of action for humans is difficult to study. Exposure to aflatoxins at high levels has been linked to liver disorders, including liver cancer in human poisonings (Campbell and Stoloff, 1979).

Torrey and Marth (1977a and 1977b) are pioneers in research on molds in home refrigerators. One of their projects involved sampling moldy refrigerated foods and taking swabs of refrigerator surfaces from 66 Wisconsin households. They found that 49% of the food isolates were penicillia and 38% were aspergilli. The researchers concluded that the most frequently isolated molds were from genera in which mycotoxins commonly have been detected and which were confirmed as producing mycotoxins in several cases at temperatures of 68F (20C), though not at refrigerator temperatures. The researchers suggested the need for a more serious attitude toward molds in home-stored foods.

Not all molds produce mycotoxins and mycotoxigenic molds are not always able to produce toxins. Current research indicates that toxin production may require specific temperature, moisture, light, air, and nutrient conditions. It is also influenced by inhibiting compounds and competing microorganisms. Mycotoxins are secondary metabolites so they do not appear during the early stages of growth. However, these toxins may remain in the food after the molds have died or the visible mold has been removed.

Mycotoxin control is by preventing fungal growth. After harvest of grains, care should be taken to sort out the visibly moldy pieces and to maintain dryness. Controlled chemical and heat treatments may be used to decrease the amounts of active mycotoxins present in animal feeds but there are no practical methods for human foods.

Since 1964, the U.S. government has had a program to regulate aflatoxins in peanuts. The FDA currently regulates aflatoxins in both human and animal food supplies as animals consuming these toxins accumulate them in their tissues or excrete via milk which humans eventually ingest.

Prevention of mycotoxin consumption at the household level involves careful examination of moldy food. Moldy food that was stored above refrigeration temperature of 45F (7C) should not be consumed since, if *Aspergillus flavus* grew, aflatoxin as well as other mycotoxin production is likely. Eating moldy refrigerated foods is controversial. Many molds that are common in foods do not produce toxins at low temperatures. However, because research is limited, the general recommendation is to discard all moldy foods other than those which are produced with molds, such as blue cheese and country-cured hams. Most researchers believe removing 1/2 to 1 1/2 inches of the moldy layer from cheese and other solid foods renders it safe to consume. Prevention of mold growth is the best household defense. Table 2.5 gives examples of householder practices that can lessen the chance of mold growth.

## Foodborne Infections

### *Escherichia coli*

One of the leading causes of infant and traveler's diarrhea in nonindustrialized countries is probably *Escherichia coli*, but victims most often rely on symptomatic treatment and rarely seek to have the exact organism identified. The role of *E. coli* in foodborne illness is complex since most *E. coli* do not cause

**Table 2.5. Methods of retarding mold growth at the household level**

| Method | Practical Application |
|---|---|
| 1. Decrease temp below optimum mold growth range | Store raisins in refrigerator, store bread in freezer |
| 2. Decrease pH below optimum mold growth range | Wrap block of cheese in vinegar-soaked cheesecloth |
| 3. Decrease humidity | Divide dried foods into small amounts so package is not repeatedly opened |
| 4. Increase air circulation to reduce condensation | Pack root cellared produce loosely |
| 5. Decrease storage time | Use refrigerated foods within several days. Freeze for long-term storage |
| 6. Decrease number of mold spores added to food | Wash cheese cutting boards well. Cover foods promptly |

illness and the harmful strains are not readily identified.

*E. coli* is part of the natural flora of the large intestine of warm-blooded animals. These bacteria routinely leave the intestine in feces and contaminate soil and water through untreated sewage, or can contaminate many foods through unwashed hands. Plants and fish grown in a sewage contaminated environment can carry *E. coli*. Improper slaughter practices may also put this gut organism in contact with the edible muscle tissue. This is a concern with the enterohemorrhagic type since cattle appear to be carriers.

Until recently, the presence of *E. coli* in food or water was considered to indicate a problem of fecal contamination but not to itself cause illness. Though most strains are not pathogenic, four different groups of *E. coli* are now known to cause illness; enteropathogenic, enteroinvasive, enterotoxigenic, and enterohemorrhagic. The enterohemorrhagic strain O157: H7 causes extremely serious illness, and is the leading cause of renal failure in children in the United States.

Some strains can invade the intestinal mucosa and others only produce enterotoxins while in the intestinal tract. An average of 26 hours after ingestion of food contaminated with $10^6$ to $10^{10}$ organisms of toxigenic *E. coli* (range 8–44 hours), diarrhea occurs. This longer time between ingestion and onset of symptoms than the interval for the food poisonings is required for the organisms to establish themselves in the upper part of the small intestine, multiply, and produce enterotoxin. The toxin causes the cells to secrete salt and water into the intestine interior, and diarrhea results. With *E. coli* 0157:H7, toxins produced in the large intestine cause blood loss in the stool. The diarrhea usually stops within 30 hours. During recovery, the organisms are cleared from the small intestine walls but may become established in the large intestine where they do not produce symptoms, but the person becomes a carrier.

Prevention of *E. coli* contamination of foods and water is a basic best defense. In the United States good control measures include washing hands after using the rest room, holding food at temperatures that prevent bacterial growth, and preventing sewage contamination of water. Heating to an internal temperature of at least 161F (72C) as in pasteurizing milk or preparing medium–well done meat will kill *E. coli* and provides a margin of safety. Lower temperatures for longer periods of time are adequate.

#### Shigellosis

Shigellosis, commonly known as bacillary dysentery, is a term for infections of *Shigella* bacteria. In 1991, there were 23,548 recorded cases in the United States. Most illnesses occur in people living in institutions and foreign travelers. Contamination of the food supply is through feces of an infected person. This most commonly is a result of poor personal hygiene, fertilizing with human waste, and consumption of uncooked shellfish or vegetables from sewage contaminated water. Poor personal hygiene in institutionalized care is responsible for afflicting many in the United States. A typical scenario involves the human carrier with contaminated (unwashed) hands handling food that is then served without further heating.

Research with human volunteers has found that consumption of as few as 10 *S. dysenteriae* organisms can cause illness. After ingestion, shigellae invade the cell walls of the colon and probably the small intestine, multiply, and destroy tissue. *S. dysenteriae* is the most virulent species. It produces a cytotoxin which increases the severity of the illness. Death of the damaged colon cells produces ulcers and the result is bloody diarrhea with pus.

The incubation period of shigellosis is 1 to 7 days. Ingestion of large numbers of the organism results in shortened incubation periods and a more severe illness. In addition to bloody diarrhea, shigellosis is characterized by abdominal pain, fever, and often vomiting.

Treatment is usually fluid and electrolyte replacement. The very young and elderly and those infected with serotypes found outside the United States may require hospitalization. Antibiotic therapy can reduce the spread by eliminating carriers; however, antibiotic-resistant strains of *Shigella* are common.

Prevention is primarily through educating food handlers on hand washing after use of the toilet, providing safe water supplies, and adequate sewage disposal.

#### Salmonellosis

Salmonellosis is a collective term for all illnesses of humans or animals caused by the *Salmonella* bacteria. This includes typhoid fever. It was recognized as a disease in the middle of the 1800s, though its presence had been recorded for centuries. Typhoid fever was a major epidemic during the U.S. Civil War. When the troops disbanded, survivors carried the bacteria to all parts of the country. Although *Salmonella typhi* and typhoid fever are no longer major problems, salmonellosis is. It is estimated to cost U.S. citizens over $1.5 billion in medical expenses annually and is a particularly serious disease for the very young and the elderly. The incidence of salmonellosis in the United States rose from 11 per 100,000 people in 1971 to 19 per 100,000 in 1991. It is probable that only about 10% of illnesses are reported. In countries with poorer waste disposal systems, the annual death toll is high.

Organisms of the genus *Salmonella* are all potentially pathogenic to humans. The most important habitat of *Salmonella* as a potential contaminant of human food is the intestinal tract of poultry, swine, and cattle. Various pets, such as freshwater aquarium snails, turtles, frogs, and children's petting zoos and public aviaries have been sources of salmonellosis outbreaks. Human-to-human transfer of the organism is also documented. Animals originally become infected through contaminated feed or from other animals. These animals may be symptomless carriers or exhibit gastroenteritis while they excrete *Salmonella* into the environment through their feces. The number of carriers and ill animals increases when animals are transported to the slaughtering plants.

Eggshells may become contaminated when they touch the nest and droppings. Salmonellae easily enter eggs with checked or cracked shells, and as the egg cools to room temperature, organisms on the surface may be drawn through the shell to the interior. *S. enteritidis* has become a recent problem world-wide due to contamination which occurs as the egg is formed. Cases in the northeastern states, caused by mishandled fresh egg products, resulted in a sharp increase in the number (27 per 100,000) of illnesses in 1985.

Although not common, plant products such as strawberries and tomatoes may become contaminated when animal waste is used for fertilizer or irrigation water is of poor quality. Washing produce with generous amounts of water will remove some surface salmonellae but cannot be relied upon completely. An outbreak of salmonellosis from cut cantaloupes resulted in the recommendation to not only wash the surface before cutting but also to limit holding to 2 hours or less in the danger zone of room temperature.

Raw meats, poultry, and their products such as eggs and unpasteurized milk may carry salmonellae into home kitchens. In the evisceration and cutting processes in slaughtering plants, the equipment is not cleansed between animals so one infected carcass can contaminate many. At poultry processing plants, 50% to 100% of the birds are often contaminated. Consumption of these products without thorough cooking may result in an infection. Illness has occurred in households that always ate very well done chicken, but were more lax about washing the countertop, where the chicken wrappings dripped, before placing fresh vegetables for a salad there. A cutting board may have been reused without thoroughly scrubbing it. Such cross contamination is responsible for many salmonellosis outbreaks.

Salmonellosis has four main syndromes: the carrier state (newly recovered or asymptomatic), enteric fever (typhoid or paratyphoid), gastroenteritis (infection of the small intestine), and septicemia (general infection through the blood). Once in the small intestine the salmonellae multiply and invade the wall. The gastroenteritis form is most common. Its incubation time is usually 20 to 48 hours, with a range of 1 to 70 hours after ingestion of infective numbers of salmonellae. The first day of the illness diarrhea, abdominal cramps, and vomiting are frequently present. Fever, headaches, chills, and prostration sometimes occur. Salmonellosis may also take the form of localized infections such as in the lymph system. Organisms may also invade the large intestine where they cause

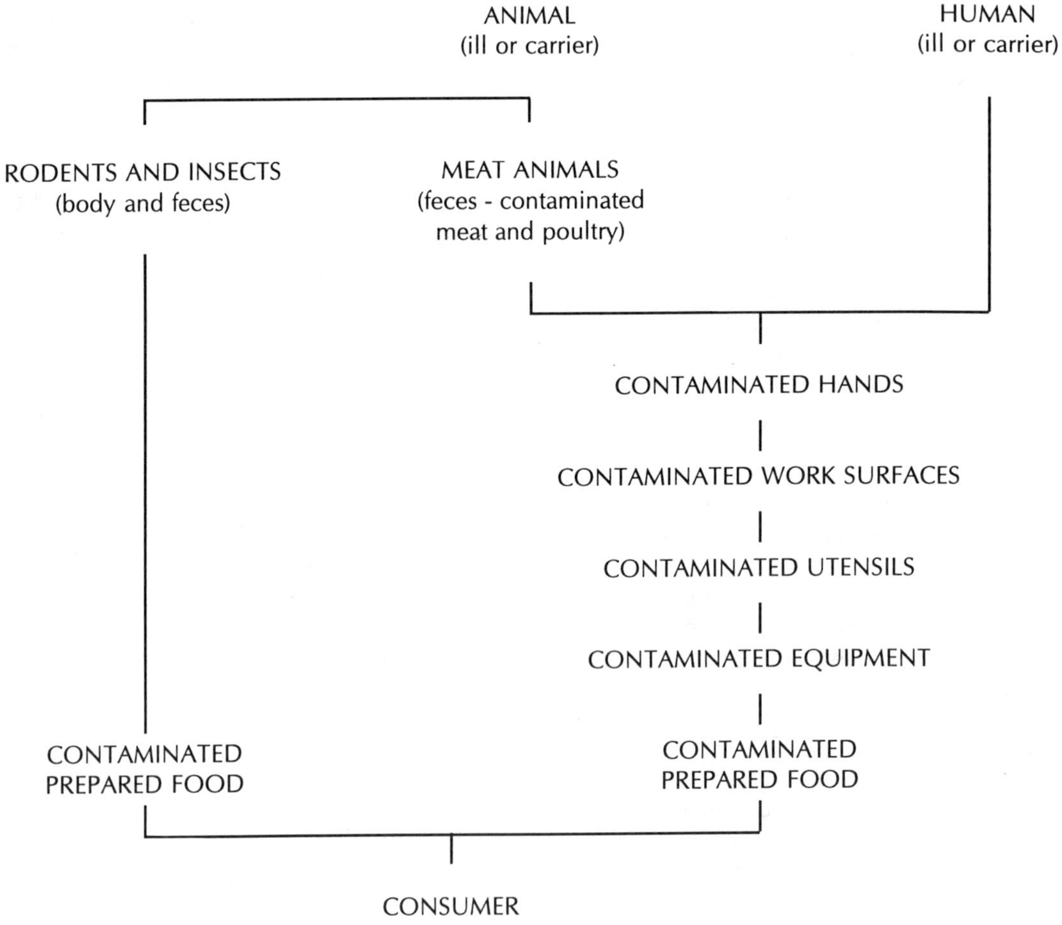


**2.3.** Sources and routes for infection due to *Salmonella, Campylobacter,* or *Escherichia coli.*

inflammation.

Afflicted persons excrete salmonellae for a while whether they exhibit symptoms or not. *Salmonella typhi* can be shed intermittently by chronic carriers for years.

In current times, the incidence of *S. typhi* has been greatly reduced, largely due to sewage treatment, testing water that shellfish are harvested from, chlorination of public drinking water, pasteurization of milk, sanitary guidelines for those processing or vending food, immunization programs, and exclusion of typhoid carriers from the food industry. However, the incidence of salmonellosis as a whole is increasing. Preventing the release of untreated sewage into bays, oceans, rivers, and lakes is needed for fish to grow free of salmonellae. There are additional guidelines many sanitarians would like to see followed such as salmonellae-free animal products in the United States.

Eradicating salmonellae from domestic animals would reduce the incidence of salmonellosis significantly. Although this goal may not be reached, practices to achieve it increase the safety of our food supply. Production of salmonellae-free feed requires stringent sanitation at the plant level after the feed is pasteurized to prevent recontamination. Increased sanitation on the farms themselves would include pest control, removal and proper disposal of waste, storing pasteurized feed so it is not contaminated, preventing farm workers from carrying salmonellae organisms among farms, and keeping visitors out.

More frequent carcass sampling at processing plants would monitor contamination through the assembly line. Sanitizing of processing equipment at intervals during the day is important for reducing animal-to-animal contamination. Such simple sanitary practices in a hog processing plant reduced salmonellae levels by 75% (Banwart, 1989). Household pets may also be carriers.

Thorough heating of meat, poultry, and their products to a minimum temperature of 160F (71C) will destroy the infective organism. Baked goods reach this temperature so are free of viable salmonellae. Although beef is less commonly contaminated with salmonellae than is poultry, it has been a source of salmonellosis, perhaps because ground or cubed beef is often consumed without thorough cooking. Eggs should be completely cooked so that the white is coagulated and the yolks hot and thick 160F (71C). Pasteurization of milk kills salmonellae and other pathogens. Children on field trips should not be given tastes of raw milk; a number of cases have occurred after such experiences. Low doses of radiation are also effective. A process has been approved for raw poultry (1990) but is not widely used in the U.S. industry. A spray or dip containing trisodium phosphate was approved in 1992 by USDA to reduce numbers of salmonellae on raw birds.

Preventing cross-contamination in the kitchen is most important. Using plastic cutting boards and washing them in the dishwasher; cleaning all surfaces with a brush and soap followed by a chlorine (bleach) solution after contact with raw meats, poultry, and seafood; and physical separation of these foods, in the refrigerator and on counters, from fruits and vegetables which are to be consumed raw will greatly reduce the chances of contaminating ready-to-eat foods with salmonellae.

Humans and household pets may also spread the illness. Simple hand washing before food handling is important. Children should be trained to not put their fingers in their mouths while playing with pets and very young children need to be supervised during play and then their hands washed. Many households do not purchase certain pets because of the likelihood they are carriers.

Large numbers of *Salmonella* must usually be present in a food for illness to occur so perishable foods should not be held between 50 to 122F (10 to 50C), an ideal growth range for *Salmonella*. To provide a margin of safety, the holding temperature should be 40F (4C) or below or 140F (60C) or above. Refrigeration temperatures prevent multiplication but *Salmonella* organisms remain viable at cold temperatures. They can also survive freezing temperatures. Ice cream made with an uncooked custard base has caused several outbreaks. Salmonellae do not grow in foods with pH below 4.6, but they may survive.

### Campylobacteriosis

An acute gastroenteritis very similar to salmonellosis in symptoms and problem foods is caused by *Campylobacter*. The most important species is *C. jejuni*. *Campylobacter* was not identified as a cause of human foodborne illness until the mid 1970s. It is now thought to be as important as salmonellae in the United States and the world. Since special incubation procedures are necessary for its growth, only recently has it been included in microbiological examination of feces or of suspect foods. As research data accumulate, *Campylobacter* will probably emerge as a very significant cause of foodborne illness.

After ingestion of even low numbers of *C. jejuni*, the organism infects the gastrointestinal tract and illness usually occurs within 3 to 5 days (range of 1 to 10 days). Predominant symptoms, which can be mild or severe, are abdominal pain, nausea and vomiting, bloody diarrhea, and fever. Although the illness may be as short as 1 day, it may persist for several weeks. It may also lead to a more generalized infection.

Like salmonellae, these bacteria are frequently a part of the bacterial flora of birds and animals, usually without causing them illness. Thus, they may be present on raw meat and poultry, and from there cross-contaminate other foods. Sewage may also cause contamination of water supplies. Growth of *Campylobacter* is unlikely in foods since it grows only within a narrow temperature range of 86–117F (30–47C) and does not compete well with other organisms. Although it does not multiply at refrigerator temperatures, *Camphylobacter* survives these well. Death occurs in frozen foods but not all will die so this cannot be relied upon.

Prevention of this foodborne illness is the same as for salmonellosis: pasteurization of milk and cooking of meat and poultry since *Campylobacter* is not heat-resistant, and the prevention of cross contamination between raw foods of animal origin and ready-to-eat foods. Humans may be carriers during an infection (even if not apparently ill) and for weeks afterward, thus instruction of food handlers in sanitation is important.

### *Clostridium perfringens* Infection

*Clostridium perfringens* is probably one of the most common types of human foodborne illnesses although reported incidents comprise only a small proportion of the total. This organism produces spores so it is present in the soil, dust, and animal (including human) feces. Since it is so widely dispersed, meats, poultry, and plants are usually contaminated with the spores. The spores survive cooking so are often present in cooked foods where competitors have been eliminated through the heating process. Cooking "heat shocks" the spores, activating them for germination, and also establishes anaerobic conditions within the cooked food. *C. perfringens* is anaerobic so its growth is then favored.

*C. perfringens* multiplies very rapidly and can double its population in 8 1/2 minutes with optimum conditions. No growth occurs below pH 5.0. The optimum temperature range for growth is 99–117F (37–47C); the minimum growth temperature is 59F (15C). The limiting $a_w$ is 0.93. Meat and vegetable foods held at warm but not hot temperatures for serving groups are especially at risk.

Food which has caused illness usually contains at least $10^6$ *C. perfringens* vegetative cells per gram. The organisms remain in the small intestine where they multiply and sporulate. Enterotoxin is released and causes ions and water to accumulate in the small intestine. Diarrhea with severe abdominal cramps and often gas results. Like *E. coli* foodborne infection where the organism must multiply and produce toxin after ingestion, *C. perfringens* foodborne illness has a fairly long time between the ingestion and appearance of symptoms. Most illnesses occur after 8 to 12 hours but the range of incubation times is 6–24 hours. The diarrhea and cramping usually do not last over 1 day.

Prevention of *Clostridium perfringens* cases is by holding hot foods above 140F (60C), cooling rapidly to below 64F (18C), and holding foods at temperatures below 40F (4C). As an additional safety measure, leftover foods may be heated to at least 160F (71C) to kill those organisms that may be present. However, householders with young children often reheat leftovers in a microwave range only to a warm serving temperature instead of "bubbling hot" and then cooling to serve. Cooked meats, poultry, beans, and gravy are the foods most frequently implicated in outbreaks.

### *Vibrio parahaemolyticus* Infection

*Vibrio parahaemolyticus* gastroenteritis is common in countries where raw seafoods are consumed. It accounts for 50% of bacterial foodborne diseases in Japan. The bacteria can grow in environments of 1–8% salt, but optimum conditions are 2–4%.

*V. parahaemolyticus* is found in marine waters, especially the warmer ones, and in fresh water containing large amounts of organic matter. Contaminated fish and shellfish which are allowed to stand at unsafe temperatures and not thoroughly cooked can cause the gastroenteritis, and cooked seafood may be cross-contaminated from the raw, such as by reusing the shellfish container. This organism is killed by heating the food thoroughly.

Ingestion of $10^5$ to $10^7$ cells of pathogenic strains produces illness in humans as the organism grows in the intestine and produces toxins. The incubation period is usually 10–20 hours, but the range is 2–48 hours after ingestion of contaminated food. The gastroenteritis is usually self limiting but it may last up to 10 days.

Since *V. parahaemolyticus* is naturally present in water, controlling the illness is best accomplished by thorough cooking of seafood, prevention of cross contamination, and icing or mechanical refrigeration to inhibit bacterial multiplication while the seafood is on the boats, at the markets, and in the household.

### Cholera

Cholera is considered by many to be a disease of the middle ages or one that presently only occurs in nonindustrialized areas after a natural disaster. However, occasional cases of cholera currently occur in the United States. The only major cholera outbreak in the Western Hemisphere this century occurred in Peru in 1991 and spread from there to other countries in Central and South America. Peru reported 169,255 probable cholera cases and 1,244 deaths. It spread rapidly due to inadequate water treatment and distribution, poor sewage disposal systems, and the popularity of a raw fish dish (ceviche). By February Ecuadorians had cholera, and by March this outbreak had reached Colombia. In April the outbreak crossed the Andes to Brazil. Ten confirmed cases from this outbreak reached the United States. Two of those afflicted were in South America when they contacted cholera through water and eating ceviche and raw oysters, and the other eight ingested noncommercially transported and improperly prepared crabmeat.

Humans are the only known natural reservoir for *Vibrio cholerae*. Both those suffering from the illness and carriers without symptoms shed the organism in their feces. Contamination occurs when feces come into contact with food, drinking water, or shellfish growing in contaminated water. There is evidence that the vibrios can establish themselves in both fresh and brackish bodies of water.

Large numbers ($10^8$–$10^{10}$) must enter the body for illness to occur, because many of the organisms are killed by acid conditions in the stomach. Some particularly virulent strains are an exception to this general rule; smaller numbers of them have caused illness. People taking antiacid medication have increased risk. The organisms that survive passage through the stomach attach to the walls of the small intestine where they multiply and produce toxin. This process takes an average of 2 days but the range is from 1 to 5 days. The cholera toxin activates adenyl cyclase which converts ATP (adenosine triphosphate) to cAMP (adenosine 3′,5′-cyclic monophosphate). The presence of the high level of cAMP causes unusually large amounts of chloride and bicarbonate to be secreted along with large amounts of water. The resulting diarrhea lasts 20 to 24 hours. It is accompanied by vomiting, but rarely by nausea or retching.

Cholera is a self-limiting disease, but requires immediate attention since 50 to 75% of patients who become dehydrated will die without treatment. Ninety-nine percent of treated dehydrated patients survive. Appropriate treatment includes fluid and electrolyte replacement either orally or intravenously. Many lives have been saved since the recognition that a solution of glucose and salts will be well absorbed. One oral rehydration solution can be made with 3 Tbsp (40 g) sucrose, 1/2 tsp (3.5 g) salt (sodium chloride), per quart (1 L) of clean water (WHO, 1985). The solution should be given at a rate equal to stool losses. In nonindustrialized countries, the mixture of dry ingredients is distributed in packets to mix with water at home. In the United States, pharmaceutical companies market premixed fluids. It is also important to provide food to the patient.

Prevention is by sanitary sewage disposal to protect water supplies, washing hands after possible contact with feces, and thoroughly cooking foods which may be contaminated, such as shellfish from areas in which there is a continuing presence of *V. cholerae* in the water. Since water quality is usually unknown, thorough cooking of all seafood is prudent. Vaccination gives a temporary 6-month immunity most of the time; however, since there is still a risk it is seldom used. Immunity is not effective for a new virulent strain which became epidemic in 1993 in India and nearby countries. Better vaccines are being sought by researchers.

### Yersiniosis

*Yersinia enterocolitica* has been identified as the cause of foodborne illness in several major outbreaks in the United States and may be a greater problem in Canada and Europe. Since only a limited number of strains of *Y. enterocolitica* are pathogenic, it is difficult to determine the extent of the problem in our food supply. A battery of tests for virulence are needed to identify such strains since harmless types are widely distributed in nature. Pigs have been shown to be a reservoir of pathogenic yersiniae as well as rodents from infected farms and slaughter houses.

Symptoms of the illness include vomiting, fever, diarrhea, and pain that mimics an appendicitis attack. Illness generally occurs 24 to 36 hours after the bacteria are consumed and lasts for 1 to 3 days. Children and young adults are the most susceptible and generally have a more severe illness.

The first foodborne outbreak identified in the United States occurred in 1976. Commercially processed chocolate milk was implicated; it apparently had been contaminated after pasteurization of the plain milk. More than 200 schoolchildren became ill and 36 of these were sick enough to be hospitalized. Appendectomies performed on 16 of the hospitalized children were identified as not necessary. A second outbreak was attributed to tofu packaged in contaminated spring water and a third to pasteurized milk apparently contaminated from the carrying cases.

*Yersinia* is a psychrotroph, thus long-term refrigeration of contaminated foods may result in increased numbers. It can grow at temperatures between 32F (0C) and 111F (44C), although growth is most rapid at 90–93F (32–34C). This organism also has a wide pH range for growth (pH 4.6–9.0) and is tolerant of 5% salt; 7% salt prevents growth. Many cured meats and some pickles reach such high salt concentrations.

Prevention is similar to that for *E. coli*. Meats and poultry should be cooked to at least the medium-done stage, milk should be pasteurized, and only tested safe water supplies used for food preparation and packaging.

### Listeriosis

The importance of *Listeria monocytogenes* in human infections has been recently recognized. *Listeria* was first known as a cause of illness in many animals and birds, such as an encephalitis of sheep called "circling disease." The sporadic nature of human and animal outbreaks suggests there are specific pathogenic serotypes involved (Kvenberg, 1988).

In 1981 a major outbreak of listeriosis associated with coleslaw was reported in Canada. The cabbage had been grown in a field fertilized with sheep manure and some of the sheep had had listeriosis. After harvesting, the cabbage had been in cold storage until used in the spring. A second outbreak occurred in Massachusetts in 1983; the cause of the contamination of the 2% and whole pasteurized milk was not identified. The most serious epidemic was in 1985 from Mexican-style cheese made in California when raw milk was added to pasteurized milk during the manufacture of the cheese. This incident resulted in 100 confirmed cases (more than 90% were infants); 40 of those afflicted died.

Listeriosis is most serious when it occurs in infants, older persons, and those with compromised (lowered) immunity. The early flulike symptoms may be ignored, but meningitis or a local systemic infection can result. Mortality rates are high, especially if patients are not promptly diagnosed and treated with appropriate antibiotics. It is especially serious for pregnant women and the fetus since, in addition to the illness of both, it may cause abortion. The fetus may be infected either via the transplacental route or during delivery. For other people, the illness is commonly a temporary, slight infection with no symptoms.

Avoidance of *Listeria monocytogenes* is difficult. The bacteria may be found in soil, water, sewage, and almost all animals, birds, insects, and on plants. Humans who are not ill as well as those who are ill, may excrete the organism in their feces. Adding to the problem is the psychrotrophic nature of the organism. It is able to multiply at temperatures as low as 32 to 37F (0 to 3C) in simplified laboratory systems (Glass and Doyle, 1989). In foods, the competition of other microorganisms and less than optimal environment may limit multiplication but refrigeration cannot be relied upon to prevent growth.

The challenge to the food and dairy industries is to ensure that all ready-to-eat foods contain no viable *L. monocytogenes*. This is a requirement of the Food and Drug Administration and, with slight modifications, of the USDA Food Safety and Inspection Service. Elimination of listeriae can be achieved through control of the endpoint temperature in pasteurization or preparation of the product since listeriae are not especially heat-resistant. The suggestion that viable bacteria within leukocytes in milk might be more resistant has been supported by most research, but the increase in resistance is small and pasteurization treatments should be effective. The second, and more difficult, control is the prevention of recontamination of the product from machinery, packaging, air, or waterborne organisms. Listeriae survive well on room surfaces, in drains, and on equipment.

For the consumer, the challenge is to cook meats, poultry, and seafood adequately and to prevent cross-contamination. Because the disease is so serious, those at high risk of illness (especially pregnant women and the immuno-compromised) are advised to reheat prepared ready-to-eat foods such as cooked seafood and deli meats. Since the organism grows most rapidly at 86–99F (30–37C), refrigeration of perishable foods is recommended even though slow growth will still occur.

#### Hepatitis A and Enteric Viruses

Hepatitis A, also known as infectious hepatitis, is caused by a virus. There were 24,387 cases in the United States in 1991. A recently identified type, hepatitis E, is also foodborne. Human and nonhuman primates are reservoirs for this virus. It is spread by the fecal-oral route so waterborne and foodborne outbreaks can occur. Raw or undercooked shellfish from sewage-contaminated water or foods contaminated by food handlers are the usual vehicle. When contamination is by infected food handlers, a variety of foods including doughnuts, raw vegetables, roast pork, orange juice, and mayonnaise have been vehicles.

The incubation period averages 30 days with a range of 10–50 days. Initially, the disease symptoms are malaise, vomiting, abdominal pain, fever, and headache. These are followed several days later by jaundice, dark-colored urine (bile in urine), and light-colored stools. The jaundice is usually most pronounced between the 7th and 14th day. It is followed by a recovery phase which may last 3–4 months. During recovery, the patient often feels tired. Mild or inapparent infections also occur during which time the virus can also be transmitted to others. People who have been exposed or might be exposed (as in the case of sharing a household with an infected person) can be given immune serum globulin. This deters secondary spread of the illness.

Prevention is by eliminating fecal contamination of food by good food handler hygiene, harvesting shellfish from nonpolluted water, and cooking it thoroughly. Infected clams added to a pot of boiling water and cooked until they opened were implicated in one outbreak. More thoroughly cooked seafood have not been implicated.

Since foods may be contaminated through an infected food handler, controlling this person is important but not easily done. The virus can be excreted in the feces for 7–10 days before there are symptoms and some people have no symptoms throughout the course of the disease. Those showing symptoms should not be allowed to work with food until they have clearly recovered. This can be determined by testing the SGOT (serum glutamic pyruvic transaminase) and SGPT (serum glutamic oxalacetic transaminase) levels in their serum.

Current research suggests that viruses do not multiply in food and tend to become inactivated during storage. Better personal hygiene and less handling of food through the use of utensils or disposable hand coverings would decrease the spread of hepatitis A and E. All food handlers should routinely wash their hands after using the rest room and again before touching foods. In some states such as Oregon, when a food handler diagnosed with hepatitis A or E is employed in a restaurant where the hand washing station is in the kitchen instead of the private rest room and the sanitation inspection scores high, the health department does not feel gamma globulin shots for all who ate there are necessary. The reasoning behind their policy is that when hand washing is supervised by all in the kitchen and other evidence is positive, hepatitis is unlikely to have been spread.

Other viruses that can cause vomiting, diarrhea, and often fever may also be transmitted in contaminated food and water (Table 2.6). Unlike many of the other causes of foodborne illness, they are unable to grow in the food. Viruses can be killed by chlorination of water and thorough cooking of food.

### Trichinosis

*Trichinella spiralis*, a roundworm, causes a painful muscle infection in humans. The disease is acquired by ingesting the larva stage of the worm in inadequately cooked infected meat of carnivores (pigs, bears, walrus). During the digestion process, the larvae are released from the meat and invade the small intestine wall where they develop into adults and mate. The female then deposits larvae into the lymph

**Table 2.6. Viruses of importance in food preparation**

| Virus | Commonly Afflicted | Incubation Period (days) | Mode of Transmission |
|---|---|---|---|
| Astrovirus | young children<br>elderly | 1–4 | food, water, f-o[a,b] |
| Calicivirus | all | 1–3 | food, water,<br>nosocomial f-o |
| Enteric adenovirus | young children | 7–8 | nosocomial f-o |
| Norwalk | older children, adults | 1.5–2 | food, water, ptp[b], f-o |
| Rotavirus A | infants, toddlers | 1–3 | food, water, ptp, f-o |
| Rotavirus B | children, adults | 2 | water, ptp, f-o |
| Rotavirus C | all | 1–2 | f-o |

[a]Viral symptoms commonly include watery diarrhea which is often accompanied by vomiting and fever.
[b]Fecal-oral route abbreviated f-o, person to person abbreviated ptp.
Adapted from Centers for Disease Control Morbidity and Mortality Weekly Reports 39 (RR-14):7. 1990.

system. The larvae are thus transported to other parts of the body where they penetrate striated muscles and encyst. *T. spiralis* goes through a similar life-stage cycle in animals.

The first symptoms of trichinosis occur from 1 to 14 days (average 7 days) from ingestion when the larvae invade the small intestine walls. Abdominal pain, diarrhea, vomiting, sweating, and anorexia are common, but symptoms may also be absent. When the newborn larvae migrate to the muscles and encyst there, muscle pain, edema, fever, fatigue, and swelling are the primary symptoms. Painful breathing, chewing, and walking; headache; constipation; myocarditis; pneumonia; central nervous system involvement; and paralysis of the diaphragm may occur if a large number of larvae are ingested. Slow recovery begins 3 to 10 weeks after the initial infection.

Trichinae in meat are killed by heating pork to 137F (58C) if this temperature is reached in conventional heating, freezing at 5F (−15C) for 20 days in pieces of pork less than 6 inches (150mm) thick, by approved curing processes, and by ionizing radiation. Pieces thicker than 6 inches require 30 days of frozen storage. Trichinae larvae have been found to remain infective up to 71 days when frozen at 21F (−6C) and in pork after some curing and short aging processes. Ionizing radiation was approved in 1985 in the United States to kill this parasite but has not yet been used.

Feeding swine only heat-processed garbage has made trichinosis in this country rare. However, isolated infested animals still are found so cooking is recommended for pork. These recommendations have been modified over the years to provide adequate, though smaller, margins of safety. Currently, the recommendation is to cook pork until it reaches a temperature of at least 155F (68C) or until there is no pink color (170F or 77C). In microwave cooking and grilling, it is especially important to be certain that all parts are done. Unfortunately, many senior citizens still cook pork to the tough, dry stage. In some cultures, uncooked pork sausages are popular; outbreaks in the United States have frequently been associated with these foods made by immigrants.

#### Other Parasitic Infections

Parasites may be present in foods throughout the world. Some cause serious illnesses but the transmission of all parasites can be prevented by simple practices. Their spread (usually through feces) is from one infected animal or human to another. When untreated human sewage contaminates the water and soil where food-animals are grown, they become infected.

Parasites are common in the digestive tracts (gut) of fish. If the fish are not gutted rapidly after harvesting, these parasites pass into the flesh. Species of *Anisakis* commonly infect herring which, when consumed raw, is a vehicle for human infection. *Gnathostoma spinigarum*, *Angiostrongylus* spp., and fluke (flatworm) infections may also be acquired through consumption of raw or undercooked fish and shellfish. Tapeworm infections can result from insufficiently cooked pork, beef, or fish. Parasites in fish can be killed by holding 7 days at ≤−4F (−20C). For beef tapeworm, 5 days at 14F (−10C) is recommended as the minimum. The endpoint cooking temperature should be 145F (63C) or higher.

Proper sewage disposal practices decrease the incidence of parasitic infections. Thorough cooking is recommended as a control measure.

Protozoa, single-celled complex microorganisms, may be transmitted from animal or human to the victim through food (food handlers to consumer) or water (via sewage). The most important are *Giardia*, amoebae, and *Cryptosporidium*. Often infections do not result in illness. Since the cysts are resistant to chlorine and other disinfectants, a filtration step is needed in water purification, both in municipal systems and in treatment of water when camping, unless water is boiled.

Toxoplasmosis is a serious infection for both mother and fetus since the parasite can be transferred through the placenta. The systemic infection which results causes congenital defects, abortions, or central nervous system problems in the infant. Pork, lamb, and perhaps beef may be infested so only meat which has been thoroughly cooked should be eaten during pregnancy. Cats are a host so the cat's litter box should be emptied by nonpregnant household members.

Since all of these parasites have a resistant cyst form, they persist in the environment. Preventing ingestion is the control measure, as eradication from the environment is usually impractical.

## Future Problems

Are there newly emerging microorganisms that are causing foodborne illnesses? Apparently not. Some of those which have been recognized recently, such as *Listeria*, have actually been known historically but not identified as a major problem before. Others have existed but not been identified earlier as causing foodborne illnesses. *Aeromonas hydrophila*, *Plesiomonas shigelloides*, and *Vibrio vulnificus* are currently being investigated to determine their roles in foodborne illnesses.

Increasing numbers of people with impaired immunity (AIDS patients and those receiving chemotherapy for cancer) have made us more aware of bacteria that normally do not cause illness. Changing food production and processing techniques and food habits may increase or decrease the magnitude of a problem. For example, brucellosis from drinking raw milk from an infected cow or goat has been eliminated as a concern in the United States because of the Federal-State Cooperative Brucellosis Control Program. In developing countries, this and other diseases remain problems.

It is important to have an active national as well as local public health program to identify changes in causes and incidence of foodborne diseases. Then control measures can be initiated (Lederberg et al., 1992).

## Prevention of Foodborne Illness

### Time and Temperature Recommendations for Safe Food Storage

Foods that have pH and $a_w$ levels within the ranges for the growth of bacteria that can cause foodborne illness are considered potentially hazardous foods. As defined by the Food and Drug Administration, these foods have adequate nutrients, their pH is above 4.6, and their $a_w$ is above 0.85.

The temperature at which these foods are held will determine the rate of bacterial growth and, for toxin producers, the level of toxin production. It has been recognized since the early part of this century that cold temperatures are bacteriostatic (static means stay the same) rather than bactericidal (kill bacteria) and many studies have been conducted to determine food storage temperatures at which pathogen growth and toxin production do not occur. There are interrelating factors that affect the temperature at which multiplication of a given pathogenic species is possible; however, in food systems the growth temperature ranges used in consumer education materials are 40–140F (4–60C), and it is recommended that perishable food not remain in this temperature range for longer than 3 hours. These temperatures represent a margin of safety compared to the 50–120F (10–49C) range for more rapid growth. The current maximum holding temperature of 45F (6C) required by most state regulatory agencies is based on the FDA standard (1976) which had a lower margin of safety than the 1993 Food Code.

There are a few exceptions to this minimum temperature for growth of pathogenic bacteria. *C. botulinum* nonproteolytic types B, E, and F have been found to grow and produce toxin at temperatures as low as 37F (3C), *Listeria monocytogenes* (Glass and Doyle, 1989), and *Yersinia enterocolitica* multiply at temperatures as low as 34F (1C) (Swaminathan et al., 1982). These organisms as well as the *C. botulinum* toxin are heat sensitive. They would not present a health hazard in foods thoroughly heated before serving. Pasteurization of dairy products eliminates *Yersinia* and *Listeria*. Heating all foods (dairy, cooked and uncooked animal foods, seafood, and vegetables) to 160F (71C) immediately before consumption eliminates bacteria. Most refrigerated foods appear spoiled before the slowly multiplying clostridia have reached hazardous numbers; the exception is smoked meats and fish where quality changes may not be noted. Therefore a maximum refrigerated time of 3 weeks is recommended for smoked meats and fish and other low-acid cooked foods. The FDAs 1993 Food Code requires a limit of 14 days of refrigerated storage for these processed foods in reduced oxygen packaging.

The expected shelf life of refrigerated foods is based upon quality changes (texture, off-flavors, off-odors) which can be brought about by nonpathogenic bacteria able to grow below 50F (10C) (Rutgers, 1971). Discards of refrigerated food due to loss of quality account for large portions of household food waste. One study found householders intentionally discarded food due to decreased quality 17% of the time, but changes they considered to be microbial, which accounted for 12% of the discards, were also largely due solely to quality changes (VanDeRiet and Woodburn, 1987). In addition to learning the difference between quality and safety changes in a food, consumers need better life expectancy guides for their refrigerated foods so there are fewer discards because of poor quality.

### Adequate Cooking

For bacteria that do not produce spores, molds, and viruses, an endpoint of 160F (71C) provides a margin of safety. Lower temperatures plus holding time may also be used. For spores, the temperature reached during cooking will allow survival so subsequent growth must be prevented by the holding temperature.

### Prevention of Cross-Contamination

Although it is impossible to exclude air and dust and surfaces will not be free of microbes, reduction of contact will increase the storage life and the safety of the food. Cross-contamination between raw foods and cooked or other ready-to-eat food is responsible for many cases of foodborne illness. Hands may transmit organisms from skin, nasal passages, and feces so need to be washed thoroughly before foods or utensils are handled. Tongs, paper squares, or other utensils should be used instead of hands. Plastic gloves are sometimes recommended but tend to give a false sense of security. They may actually become almost as contaminated as skin.

Industry has applied a system, called Hazard Analysis and Critical Control Points (HACCP), to the prevention of foodborne illness. Some have termed this "applying what we know to reduce risk." The production of a food from beginning to end is observed, and the steps which are critical in the prevention of foodborne illness are identified. For each step, the necessary safeguards and monitoring steps are identified. This same process can be applied in home food preparation and preservation as guidelines are developed (USDA Food Safety and Inspection Service, 1989).

## ■ Summary

The safety of foods depends upon the use of naturally safe products, the choice of safe additives and the prevention of contamination with harmful chemicals, and the control of

**Table 2.7. Microorganisms associated with illness from eating raw or undercooked products**

| Pathogenic organism | Fish | Poultry[a] | Beef | Pork[b] | Eggs | Milk |
|---|---|---|---|---|---|---|
| *Campylobacter* | | ++ | + | + | | + |
| *E. coli* | + | + | ++ | + | | + |
| *Listeria* | + | + | + | + | | + |
| *Salmonella* | + | ++ | + | + | ++ | ++ |
| *Vibrio* | + | | | | | |
| *Yersinia* | | | | ++ | | + |

[a]Chicken, Turkey, Duck, Rabbit.
[b]Trichina is an additional risk but is very rare in domestically raised pork in the United States today.

microbial hazards. The last is the most important measure in our current food supply. The human pathogens and their sometimes unusual entries into food can be alarming to consumers. This need not be. It is important to focus on the methods of prevention which are easily followed and inexpensive at the household level.

**1.** EXAMINE FOOD. Off-odors or visible slime should be reasons for discard. This decreases the chance of large numbers of organisms in the food which may but often does not include those which can cause illness.

**2.** KEEP HOT FOODS HOT AND COLD FOODS COLD. Warm is an unsafe temperature. Hold food below 40F (4C) or above 140F (60C). Use hot plates, chafing dishes, and beds of ice to hold foods at a safe temperature for serving in buffet lines. Put out only a small amount of food at a time and replace with fresh bowls as needed. Do not combine the leftover with the fresh food. Know the temperature at which the food has been held and the time it has been in the rapid growth zone (50–120F, 10–49C). If enterotoxin production by *S. aureus* is a potential hazard, perishable food which has been in the danger zone more than 3 hours must be discarded.

**3.** COOK RAW MEATS, POULTRY, AND FISH AND OTHER SEAFOOD THOROUGHLY. This kills microorganisms that cause foodborne illness, except for spores of *Clostridium* and *Bacillus*. Serve only pasteurized milk (not raw). A general rule is to heat to a minimum of 160F (71C). Eggs need a minimum heat treatment of 140F (40C) for 3.5 minutes, or reaching an internal temperature of 160F (71C) to kill any salmonellae present. For palatability, poultry is generally cooked to 180F (82C), but 165F (74C) is adequate for safety. Barbecuing often results in undercooked poultry and must be carefully checked. Microwave ranges usually have uneven heating so meats should be cooked to a slightly higher temperature, with longer time on medium power, and/or in a heat-resistant plastic bag. A stand-by period after cooking and before serving is also important for microwaved meats and poultry. Heat-resistant spores of clostridia and bacilli which are not killed during thorough cooking can be controlled by keeping hot foods hot (140F, 60C) and cold foods cold (40F, 4C).

**4.** COVER AND REFRIGERATE LEFTOVERS PROMPTLY. Large quantities of food or casseroles cool more rapidly when placed in a metal container in a sink of cold water than they do surrounded by air on a countertop. Hot foods may be refrigerated. When hot food was added to iceboxes, the ice rapidly melted and cooling properties were lost. The increase in energy use will be slight. For quick cooling in the refrigerator, hot foods should be less than 3 inches deep in flat pans or in 1/2-gallon or smaller jars.

**5.** WASH HANDS AND WORK SURFACES WELL. Routine, thorough hand washing with soap and lots of water reduces fecal and most hand-nasal contamination. Cutting boards, countertops, and other surfaces should be scrubbed with soap and water and rinsed with a bleach and water solution (1 Tb bleach to 1 gal water gives 200 ppm available chlorine) after they have come in contact with raw meat, fish, or poultry. Recent research has demonstrated that plastic boards need to be replaced when scarred and are at least as difficult to clean as wooden boards with ordinary scrubbing (Ak et al., 1994). However, since wooden boards are porous and cannot go into the dishwasher, they are not recommended for use with raw foods.

**6.** PREVENT CROSS-CONTAMINATION between raw and ready to serve foods. In the refrigerator protect foods from dripping meat and poultry packages. Do not contaminate with re-used shopping bags or egg cartons. Washing poultry before using is a good way to splatter salmonellae over the work area and does not remove an important number of bacteria. In barbecuing or preparing fondue, cooked food should not be put on the plate that had held the raw. Marinades should not be used for raw meat and poultry and then for basting the meat near the end of cooking. Boil marinades if they are also to be served with the meat.

**7.** REHEAT LEFTOVERS THOROUGHLY. Heating until "bubbling hot" inactivates any heat labile toxins or microorganisms that entered after the initial cooking. This does not give assurance that an unsafe food is rendered safe to consume as heat-resistant toxins, including the very common *S. aureus* enterotoxins, are not affected; but it may decrease the occurrence of foodborne illnesses from *C. perfringens*, *C. botulinum*, and *Listeria*, especially. Thus, it may be wise advice to give to someone at higher risk of foodborne illness.

**8.** CARE LABELS are on some perishable foods. Read and follow the label. The old adage "when in doubt, throw it out" still applies, but the informed householder does not encounter many situations in which there is doubt.

Ethnic groups in the United States may follow preparation practices for familiar foods which have a risk of foodborne illness. For example some nationalities like sausages made from raw pork which are served without cooking. Cultures in which raw fish is routinely consumed do not view it as a potential hazard and cross-contamination from raw fish to cooked foods in those kitchens is common. Asian families often have hunted mushrooms in their native country, only to find and use poisonous ones here. However, the safe practices outlined above are a good guide in evaluating the safety of foods from other cultures throughout the world.

## References


Ak, V.O., Cliver, D.O., and Kaspar, C.W. 1994. Decontamination of plastic and wooden cutting boards for kitchen use. J. Food Prot. 57:23.

Angelotti, R., Wilson, E., Foter, M.J., and Lewis, K.H. 1959. Time-temperature effects on salmonellae and staphylococci in foods. I. Behavior in broth cultures and refrigerated foods. Cincinnati: Robert A. Taft, Sanitary Engineering Center, U.S. DHEW.

Banwart, G.J. 1989. Basic food microbiology. 2d ed. New York: Van Nostrand Reinhold.

Bryan, F.L. 1988. Risks of practices, procedures and processes that lead to outbreaks of foodborne diseases. J. Food Prot. 51:663.

_____. 1988. Risks associated with vehicles of foodborne pathogens and toxins. J. Food Prot. 51:498.

_____. 1979. Foodborne disease. Atlanta: Center for Disease Control.

Butler, W.H. 1974. Aflatoxin. *In* I.F.H. Purchase, ed. Mycotoxins. New York: Elsevier.

Campbell, T.C. and Stoloff, L. 1979. Implication of mycotoxins for human health. J. Ag. Food Chem. 22:1006.

CDC. 1988a. Morbidity and mortality weekly report. Vol. 37. Centers for Disease Control, Atlanta.

Cliver, D.O., ed. 1990. Foodborne diseases. New York: Academic Press.

Cliver, D.O. 1994 Epidemiology of viral foodborne disease. J. Food Prot. 57:263.

Cody, M. Mcl. and Keith, M. 1991. Food safety for professionals: A reference and study guide. Chicago: The American Dietetic Association.

Derr, D.D. 1993. International regulatory status and harmonization of food irradiation. J. Food Prot. 56:882.

Doyle, M.P., ed. 1989. Comprehensive reviews of microbial food safety. Foodborne bacterial pathogens. New York: Marcel Dekker.

Food and Drug Administration. 1993. Food code 1993. Washington, DC: U.S. Dept. Health and Human Services, U.S. Public Health Service.

Food and Drug Administration. 1976. Food service sanitation manual. Washington, DC: U.S. Dept. Health Education and Welfare, Public Health Service.

Food Safety and Inspection Service. 1990. Preventing foodborne illness. Home and Garden Bulletin 247, U.S. Dept. Agr., Washington, DC.

Frazier, W.C. and Westhoff, D.C. 1988. Food microbiology. 4th ed. New York: McGraw-Hill.

Genigeorgis, C. 1975. Public health importance of *Clostridium perfringens*. J. Am. Vet. Asso. 167:821.

Glass, K.A. and Doyle, M.P. 1989. Fate of *Listeria monocytogenes* in processed meat products during refrigerated storage. Appl. Env. Microbiol. 55: 1565.

Hirschborn, N. and Greenough, W.B. 1991. Progress in oral rehydration therapy. Sci. Am. 264 (5):50.

Holmberg, S.D. and Blake, P.A. 1984. Staphylococcal food poisoning in the U.S. J. Am. Med. Asso. 251:487.

International Commission on Microbiological Specifications for Foods. 1978. Microorganisms in Foods I. 2d. ed. Buffalo: Author.

Jarvis, A.W., Lawrence, R.C. and Pritchard, G.G. 1973. Production of staphylococcal enterotoxins A, B, and C under conditions of controlled pH and aeration. Infect. Imm. 7:847.

Jay, J.M. 1992. Modern food microbiology. 4th ed. New York: Van Nostrand Reinhold.

Jones, J.M. 1992. Food safety. St. Paul: Eagan Press.

Kopanic, R.J., Jr., Sheldon, B.W., and Wright, C.G. 1994. Cockroaches as vectors of *Salmonella:* Laboratory and field trials. J. Food Prot. 57:125.

Kvenberg, J.E. 1988. Outbreaks of listeriosis/*Listeria*-contaminated foods. Microbiol. Sci. 5:355.

MacDonald, J.F., ed. 1990. Agricultural biotechnology: Food safety and nutritional quality for the consumer. National Agr. Biotechnol. Council Report 2. Ithaca: Cornell Univ.

Minor, T.E. and Marth, E.H. 1976. Staphylococci and their significance in foods. New York: Elsevier.
Notermans, S. and Heuvelman, C.J. 1983. Combined effect of water activity, pH and sub-optimal temperature on growth and enterotoxin production of *Staphylococcus aureus*. J. Food Sci. 48:1832.
Popovic, T., Olsvik, O., Blake, P.A., Wachsmuth, K. 1993. Cholera in the Americas: Foodborne aspects. J. Food Prot. 56:811.
Roberts, T., 1989. Human illness costs of foodborne bacteria. Am. J. Agr. Econ. 71:468.
Rose, J.B. and Sobsey, M.D. 1993. Quantitative risk assessment for viral contamination of shellfish and coastal waters. J. Food Prot. 56:1043.
Rutgers University Food Science Department. 1971. Food Stability Survey I & II. Washington, DC: U.S. Government Printing Office.
Schultz, H.W. 1981. Food law handbook. Westport, CT: AVI Publ.
Smith, J.L. 1993. *Cryptosporidium* and *Giardia* as agents of foodborne disease. J. Food Prot. 56:451.
Swaminathan, B., Harmon, M.C., and Mehlman, I.J. 1982. A review: *Yersinia enterocolitica*. J. Appl. Bact. 52:151.
Torrey, G.S. and Marth, E.H. 1977a. Isolation and toxicity of molds from foods stored in homes. J. Food Prot. 40:187.
Torrey, G.S. and Marth, E.H. 1977b. Temperatures in home refrigerators and mold growth at refrigeration temperatures. J. Food Prot. 40:393.
Tyler, V.E. 1993. Honest herbal. 3rd. ed. Binghamton, New York: Haworth Press.
U.S.D.A. Food Safety and Inspection Service. A margin of safety: the HACCP approach to food safety education. 311-374/40697. 1989. Washington: U.S. Govt. Printing Office.
U.S. General Accounting Office. 1990. Food safety and quality: Who does what in the federal government. Washington, DC: GAO/RCED-91-19B.
VanDeRiet, S.J. and Woodburn, M.J. 1987. Food discard practices of householders. J. Am. Dietetic Asso. 87:322.
Vanderzant, C. and Splittstoesser, D. ed. 1992. Compendium of methods for the microbiological examination of foods. 3d ed. Washington: American Public Health Asso.
Woodburn, M.J. and VanDeRiet, S.J. 1985. Safe food: Care labeling for perishable foods. Home Ec. Res. J. 14:3.
World Health Organization. 1988. Food irradiation: A technique for preserving and improving the safety of food. Albany: WHO Pub. Center.
World Health Organization. 1985. Treatment and prevention of acute diarrhea. Albany: WHO Pub. Center.

VIDEOTAPES

The Danger Zone—A Food Safety Program for Teens. U.S. Department of Agriculture. Food Safety and Inspection Service. (To order: Modern Talking Picture Service, 5000 Park Street North, St. Petersburg, FL 33709)
Eating Defensively: Food Safety Advice for Persons with AIDS. FDA, Office of Public Affairs, 8255-H. Patuxent Range Rd., Jessup, MD 20794.
Food Safety is no Mystery. U.S. Dept. Agriculture. Food Safety and Inspection Service. (To order: address above)

SLIDE SET

Oregon/North Carolina Home Economics Extension. Dr. Sal Monella's Safe Food Prescriptions. OSU Extension Service, Corvallis, OR.

## CONSUMER QUESTIONS

**Q.** *When should food be thrown away?*

**A.** Accurately determining which changes in a food are only quality changes and which changes indicate possible microbial spoilage by pathogenic bacteria is difficult for many consumers. A few simple criteria help identify common characteristics of food spoilage due to microbial growth, PATHOGENS MAY BE PRESENT, FOOD SHOULD BE DISCARDED (garbage disposal, animal-proof garbage can).

1. Foul odor is often due to microbial growth. (Exception is rancid oil and fat.) Putrefaction may have occurred in protein foods with accompanying sewagelike odors.
2. Surface slime is often, but not always, apparent when total counts of bacteria are high.
3. Foods that have molded during storage, especially at room temperature, may have mold toxins. Exceptions are old-fashioned cured hams and cheeses which are mold-ripened.
4. Canned foods in which gas has been produced causing a loss of seal or bulging of the lid. In acid foods, however, the gas may result from a chemical reaction.
5. Cloudy liquid and softer-than-usual texture in a canned food after storage may indicate bacterial growth.
6. For many perishable foods, the time-temperature history may be the only guide. If such foods have been in the danger zone of 50F (10C)–120F (50C) for 2 to 3 hours, then they may be unsafe even though the food

appears unchanged. For a wide margin of safety in this temperature recommendation, 40F (4C)–140F (60C) is commonly given.

QUALITY CHANGES: IF NONE OF THE CHARACTERISTICS OF MICROBIALLY SPOILED FOOD ARE PRESENT AND THE HOUSEHOLD FINDS THE FOOD STILL ACCEPTABLE, IT CAN BE CONSUMED WITHOUT THE THREAT OF FOODBORNE ILLNESS. It is assumed for all of the following examples that the refrigerator temperature is at or below 40F, and that leftovers have been handled in an otherwise safe manner (refrigerated promptly); temperature abuse has not occurred.

1. Food separated into layers. Emulsions such as sour cream and salad dressings break over time. The food can simply be stirred or shaken, and served. Cottage cheese or yogurt that acquires a clear liquid layer after several days refrigerated storage, but has no off-odor is safe. Soups and gravies thickened with flour or cornstarch often separate during several days of refrigerated storage.

2. Uncooked fruits and vegetables shriveled. This is only water loss and not a microbial change. The vitamin C content may be reduced as much as 90% with noticeable shrivel, but the produce is safe to consume. The center rings of cut onions may appear to "grow" in the refrigerator due to excess water in those portions which enlarges them and they expand outward—only a physical change and safe to consume.

3. Improperly covered items dry out in the refrigerator, or nonperishables such as bread dry out on the counter. Loss of water is responsible, not microorganisms.

4. Lettuce acquires brown, rusty-looking edges. They may even be slimy. This may be the result of microorganisms, which are plant pathogens, not human pathogens. Usually browning is due to death of cells and enzyme action. The brown areas should be removed for best appearance, but all can be safely served.

5. Foods in the refrigerator that freeze. Adjust refrigerator setting to correct problem; use thermometer to monitor the adjustment so refrigerator stays under 40F. These foods are safe to consume. Water expands as it freezes, and in doing so, it has probably ruptured the food structure. After thawing, the quality may be poor and there may be excessive liquid with the food. As long as thawing of perishable foods occurs at temperatures under 40F so that pathogens do not multiply, the food is as safe to eat after thawing as it was before freezing.

**Q.** *What should be done with packages that have been stored at the proper temperature, but for longer periods than recommended by the manufacturer?*
**A.** "Best if used by" dates are a storage guide for consumers. They contain a large margin of time in which quality remains acceptable and the food remains safe to eat. Examine the food for quality acceptability, but do not assume microbial spoilage has occurred if there are no signs of it. "Sell by" dates are pull dates because they are a guideline for the retailer to pull the items off of the shelves for best quality.

**Q.** *When is food left unrefrigerated safe to eat?*
**A.** First, determine if the food is highly perishable or not. Perishable food held in the danger zone (50 to 120F, 10 to 49C) for over 3 hours (2 hours at very warm temperatures) should be discarded because pathogens may have multiplied.

Examples of foods that will have only quality spoilage are those with sugar content so high that microorganism growth is inhibited such as jams, jellies, frostings (including seven minute), syrups, and cakes. This does not include all sweet foods. Some desserts still have a concentration of milk, starch, or eggs high enough for pathogen growth such as soft meringues, cornstarch and tapioca puddings, custard pies, and custard-filled pastries.

Examples of foods that are less perishable due to their low moisture include bread, jerky, dried fruits and vegetables, dried pasta, and hard meringues.

Examples of foods that will have only quality spoilage due to their high acid content include salad dressings, mayonnaise, cream cheese frosting, vegetable and fruit pickles, fruit juices, and cooked fruits in juice or syrup (such as opened jars of canned fruits). Fruit pies are both low in pH and contain sugar. These foods will spoil in time, but the spoilage is by non-pathogens or by visible mold. Combination foods that contain both potentially hazardous and less perishable items should be handled as perishable. Examples include mayonnaise and macaroni salad, mayonnaise and tuna mixtures, and meat loaf sandwiches.

Some foods are nonperishable when purchased, but after cooking and/or rehydration become perishable such as rice, macaroni, and dried soup mixes.

Raw foods are usually not a good environment for pathogens to grow. Fresh fruits and vegetables are safe to consume unless grown in contaminated soil or with unsafe water. Raw meats, fish, and poultry are a special case. It is advisable to thaw raw meats, fish, and poultry at temperatures below 40F (4C), and to store them in an ice chest if the trip home from the market is long or the weather particularly hot because growth of pathogens is possible. This increases the risk of cross-contamination during its preparation for cooking. However, when these recommendations are not followed, the food is still safe to consume after cooking well because most pathogens, such as *S. aureus,* are poor competitors with other microorganisms on raw foods and also because thorough cooking kills the bacteria that cause foodborne illness. A notable exception is scombroid poisoning caused by spoilage of raw fish.

# 3 Drying

Naturally dry foods such as nuts, dried legumes, and grains were staple foods throughout the year for many early cultures. Decreasing the moisture content of fresh foods to make them less perishable was a simple extension of the natural process. It easily provided a greater variety of fruits and vegetables outside of growing seasons and meat for journeys. For much of human history, sun drying was used to decrease the moisture content. Ancient Egyptian hieroglyphics show outside production of salted, dried fish, and the Incas sun-dried food in summer for use during harsh winters more than 3,000 years ago. Tribes of the Nile Valley dried fruits and meat by burying them just beneath the surface of the hot desert sand and then stored them in covered earthenware, and the Bible mentions dried grapes and figs. The first use of fire to speed drying is unrecorded, but was prevalent in early meat and fish drying. Perhaps the smoke flavor helped mask some undesirable off-odors and -flavors that occurred with room temperature meat storage (Labuza, 1976). Drying was still an important preservation method for Europeans settling the North American continent; the original Massachusetts colonists produced a dried fish product that was the first colonial export to England.

When armies needed stable food for war, artificial heating techniques were used instead of relying on the more unpredictable sun in nondesert climates. The World War II time period saw the greatest strides in food-drying knowledge, but the advantages of low-cost, low-weight, nonperishable foods were seen by military minds much earlier. Marco Polo reported in his journal that the Mongols prepared a sun-dried milk powder they used on military expeditions. In England in 1780, the first patent on drying was issued to J. Grafer (Labuza, 1976). He scalded vegetables in salted water and dried them in a heated room for 20 to 30 hours. These vegetables had a very limited shelf life. In 1795 the French improved upon this process by thinly slicing vegetables and drying them in forced air. The dry vegetables were then compressed and sealed in tinfoil and cans. Carrots and potatoes dried by this process were used in the Crimean War (1854–1856) to treat scurvy; unfortunately oxidation of the vitamin C rendered them almost useless for this purpose. In another effort to improve nutrient intake using dried vegetables, unblanched, sun-dried onions from California were shipped to Alaska during the Klondike gold rush to combat scurvy. Dried vegetables were used in limited amounts due to their poor quality during the U.S. Civil War and by the British in the Boer War (1899–1902). Because of few production plants and unacceptable products, the U.S. troops in World War I coped with the same poor quality dried foods that soldiers a century before had been served, but this war stimulated research interest in drying. By World War II dried vegetables were a major focus. In 1946 over 375 companies which dried vegetables were in operation and the quality was greatly improved due to re-engineered dryers and sulfite research. The U.S. army now uses freeze-dried combination dinners that come in olive-green, camouflage, or desert-sand colored pouches. Hot water is added directly into the pouch and after a few minutes for absorption, a warm meal is ready.

Today, drying is used extensively in cultures where electrical energy is expensive or nonexistent in order to eliminate long-term, low-temperature storage. For the modern North American home food preserver, drying is a way to produce special foods for sack lunches, backpacking, snacks, and gifts.

In drying, water is transferred from the food to air. This air is then allowed to escape to the atmosphere (usually directed out of the dryer by a fan) and replaced by air with lower humidity. In most home dryers this is a continual process. The actual mechanism of drying involves energy, usually in the form of heat, being supplied to the water in the food. The vapor pressure of this water is raised until it evaporates from the surface of the food. As this surface water is removed, it is replaced with water from deeper within the food by diffusion, convection, capillary flow, shrinkage, and other means (Troller and Christian, 1978).

The solid cellulose skeleton of plant material is composed of many cells joined together forming a network of small spaces. During drying, moisture first leaves the larger spaces and then starts to evaporate from some of the small ones. Next, some of the cellular moisture held osmotically by the semipermeable cell walls is removed. The most tightly held water is bonded chemically and cannot be removed without changing the product greatly.

As the food dries, different types of moisture movement are predominant at different times. Initially, the pores—small spaces—are full and as their water is lost at the surface, more water moves in to fill them from other parts of the food through diffusion (Fig. 3.1a), but gradually air pockets appear and the water is withdrawn from the wide center portions of the pores (Fig. 3.1b). In this situation, moisture can migrate either by creeping along the walls (capillary flow) or by successive evaporation and condensation between liquid bridges (liquid-assisted vapor transfer) (Fig. 3.1c). With further drying, these liquid bridges evaporate entirely and moisture moves by unhindered diffusion of vapor (Fig. 3.1d). The final stage is one of desorption-adsorption where any moisture that vaporizes is condensed and the food is in equilibrium with the environment (Keey, 1972). The environment is usually provided by

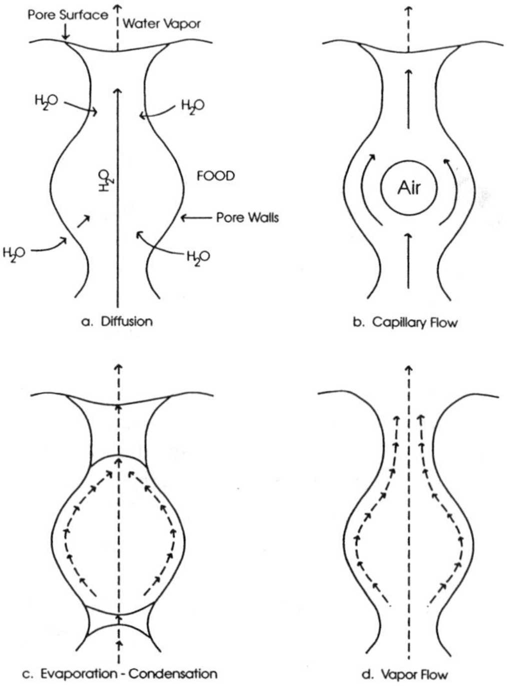


**3.1.** Types of water movement in a food during the drying process. Solid lines represent movement of liquid water. Dotted lines show vapor molecules. Adapted from Keey, 1972.

a sealed plastic sack in the storage of home dried food.

In foods, water is present as unbound water and bound water. Unbound water is within the pores and spaces between plant cells. This unbound water has the same characteristics as pure water. It exerts the same vapor pressure (tendency to become a gas) and has the same latent heat of vaporization (amount of energy required to change state from liquid to vapor). Unbound water is easily removed during household food drying.

Bound water is held on the surfaces of solid compounds, such as the hemicellulose in the cell wall; by molecular interactions between water and the solid; or between several layers of water molecules and the solid molecule forming a multilayer of water molecules. Bound water exerts a vapor pressure less than that of pure water, it does not evaporate easily, and the latent heat of vaporization is greater than that of pure water at a given temperature. It requires more heat to release this water, therefore, bound water is not usually removed during drying.

For a given food, the total amount of moisture that can be lost will vary with the humidity in the air. The water vapor concentration in the air and the temperature are important. Household drying should be done on warm days when the humidity is low.

As water leaves and tissue shrinks, the flavors (acids, salts) are concentrated. As water migrates out during drying, dissolved salts are carried along to the surface. Here, water evaporates into the air but the salts do not. The salts concentrate and may even precipitate at the surface.

In a food system, the water removal may be restrained by the drying process itself. Food tissue often shrinks as it loses moisture and the structure may change, blocking the exit of water. Intensive drying can also cause chemical changes. Case hardening is a term used to describe a tough outer surface brought about by rapid moisture loss there. This hard surface is more impermeable to water and a moist interior remains that is susceptible to microbial spoilage (Fig. 3.2). For this reason home food drying instructions start with the dryer at a low

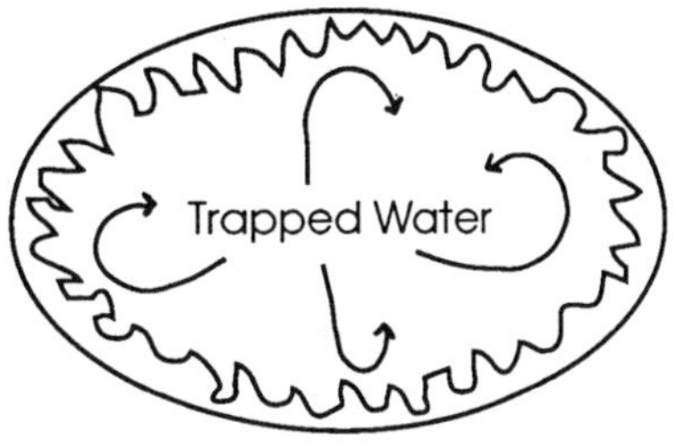


**3.2.** Case hardening traps interior water.

or medium setting. Even with slow drying, the surface of home dried food does tend to be harder than the interior at the completion of drying. To help alleviate this, the last step of most recipes recommends placing the dry food in a sealed container for several days so an equilibrium with the moister inside is reached. The product feels less dry to the touch after this process, called conditioning, and it is an overall softer dehydrated food. It also is the time to check to be certain that the food is dry enough to store.

## Water Activity

Fresh foods become less perishable when their water content is drastically decreased but this reduction in perishability is associated with different water contents for different foods. Potato starch containing 24% moisture on a dry basis is stable while crystalline white sugar may spoil at only 4% moisture (Van den Berg and Bruin, 1981). The nature or state of the constituent water—bound or unbound—not the actual water content, determines perishability. This is called the availability of water for deterioration reactions in foods. In 1953, the term water activity, denoted $a_w$, was first used to describe this phenomena. Pure water has an $a_w$ of 1.00. As water content decreases, so does the $a_w$ value. As important as $a_w$ is for both microbial and quality aspects of foods, its reliable measurement is limited (Prior, 1979); the dew point method, electronic hygrometers, hair hygrometers, psychrometers, vapor pressure manometers, and other instruments may be used. As expected, the use of different instruments and methods gives varying $a_w$ readings.

Many foods are multicomponent multiphase systems, partially dissolved in a gel condition. For example, fruits and vegetables have fibrous parts that are solid, parts that are water with dissolved sugars and salts in various concentrations, and oil or fat droplets may also be present. A single activity of water does not exist throughout such a system, but for simplicity, an average $a_w$ is given for the total food.

The optimum $a_w$ for dried foods differs for the specific food but is generally between 0.20 and 0.30 for protein foods, lower for vegetables, and higher for fruits (up to 0.60). An $a_w$ below 0.60 is adequate to prevent microbial growth but chemical and enzymatic reactions may continue at an unacceptable rate.

A decrease in $a_w$ slows chemical reactions and microbial growth until most reactions are inhibited except for lipid oxidation which is favored by further decreases (Fig. 3.3). Molds and yeasts are able to multiply at lower $a_w$'s than bacteria. When home dried foods show microbial growth, it is usually molds. At $a_w$'s of about 0.7–0.85, reaction rates are high due to concentrating the reactants since as compounds come closer together, they are more likely to be involved in reactions. With further reductions in $a_w$, the reaction rate slows, probably because too little water is available to participate directly in the reactions and the reactants are less able to move close to each other as the food system dries.

## Microbial Growth

Though dried foods are often tasty snacks and their lighter weight and reduced mass make them appropriate for recreational backpacking and air transport of foods in emergency situations, the primary reason most foods are dried is to reduce spoilage due to bacteria, yeasts, and molds. As the $a_w$ is decreased from $a_w$ 1.0 to $a_w$ 0.99 the rate of bacterial growth usually increases (Fig. 3.3). The maximum growth rate for many bacteria is at an $a_w$ of 0.990–0.995. As the $a_w$ is reduced below 0.99–0.95 (the specific level differs with the genus), bacteria experience an increased period of adjustment or adaptation (lag phase), the overall growth rate is decreased, and the total cell count is reduced. This inhibition of bacteria may occur with little change in $a_w$. The growth of *Staphylococcus aureus* at $a_w$ 0.90 is 10% of its maximum growth rate (Sperber, 1983). When the $a_w$ is decreased below the minimum required for growth, the bacterial cells either remain dormant or die.

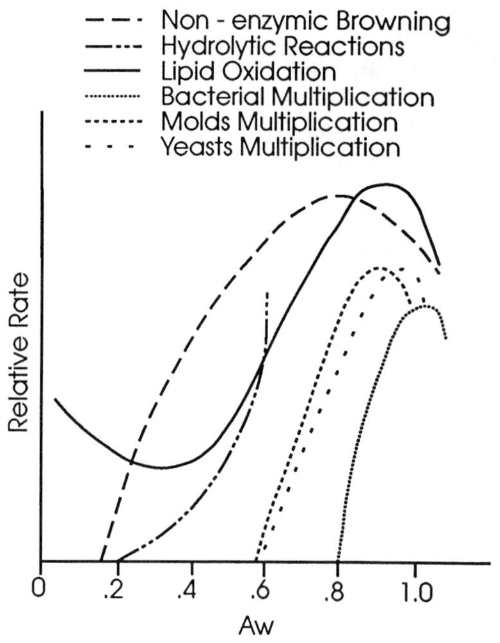


**3.3.** Relationship between available water ($a_w$) and changes in food qualities at room temperature.

Most microorganisms cannot grow at an $a_w$ below 0.85. Although halophilic bacteria will grow at 0.75, they pose no known health problems and are not common in foods. Molds are the most tolerant of a low $a_w$. The minimum $a_w$ for toxin production by molds is approximately 0.78 (Beuchat, 1981). Mold growth is prevented below 0.70. A few species of yeasts are able to grow in the 0.61–0.70 range (Beuchat, 1981). The pH, oxidation-reduction potential, nutrients, type of solutes present, oxygen and carbon dioxide levels of the food, temperature, and the addition of chemical preservatives and antimicrobial substances will influence the minimum $a_w$ for growth and survival. When these other environmental conditions are not optimum, the minimum $a_w$ is usually increased (Troller and Christian, 1978). For instance, though fresh fruits have an $a_w$ above 0.95 (Table 3.1), they usually have a pH low enough to inhibit the growth of many species.

The growth of bacteria is readily controlled by drying. Excluding *S. aureus*, the minimum $a_w$ for growth and toxin production for pathogenic bacteria is 0.91. For growth of *Staphylococcus aureus* it is 0.86, but the minimum $a_w$ enabling toxin production is slightly higher. Generally, the minimum $a_w$ for growth of a particular bacteria is the same as the minimum $a_w$ for spore formation (Jakobsen and Murrell, 1977); however spores, once produced, can usually germinate at an $a_w$ below that required for growth (Chyr et al., 1977 and Pitt and Christian, 1968). At $a_w$ values above 0.85, molds and yeasts are not usually the predominant early spoilage microorganisms, because they do not compete well with bacteria and are quickly outgrown by them.

Drying cannot be counted on to render an unsafe food safe. Even vegetative cells of microorganisms often survive drying well. *Salmonella, Enterococcus faecalis*, and *Escherichia coli* die more rapidly at $a_w$ levels of 0.71–0.53 than they do at $a_w$'s of 0.34–0.11 because these bacteria are not in an active state at the lower

**Table 3.1. $a_w$ parameters for foods and microorganisms**

| $a_w$ Range | Inhibited by Lower $a_w$'s | Foods in Range |
|---|---|---|
| 1.00–0.95 | *Bacillus, Clostridium perfringens, Escherichia, Klebsiella, Pseudomonas, Proteus,* some yeasts | fresh vegetables, fruit, meat, fish, milk, bread, foods containing up to approximately 40% sugar or 7% salt |
| 0.95–0.91 | *Clostridium botulinum, Salmonella, Vibrio parahaemolyticus,* some molds and yeasts | cheddar and Swiss cheese, cured meats, foods containing 55% sugar or 12% salt |
| 0.91–0.87 | many yeasts (Candida) | fermented sausage (salami), dry cheeses, margarine, sponge cakes, foods containing 65% sugar (saturated) or 15% salt |
| 0.87–0.80 | *Staphylococcus aureus,* most molds, mycotoxigenic penicillia | fruit syrups, uncooked rice, flour, fruit cake, country ham, fondant, high-ratio cakes |
| 0.80–0.75 | most halophilic bacteria, mycotoxigenic aspergilli | jam, marmalade, glace fruits, marzipan |
| 0.75–0.65 | xerophilic molds | rolled oats, fudge, jelly, molasses, some dried fruits, nuts |
| 0.65–0.60 | osmophilic yeasts, few molds | honey, caramels, dried fruits, fruits containing 15–20% moisture |
| 0.60–0.20 | no microbial proliferation | pasta, spices, whole egg powder, cookies, crackers, whole milk powder, dried vegetables |

Adapted from Beuchat, 1981 and from Troller and Christian, 1978.
Used by permission from Fennema, O.R. (1985), p. 57.

$a_w$ range. Such bacteria can survive for long periods in dried foods (Sperber, 1983). The exact survival mechanism some bacteria possess is a topic of much current research but the most probable theory is that the bacterial cell changes its internal $a_w$. During drying when the $a_w$ of the surrounding food is decreased from 0.995–0.950, water also leaves the bacterial cell. The cell loses its shape (turgor) and is subject to breaking apart (plasmolysis) but some will survive. During drying, *S. aureus* in a food loses approximately 50% of its intracellular water and *Salmonella* loses approximately 44% (Sperber, 1983). Bacterial cells do not grow when they are so severely stressed; but some types survive in a dormant state. In order to grow again, the surviving bacteria must regain their turgor by taking up water to replace what was lost. Rehydration is a complicated process involving a series of enzymes and sometimes changing potassium levels within the cell, but water does re-enter the cell and growth slowly resumes.

## Chemical Reactions

Prunes, dark raisins, and some varieties of figs are acceptable to consumers as browned dried foods. Other brown dehydrated fruits and vegetables are usually rejected due to color and flavor changes. Two main categories of browning reactions may be involved: enzyme catalyzed (enzymatic) and nonenzymatic.

Enzymatic reactions are usually the first to affect fruits. Enzymatic browning can occur when the natural fruit structure separating the enzymes from the substrates is broken. Dark colored bruises are the result of these structures being broken by mechanical stress such as dropping or transportation damage. A knife cut also brings enzymes and substrates into contact. Use of a dull knife will result in more tissue damage and ultimately more browning than a sharp knife. Iron knives may also react with the food to produce dark colors.

Oxygen plays a part in these enzymatic reactions. Decreasing the exposure to oxygen by immersing in water reduces the discoloration. Enzymatic browning involves several oxidation steps in which colorless compounds naturally present in fruit become brown colored compounds. One of the intermediary products (o-quinones) in this series of reactions can react with ascorbic acid to produce dark colored substances during drying. For this reason, ascorbic acid treated foods may be darker than untreated dehydrated foods.

The nonenzymatic group of browning reactions includes Maillard reactions and browning due to oxidation without enzyme catalysis. Maillard is the main nonenzymatic browning reaction in dehydration. This chemical reaction with many steps is between amine groups of a protein and the carbonyl groups of a reducing sugar or other organic compound. The result is yellow or brown products and off-flavors. Only small amounts of reactants are needed. The proteins which make up cell membranes and other cell components in plants can result in an unacceptably discolored food. The Maillard reaction is accelerated by heat and occurs optimally at moisture levels between 1% and 30%. Low moisture levels accelerate the Maillard reaction since, as water leaves, the protein and sugar groups are closer together to react more easily. This is why browning is such a problem in dried foods. Lower pH, lowering the temperature, or removing the sugar or protein substrates can reduce Maillard browning.

If a food has undergone Maillard browning, some loss of the amino acids from a nutritional standpoint has occurred. This loss of nutritive value can occur before the colored pigments appear. Basic amino acids such as L-arginine and L-histidine are most susceptible, but the loss of essential amino acids also occurs. Loss of lysine in grains is of great importance if dietary intake is limited. Maillard browning also has nutritional significance in commercially dried milk and egg products. In addition, browned foods will not absorb water easily; dried milk that has undergone Maillard browning during storage will not rehydrate to a smooth beverage.

### Pretreatments

The reactions that cause quality to deteriorate function only in specific ranges of pH, temperature, and oxygen availability; all factors which can be altered by householders (Table 3.2). Additionally, substances that bond chemically to compounds in browning reactions and hold them in a colorless state can be included in dips or smoke.

Pretreatment is necessary for most top quality dried fruits and vegetables. Sulfuring produces the highest quality, followed by sulfiting. Syrup blanching adds sugar to the dried fruit product to inhibit enzymes but the finished product is very good. Ascorbic acid, lemon juice, or salt water dips may be used for holding the cut fruit before drying but not as the sole pretreatment for top quality. Pineapple juice has been effective (Lozano-de-Gonzalez et al., 1993).

Sulfuring—exposing to sulfur fumes—most effectively maintains the quality and nutrients of foods during drying and storage. The sulfuring process involves stacking drying trays at least 10 inches above the ground, with air space in between, all covered by a vented container such as a wooden or cardboard box. Sulfur (flowers of sulfur) is burned in a small container at ground level and the sulfur dioxide fumes rise throughout the trays. Sulfuring treatments can penetrate the fruit 1–2 inches which is very effective. The amount of sulfur to ignite varies with the airtightness of the covering box. Generally, 1 tablespoon per pound of fruit is used with nonairtight cardboard boxes and 1.5 teaspoons. per pound of fruit with wooden boxes especially made for sulfuring. Sulfuring times vary from 45 minutes to 2 hours, depending upon the type and quality of fruit and the thickness of slices. A reputable recipe should be consulted for specific times. Sulfuring must be done outside away from plants, children, and pets because the fumes are toxic. Sulfured foods should not be dried indoors due to objectionable fumes (respiratory irritants) released during dehydration. Those doing a lot of drying may find sulfuring worth the effort to set up, but the majority of householders simply use sulfite dips and settle for a lesser quality finished product.

Solution dip treatments are most frequently used in household drying to inhibit darkening because they are inexpensive, can be done indoors, and they avoid the danger of burns, fire, and chemical irritation to the respiratory system that is a concern with sulfur dioxide. The type of sulfite available to the home food dryer varies with locality. One part sodium bisulfite is equal to 4 parts metabisulfite. Food grade quality can often be purchased from canning supply sections of retail stores, pharmacies, and wine making supply stores. The concentration of sulfite solution needed will vary with the size of the fruit pieces (sulfite must penetrate 1–2 mm to be effective), the ripeness and variety, and the length of storage desired. Larger pieces, immature fruit, and fruit to be stored longer than 6 months or above

**Table 3.2. Pretreatments decrease browning in different ways**

| | Method of Inhibition | | | | |
|---|---|---|---|---|---|
| Pretreatment | Chemical Bonding | Heat[a] | Acid | Oxygen Depletion | Ionic Strength |
| Sulfuring | x | | | | |
| Sulfiting | x | | | | x |
| Lemon juice | | | x | x | |
| Salt water | | | | x | x |
| Vitamin C[b] | | | | x | |
| Steam blanch | | x | | | |
| Sugar blanch | | x | | | x |
| Water blanch | | x | | | |
| Water dip | | | | x | |

[a]Heat for enzyme inactivation.

[b]The new compound formed from the reaction of ascorbic acid can also become dark colored during drying and storage.

60F (15C) require the upper range of penetration and hence stronger sulfite concentrations should be used. A strength of 1–2 tablespoons of sodium bisulfite per gallon of water is generally used for pretreatment dips of 5 minutes for thin slices and 15 minutes for larger pieces. During the dip, water is absorbed. This lengthens drying times by 15–20%. After dipping, the fruit may be rinsed briefly under cold tap water to improve flavor; however, this does lower the sulfite content slightly. Some people are allergic to sulfites and suffer difficulty in breathing when they consume them. Labeling packages of home prepared, sulfited foods used as gifts is recommended.

Enzymes which catalyze browning reactions during drying can be inactivated by heat. Steam blanching is the preferred pretreatment for vegetables, since in water blanching a greater part of the water soluble vitamins are lost. However with steam blanching large amounts of vitamin A and C still may be destroyed (Maggard, 1982). Increased blanching time and temperature results in increased chlorophyll retention and increased color quality in dried green beans, perhaps due to inactivation of a heat-labile factor involved in the conversion of chlorophyll to pheophytin which is an olive green compound (Foda et al., 1969). Adding sodium bisulfite to the water used for steam blanching is recommended to improve the keeping quality of corn, green beans, potatoes, and mushrooms by also decreasing the Maillard reaction.

### Oxidative Rancidity

Oxidative rancidity is a chemical reaction involving an unsaturated, especially polyunsaturated, fat. This reaction is favored by the presence of light, heat, some trace metals, and oxygen. The products are responsible for off-odors and off-flavors in stored foods. This occurs even in vegetables which contain only trace amounts of lipids.

Oxidative rancidity of fats can be a problem in dried foods due to the length of storage and their low moisture content (Fig 3.3). Flesh from wild game animals typically contains more polyunsaturated fats. This is a particular problem with low-altitude game animals that have been feeding in abundant grain fields, as they will accumulate adipose tissue and lay down more unsaturated fats than domestic beef.

In preparing meats for drying, trimming the fat is recommended to decrease rancid odors and flavors. Long-term storage of home dried meats is recommended at 0F (−18C) to retard this process. This is important for beef but especially for venison jerky.

## Nutrient Loss

Drying is associated with some decrease in vitamin content for both commercial and home dehydrating processes. In a study of fruits and vegetables dried in a food dryer designed for home use and prepared following recommended home drying methods, the vitamin retention was found to be particularly low for the heat-labile vitamin C (Holmes et al., 1979). Especially high vitamin C loss occurs in foods with long drying times such as green beans (6 hr) and tomato puree (4 hr).

## Color Changes in Dried Food

Drying changes the physical and sometimes the chemical properties of a food. This in turn changes the food's ability to reflect, transmit, absorb, and scatter light; thus changing the color we perceive. Anthocyanin pigments can be in a blue (blueberries) or a red (raspberries) form. They may be changed during drying and sulfur treatments tend to bleach them; however, the appearance of these dried foods is usually acceptable. Carotenoid structure may be altered during drying, but orange foods generally hold their color well. Lycopene pigment (red in tomatoes) is usually found to stay acceptably red. Chlorophyll a and b molecules may lose their magnesium and be converted to olive green pheophytin. This conversion is accelerated by the presence of acid which can be released from the plant tissues themselves as their internal structure breaks during drying. Darkening of fruit and vegetable tissue tends to be a more significant factor in appearance than other changes in plant pigments.

## Hygroscopicity

Hygroscopic foods draw moisture from the air into the food. The compounds, including carbohydrates, that are responsible for this water uptake are called hydrophilic. They have many hydroxyl (-OH) groups to bond with water. It is important that dried foods, because of their hydrophilic nature, be stored well sealed to avoid the uptake of water.

The structure of the carbohydrate molecule affects the rate of water binding and the amount of water bound. Sugars are more hydrophilic than starch, and fructose is the most hydrophilic of sugars. Fruit leathers sweetened with honey, which contains fructose, instead of table sugar (sucrose) will become sticky upon exposure to air. Sometimes the cause of the problem is not so easily detected. A small fruit leather production plant in Washington State specialized in selling its apple leather to tourists. Their third year of operation they encountered a major problem with spoilage. They dried the leather to the usual endpoint, then cut, and rolled it the same as previous years. This year, however, the product was weepy. They consulted the local county Home Economics Extension agent who questioned them about any changes in the sweetening agents added. The company adamantly declared that their natural leathers contained "no sugar." Further investigation revealed they had started adding a new apple cultivar (variety) into their pulp mixture. This new cultivar had a slightly higher fructose content and was responsible for the water uptake problem.

## Home Food-Drying Methods

Home dried foods were very common in the United States until the 1940s when household canning supplies became affordable. Many senior citizens remember dried apples being the only fruit available in wintertime during the Great Depression. Dried foods have gone from being off-season staples 100 years ago to specialty treats.

Drying foods at home involves many variables such as the amount of moisture in the air, quality and variety of produce, preparation, and dryer efficiency in the process itself. In addition, packaging and storage conditions affect the final product.

Clean, ripe unblemished produce should be selected. Peak ripeness is more important in drying than in canning or freezing (Maggard, 1982), but slightly overripe fruit can be used satisfactorily in leathers. Choice of cultivar is also responsible for large quality variations. Rarely do recipes state the cultivar recommended so trial and error is used.

Most fruits and vegetables must be peeled as their skins are more resistant to water movement and peels result in gritty leathers. All but berries and very small fruits need to be cut into pieces or divided into sections (apple slices instead of quarters) small enough to dry in a reasonable time without case hardening. Slicing to increase surface area to volume ratios speeds drying. Most instructions suggest a starting temperature of 140F (60C) for 1–3 hours and then heating at 90F–130F (32–54C) until dry. Many foods are first blanched to inactivate enzymes and treated with antioxidants to decrease discoloration. These pretreatments are optional from a food safety standpoint, but the quality in dried preservation deteriorates so badly without them that the food is usually deemed inedible.

Home drying is done in kitchen ovens, by the sun (solar drying), in warm dry places such as herbs hung in attics, and with home food dehydrators. Forced air food dryers usually have a heat source at the bottom and perforated trays (for air circulation) layered above. Lining the trays with foil or plastic for easier cleanup is not a good practice since it restricts air flow; the exception is fruit leathers that must be on a solid surface. Trays for some dryers are made to be dishwasher safe. Galvanized screens should not be used on drying trays as there is a reaction with acid that darkens food and zinc dissolved in foods causes illness. Fiberglass, vinyl, and copper screening are also not recommended because their components may be picked up by the food. Wood slats around the edges of screens are an excellent nonreacting material; covering them with cheesecloth before drying may aid in cleaning.

Plans for homebuilt wooden dryers are available from some Cooperative Extension Services. This is a large wooden forced air unit which uses light bulbs as its heat source. The construction costs depend upon the amount of materials that the builder can salvage from used appliances and whether the screens are purchased premade. Smaller dehydrators commercially manufactured for home use are widely available in retail stores.

Several kinds of fruits or vegetables can be dried at the same time as long as strong-smelling vegetables are not included. It is important that only single layers of food are on each tray and there is space for air circulation between them. The time required for drying and also the quality of the product will be affected by the velocity of air in the dryer, the relative humidity of the air, the temperature during drying, any pretreatment given the food, and the type of food itself. Food on lower shelves near the heat source may dry most rapidly. Each food should be removed as it reaches its endpoint. Some dryers are designed so that checking and removing partial lots is a simple process. As soon as the foods are cool, they should be packaged to avoid uptake of moisture from the air.

The desired endpoint for dryness in home dried food is an $a_w$ below 0.60 to inhibit microbial spoilage. Drying is rapid in the home dryer to an $a_w$ of approximately 0.70 (35% moisture) and then progresses slowly to 0.60 (Sullivan and Weber, 1982). Most vegetables are brittle when they are dried to 0.60 $a_w$. Fruits dried to a proper endpoint feel pliable and leatherlike and contain no pockets of moisture after they have cooled to 72F (22C) and equilibrated.

Since each fruit and vegetable will appear different when dry, inexperienced food driers may wish to calculate the amount of water lost to determine drying endpoints. When 80% of the water has been removed, the product is definitely dry enough for safe storage. A drawback to this calculation method is that some foods may be slightly overdried by the time 80% of the water is lost; for example top quality dried peaches average 25% water (ASHRA, 1986).

For calculations, use Table 7.2 in the freezing chapter to determine beginning water content. Weigh all pieces of a prepared fruit or vegetable that will be placed on one tray (example: 5 lb peaches). Calculate the weight of water in the sample (5 lb × 89% water in peaches = 4.5 lb total water content). Calculate the amount of water to remove (4.45 lb × 75% = 3.3 lb weight loss needed for top quality peaches). Use 80% for most other fruits. Calculate the end weight of the sample (5 lb fresh wt. − 3.3 lb weight loss needed = 1.7 lb dry weight). Visual determination of endpoints by experienced dryers takes into account individual differences on the tray, such as edge pieces drying first, that calculations do not.

Foods tend to dry unevenly so conditioning in a tightly closed container for 4–10 days to distribute residual moisture is usually necessary. Condensation during conditioning indicates longer drying is required; if so proceed as above for drying.

Sun-dried foods should be pasteurized to kill insects. After conditioning, these foods may be placed in a 175F (80C) oven for 15 minutes or 160F (71C) oven for 30 minutes to pasteurize, or they may be packaged and frozen for at least 4 hours before storing. To be certain that all portions have frozen, 48-hour storage in the freezer is recommended to kill insects and their eggs. Heat-labile vitamin loss can be high with the oven pasteurizing method (Maggard, 1982).

Dried foods should be stored in containers that exclude air, light, and moisture. The containers should be in a dry, dark place with optimum temperature 60F (15C) or below to retard oxidative rancidity reactions. Food should be packaged in amounts to be used at one time so moisture is not added with several openings. Storage life is usually 6 months to 1 year. With longer times, quality can be expected to decrease, but the food is safe as long as it remains dry.

Commercial drying is conducted in much the same way as drying on a small scale at the household level; warm air is circulated over the food. Explosive puffing is used for its special effect and freeze drying for top quality. Industry does have better packaging methods available. Often their dried foods are packed in nitrogen to reduce the oxygen available for reactions such as oxidative rancidity. Commercial packages also may be opaque to eliminate light exposure which accelerates rancidity. The quality of commercially dried foods as they exit the dryer is often not superior to home dried, but during storage the commercially dried have an advantage.

## Drying Vegetables

Some vegetables do not need anti-browning pretreatments, but most need to be blanched to destroy enzymes. These enzymes can quickly cause quality loss to the point that the food is considered spoiled. Even low-fat foods such as green beans and peas contain enough lipid in their cell structures to spoil by oxidative rancidity during only a few weeks storage if dried unblanched. Green peppers are an exception to this general rule because they lack the enzyme(s) responsible for quality loss. In addition to enzyme inactivation, blanching softens plant structure for shorter drying times and may partially cook the vegetable, which shortens preparation time of cooked dishes.

Temperatures near 212F (100C) inactivate enzymes within 1 minute of exposure under moist conditions. Steam and water blanching are often used in households to inactivate enzymes before a food is dried. The enzymes catalase and peroxidase are generally used as indicators of residual enzyme activity in foods, but catalase is less resistant to heat than is peroxidase. Pretreatment and blanching can be combined. One teaspoon of sodium bisulfite per cup of water in steam blanching will help preserve the color of some vegetables during drying and storage and decrease the loss of vitamins A and C through oxidation. However, since some consumers cannot tolerate sulfite, such products should be labeled.

Strong-flavored vegetables such as hot peppers should not be dried inside as they give off strong odors and the fumes may cause eye irritations. Onion and garlic are susceptible to quality loss when dried. The Maillard reaction is particularly noticeable due to their light color, and the enzyme reaction responsible for their fresh flavors and odors continues during the slicing and drying; the products volatilize and the dried food lacks the typical onion or garlic taste and odor.

Dried vegetables lose more of their water content than dried fruits because they lack the large amounts of sugars which bind and hold water and consequently vegetables rehydrate more slowly. Usually the end product is best if the vegetables are barely covered with boiling water and allowed to stand at room temperature 15 minutes to 2 hours until rehydrated. If longer rehydration time is needed, the vegetables and their soaking water should be placed in the refrigerator to prevent bacterial spoilage. The rehydration water is best used for the cooking water as some nutrients will have leached into it. Rehydrated vegetables should be simmered gently for best quality. Fully rehydrated vegetables take only slightly longer to cook than the same vegetable frozen. The appearance of cooked, dried vegetables may be inferior so they are often used in soups or with sauces. Drying is not known as a method that preserves foods similar to their original state, but in some instances a low-cost, low-technology method that produces lightweight food is applicable.

## Spoilage of Dried Foods

Products may become darker during storage and take up less water when rehydrated. These changes are slower at low temperature so quality is maintained longer. If the drying process was too short or packaging was not airtight and the moisture content increased, molds may grow. Some species of molds have been found to produce toxic compounds. Since there is little research on specific products, it is safest to discard any food which becomes moldy.

## Summary

Dried foods (Note: Cured meats are not considered dried unless dried as jerky.)

**SAFE IF:**

Dried sufficiently to prevent molding and stored so food will remain dry and free of insects.

**TOP QUALITY IF:**

**a.** Best maturity or quality for eating.

**b.** Blanched if vegetable other than green pepper, onions, zucchini, herbs, following directions given in reliable source.

**c.** Pretreated as recommended if light-colored fruit and the light color is desired.

**d.** Cooked as directed for meat products.

**e.** Drying rate controlled so neither too slow nor too rapid.

**f.** Packaged in moisture vapor-resistant packaging.

**g.** Used within times recommended.

## References


ASHRA, 1986. ASHRAE Handbook, refrigeration systems and applications. Atlanta: ASHRAE Publ.

Banwart, G.J. 1988. Basic food microbiology. Westport: AVI Publ.

Beuchat, L.R. 1981. Microbial stability as affected by water activity. Cereal Foods World, 26:345.

Briggs, R. and Richardson, D.G. 1976. Unpublished paper. Anti-browning solution dips compared in home type fruit dehydration. Oregon State Univ., Corvallis, OR.

Chung, D.S. and Chang, D.I. 1982. Principles of food dehydration. J. Food Prot. 45:475.

Chyr, C.Y.L., Walker, H.W. and Hinz, P. 1977. Influence of pH, temperature, curing agents, and water activity on germination of PA 3679 spores. J. Food Prot. 40:369.

DellaMonica, E.S. and McDowell P.E. 1965. Comparison of beta-carotene content of dried carrots prepared by three dehydration processes. Food Tech. 10:1597.

Desrosier, N.W. and Desrosier, J. 1977. The technology of food preservation, 4th ed. Westport: AVI Publ. Co.

Eichner, K. and Ciner-Doruk, M. 1981. *In:* Water activity: Influences on food quality, eds. L.B. Rockland and G.F. Stewart, 567. New York: Academic Press.

Fennema, O.R. 1985. *In:* Food chemistry, ed. O.R. Fennema. New York: Marcel Dekker.

Foda, Y.H., EL-Waraki, A. and Zaid, M.A. 1969. Effect of blanching and dehydration on the conversion of chlorophyll to pheophytin in green beans. Food Tech. 22:233.

Fry, R.M. and Greaves, R.I.N. 1951. The survival of bacteria during and after drying. J. Hygiene. 49:220.

Holmes, A.A., Miller, L., Edwards, M. and Benson, E. 1979. Vitamin retention during home drying of vegetables and fruits. Home Econ. Research J. 7:258.

Jakobsen, M. and Murrell, W.G. 1977. The effect of water activity and $a_w$-controlling solute on sporulation of *Bacillus cereus* T.J. App. Bact. 43:239.

Karel, M. and Yong, S. 1981. *In:* Water activity: Influences on food quality, eds. L.B. Rockland and G.F. Stewart, 511. New York: Academic Press.

Keey, R.B. 1972. Drying: Principles and practice. Elmsford: Pergamon Press.

Labuza, T.P. 1980. The effect of water activity on reaction kinetics of food deterioration. Food Tech. 34(1):36.

Labuza, T.P. 1976. Drying food: Technology improves on the sun. Food Tech. 30(1):37.

Labuza, T.P. and Saltmarch, M. 1981. *In:* Water activity: Influences on food quality, eds. L.B. Rockland and G.F. Stewart, 605. New York: Academic Press.

Leistner, L., Rodel, W. and Krispien, K. 1981. *In:* Water activity: Influences on food quality, eds. L.B. Rockland and G.F. Stewart, 855. New York: Academic Press.

Lozano-de-Gonzalez, P.G., Barrett, D.M., Wrolstad, R.E., and Durst, R.W. 1993. Enzymatic browning inhibited in fresh and dried apple rings by pineapple juice. J. Food Science. 58:399.

Maggard, P.D. 1982. Practical approaches to home food dehydration. J. Food Prot. 45:492.

Mathew, A.G. and Parpia, H.A.B. 1971. Food browning as a polyphenol reaction. Adv. Food Res. 19:75.

Pitt, J.I. and Christian, J.H.B. 1968. Water relations in xerophilic fungi isolated from prunes. Appl. Microbiol. 16:1853.

Powrie, W.D., Wu, C.H., Rosin, M.P. and Stich, H.F. 1981. Clastogenic and mutagenic activities of Maillard reaction model systems. J. Food Sci. 46:1433.

Prior, B.A. 1979. Measurement of water activity in foods: A review. J. Food Prot. 42:668.

Sperber, W.H. 1983. Influence of water activity on foodborne bacteria: A review. J. Food Prot. 40: 369.

Sullivan, J.F. and Weber, H. 1982. Home dehydrators for food preservation. Home Econ. Research J. 10:411.

Troller, J. 1983. Methods to measure water activity. J.

Food Prot. 46:129.

Troller, J.A. and Christian, J.H.B. 1978. *In:* Water activity and food. New York: Academic Press.

Van den Berg, C. and Bruin, S. 1981. *In:* Water activity: Influences on food quality, eds. L.B. Rockland and G.F. Stewart. New York: Academic Press.

Vaughn, R.H. 1951. The microbiology of dehydrated vegetables. Food Res. 16:429.

## CONSUMER QUESTIONS

**Q.** *How can I prevent my home dried food from turning unacceptably dark?*
**A.** This browning can be retarded by the use of antioxidants and inhibitors of the Maillard reaction. Sulfite dips before drying are the most effective for consumers not allergic to them. The ascorbic acid molecule can participate in Maillard browning reactions and form dark compounds itself so it is not appropriate for use in drying. Steam or syrup blanching may prevent browning in fruits due to enzymes. Exposure to heat and light during storage accelerates browning.

**Q.** *Why do dried foods sometimes mold during storage?*
**A.** Invisible mold spores are in the air. They contaminate dried foods; however, food dried sufficiently and packaged to keep atmospheric moisture out does not have enough available water—an $a_w$ high enough—to support mold growth. Prevent by drying completely, conditioning, and then airtight packaging.

If food is moldy, it should be discarded because some mold species have been shown to produce harmful chemicals during storage at room temperature.

**Q.** *My fruit leather is too tart. What can I do next time?*
**A.** Sugars can be added to the leather puree. Sucrose (table sugar) is best as honey contains fructose which attracts water and tends to shorten the shelf life. Rolling the leather in coconut or spreading with a sweet filling (jam, chocolate, marshmallow creme) just before serving will tend to decrease the tartness.

**Q.** *Why does fruit leather crack when I try to roll it for storage?*
**A.** Overdried. This is especially a problem along the edges which may have been spread thinner. The crisp parts can be removed and served separately as snack chips. Slightly overdry edges are common as the center dries last. In these instances, the moisture will equalize with storage.

**Q.** *What causes some fruit leather to be grainy?*
**A.** Fruit skins tend to be coarser textured even though pureed in a blender. Peeling will give a more uniform consistency. Pear flesh contains stone cells which give it a grainy texture fresh or dried.

**Q.** *Why does my beef jerky have an off-odor after 2 months storage?*
**A.** Oxidative rancidity can produce off-odors and off-flavors if jerky is stored at room temperature. Proteolytic bacteria can grow if the moisture level is too high. Proteolytic bacteria produce sewagelike odors and are reason for discard. Storage at 0F (−18C) will greatly slow the rancidity reaction and eliminate bacterial growth.

**Q.** *Why was my last batch of jerky crumbly?*
**A.** Drying at too high temperatures can produce hard, brittle jerky. Cutting the raw meat across its grain leaves it in natural sections that tend to separate after dehydration.

**Q.** *Is it possible to make a low-salt jerky?*
**A.** Season the raw meat to your personal preference. If you choose to marinate for several hours or overnight, do so under refrigeration.

**Q.** *Why does the dried vegetable soup mixture have some vegetables that do not cook easily?*
**A.** Perhaps improper rehydration. Before cooking, soak the vegetables. Some vegetables require 2 hours soaking time after boiling water has been poured over them. Smaller pieces and some vegetables may rehydrate more quickly. A long blanching time may partially cook some of the vegetables which otherwise require longer cooking.

**Q.** *How do I tell when foods I am drying are*

*dry enough?*

**A.** Vegetables are generally brittle; fruits are generally tough, leathery, and pliable; jerky will bend with cracking on the surface only. There should be no moist areas in any of the dried foods though some fruits and leathers will be slightly sticky from their high sugar content. Test a sample that has cooled to room temperature as they will seem more pliable when warm. After equilibrating, if there are moisture droplets visible, the food should be immediately redried.

**Q.** *I want to use a recipe book that gives different temperatures for various foods. How do I regulate dryer temperature?*

**A.** This is important for food safety, to avoid case hardening (the outside forming a dry hard crust and the inside remaining moist), and to avoid actually cooking the food. In most domestic food dehydrators, the heat source is at the bottom. A fan helps distribute the heat, but the bottom trays usually are warmest. Testing the temperature at several tray levels, including the lowest tray, and also rotating the trays helps. Holding moist food at temperatures between 40F (4C) and 140F (60C) for over 3 hours risks foodborne illness for the consumer, so dryer temperature should be adjusted accordingly.

**Q.** *I misjudged the drying endpoint of a perishable food (vegetable and meat). The product was packaged and left at room temperature to condition overnight. The next morning I discovered it was not dry enough. What do I do now?*

**A.** Repeat drying process at once since the moisture was noticed during conditioning. If excessive moisture is noticed during storage; discard for safety.

# 4 Quick Pickles

As early as 1,000 B.C. in the Middle East, there is evidence of people preserving crab apples, pears, plums, onions, and raw walnuts in vinegar and spices. Such quick pickles, also called fresh pickles, are raw vegetables or fruits which are covered with a vinegar-water solution and then usually processed at boiling-water temperature to destroy enzymes and give a seal. They differ from fermented types in method of preparation, flavor, texture, and appearance. The quick or fresh pack pickling method does not result in as crisp a pickle as does a longer brining process but it has become popular due to its shorter preparation time.

## Microbiology

Quick pickles are preserved by adding acid which lowers the pH to the point where it restricts the growth of microorganisms, or by a combination of low pH and low $a_w$, and also by having a vacuum seal on the jar to prevent future contamination and give anaerobic conditions. Anaerobic conditions are important to prevent the growth of molds, which require air, because many molds can grow at the low pH of pickles. It is always a concern that molds may use some of the acids in their metabolic activity, thus raising the pH to levels where yeasts and even bacteria can grow, making the product potentially unsafe to consume. Heating to boiling in a jar with a lid eliminates oxygen and inactivates softening enzymes which can destroy pickle texture during storage, but the processing times are not long enough to kill all bacteria and their spores. Low pH is the main preservation factor for human pathogens in pickles, not canning.

After the acid brine and fresh fruit or vegetable are combined, an equilibration period follows. During this time the acid and seasonings penetrate all parts of the plant tissue; the brine becomes less acid and the produce more acid. Quick pickles usually are not consumed for at least 3 weeks after preparation to allow for this equilibration.

The United States Department of Agriculture (USDA) recommends that homemade pickles contain at least 0.70% acetic acid (maximum pH 4.0), when the cover solution has fully equilibrated with the vegetable. Their recipes and those from reliable sources provide this level of acidity. There is a safety concern when fresh pack pickles are not sufficiently acidified because at pH above 4.6, low-acid vegetables such as cucumbers, green beans, cauliflower, and corn can support growth and toxin production of human pathogens. Bacteria present on the vegetables in a spore state such as species of *Clostridium* are not killed in the processing in the boiling-water canner; actually the heat shock they receive may initiate spore germination. Pathogenic bacteria may grow and produce toxin without accompanying quality changes in the food and thus be undetected by the consumer. In a food acidified to at least 4.6, the vegetative state of these spores cannot grow and produce toxin. The pH of properly prepared pickles ranges from 2.6 to 4.0 varying with the type: sours 2.6, sweets 2.8, hamburger dills 3.2, fresh pack sweets 3.6, fresh pack dills 3.8, and refrigerated dills 4.0 (Fleming, 1981). The sweet pickles do not contain less acid; higher amounts of sugar give them their sweet taste.

Most quick pickle recipes are safe since in order to have good pickle flavor, the pH is almost always below pH 4.5, but unfortunately errors in choosing vinegars and adding correct amounts occur. One such rare problem arose when a senior, no longer able to do her own grocery shopping, wished to make her traditional spiced pickled beets. A neighbor was given the market list and brought back the ingredients. Months later, when the jar of beets was opened, the pickler complained that they did not taste as sour this year as in the past. Checking of the home cupboards revealed a partially full gallon of apple cider which had been mistakenly purchased and then used instead of apple cider vinegar.

Due to possible errors food preparers may make in vinegar addition, the state of Oregon allows only quick pickled fruits with a pH below 4.5 to be produced in approved home kitchens for commercial sale. The few food banks that accept home preserved foods also ban vacuum-sealed vegetables, including vegetable pickles for this same reason. Householder errors are infrequent, but when they occur questions are always raised about the safety of home vacuum-sealed foods of any type. Sound recipes such as those tested and published by the USDA, and those having at least a 1:1 vinegar:water ratio, properly prepared with commercially manufactured vinegar, result in products with a very wide margin of safety.

The processing step is designed to pasteurize, which prevents spoilage by killing bacteria capable of fermentation and those which metabolize acids. The internal whole cucumber temperature should be at least 165F (74C) for 15 minutes to protect the product from spoilage during a 6 month storage period (Fellers and Pflug, 1965 and Monroe et al., 1969). The USDA times in a boiling-water canner are calculated to provide this heat treatment plus a margin of safety.

As with canning, heat penetration inside the jars is important. A tight pack (25% brine and 75% cucumbers) results in underheating and increased spoilage. A loose pack (45% brine and 55% pickles) results in overheating of the cucumbers which adequately pasteurizes them but the final levels of salt, acid, and spices in the flesh are increased above a desirable range (Monroe et al., 1969).

The tightness of pack, kind and amount of acid, buffering action of the vegetables, and amount of salt all need to be considered for safety when recipes are formulated (Fleming, 1981). This is best done by food scientists, although adjusting the spices is a safe outlet for householder creativity in pickling.

Storage of the sealed jars in a cool dark place is recommended; however, pickles are not overly sensitive to heat changes. Pickles can be stored at 32F (0C) and 72F (22C) without showing quality changes for a 12-month period. After storage at 100F (38C) for 4 months pickles start to loose quality (Boggess et al., 1974).

## Vinegars

The added acid in quick pickles is usually cider vinegar because it is inexpensive and readily available. Vinegar is a mixture of acetic acid and water. Commercial brands are approximately 5% acetic acid and pickling recipes are formulated for this acidity. Grain strength is a term used to characterize commercial vinegars instead of percent acetic acid. The percent acid times 10 is the strength of a vinegar in grains.

Four percent acetic acid vinegar is 40 grain. Homemade vinegars will vary in acid content so are not recommended for pickling. They are fine products for use in salad dressings and sauces, but cannot be relied on when acid is used to inhibit bacteria.

Assorted vinegars are available in retail stores. Selecting the appropriate one will influence the quality of the pickles. Distilled white vinegar is made from acetic acid and water. The acetic acid is from fermentation of a dilute solution of alcohol. It has a mellow aroma, but a sharp acid taste. It gives a delicate flavor to pickled products and does not darken them. Cider vinegar is produced from fermented apple juice. It is basically a dilute solution of acetic acid in water but contains fruit acids, esters, inorganic salts, and extracted substances. It has a mellow acid taste and its fruity flavor blends well. However, it will darken light fruits and vegetables.

The vinegar-water solution surrounding the pickled product needs to be a minimum of 0.7% acid for safety (USDA, 1965). Quick pickles average about 1.2% acid with the brine being slightly more acid than the pickled tissue. This acid content does not change during storage (USDA, 1988).

## Ionic Concentration

Dissolved sugar and salt can act with vinegar to inhibit microorganisms in pickles. Most home recipes contain enough acid to inhibit bacteria solely through low pH. These recipes have at least a 1:1 ratio of vinegar:water in the solution poured over the pickles. Sugar and salt are included for flavor and texture (quality characteristics), and the amounts can be adjusted for personal preference.

The Morton Company's Lite Salt is a blend of equal parts of potassium chloride and sodium chloride. The bitter taste of potassium chloride is reduced in this product since it is a blend. It is used in some commercial fresh pack dill and sweet cucumber pickles. Potassium chloride can often be entirely substituted for sodium chloride in sweet homemade pickles, but tends to be bitter in dills.

In cases where there is less vinegar than water, higher concentrations of salt and/or sugar are needed for safety. Calculating appropriate amounts of these three ingredients to provide safe storage of perishable foods (such as pickled vegetables) is complicated and should not be undertaken by householders or by nonmicrobiologist cookbook writers. Recipes from reliable sources such as the USDA or some canning supply firms which have their own laboratories have tested the ionic concentration for safety.

## Texture

Crispness is a significant factor in pickle quality. Selecting top quality produce to pickle, using a recipe with ionic concentrations that do not result in toughening or softening, and inhibition of softening enzymes are all important.

Cucumbers which have not had enough water during growth tend to have poor texture whether eaten fresh or pickled. Drought years are accompanied by many complaints of poor pickle quality. Overripe or undermature produce also does not have the initial structural network needed to tolerate the pickling processes.

High-salt brines may draw moisture out of vegetable tissue, leaving it tough and shriveled. High-sugar brines may increase the firmness through changes in plant pectins. In sweet pickled peaches, both the viscosity of the brine—also called a syrup with large amounts of sugar—and flesh firmness will increase with increased sugar concentrations.

Heat exposure decreases firmness. When the fruit or vegetable is heated in the brine or held in a hot brine for long periods of time, the tissue cooks and crispness decreases. Some exposure to heat is necessary, but having all the equipment ready and working quickly during pickling eliminates excessive heat exposure. Pickle recipes in which cucumber chunks are heated with a brine in a saucepan before filling jars usually specify the length of time to heat, either just until thoroughly hot or in minutes of simmering. Hot brines are usually poured over the vegetable (hot pack) because they have better penetration than cold brines, and the acids and spices need to penetrate well. Hot packing is also recommended for those pickles which are boiling-water processed for storage. These limited exposures to heat still result in good texture. Processing at increased temperatures through use of a pressure canner will result in softer flesh (Boggess et al., 1974).

At temperatures above 160F (71C) the intercellular pectin is degraded at an acid pH, which results in a softer pickle. The degradation rate increases with increasing temperature. This is the reasoning behind some recommendations to process pickled products in water baths of temperatures ranging from 170–190F (77–88C). However, in a household situation, 212F (100C) can be easily visually determined so it is generally used.

Enzymes which soften plant tissue can cause quality spoilage during storage if they are not inactivated by heat processing in a boiling-water canner. Several such enzymes can break down pectin which is a major structural component of plant tissue (see fermentation in Chapter 5). Peroxidases are a group of enzymes found in all higher plants. They are the most heat-resistant plant enzymes so are used as an indicator of adequate processing; if peroxidase is not present, it is assumed the other enzymes which may cause quality changes are also destroyed. Pickle processing times are based on peroxidase tests, since the vegetable tissue is preserved by the brine and heat mainly provides an airtight seal and inactivates enzymes; at enzyme inactivation temperatures, vegetative bacterial cells are also killed.

Chemical additives are also available to householders to increase pickle crispness. Lime and alum are sometimes included in recipes for this purpose. Some decades, these additives have been omitted from USDA recipes, but favorite family recipes tend to be passed through the generations unchanged and questions arise about the function and safety of these ingredients.

Calcium oxide is the form of lime used in pickling; it is also called slaked or builder's lime. Slaked lime dissolves readily and the calcium is absorbed into the plant tissue where it combines with pectin to form calcium pectate which results in a firmer pickle. Unslaked lime is not as soluble so it has less calcium available as a firming agent. Its use is not recommended. Food grade lime can usually be purchased at a pharmacy or canning supply outlet. Nonfood grade lime is not as pure and may contain harmful substances if ingested.

Alum is an aluminum compound. Like calcium, aluminum has a firming effect on plant tissue, but unlike calcium, there are some health concerns with aluminum in the diet. For this reason, it is no longer recommended that alum be included in pickles. There are no direct links with aluminum poisoning and pickle consumption since most people consume small amounts of pickles, however digestive irritation may occur. Simply leaving alum out of a favorite recipe usually results in a softer pickle. New recipes have been formulated to make crisp pickles without alum. To rewrite a favorite recipe to exclude alum, start with a similar current USDA recipe and adjust the spices. Do not adjust the acid:water ratios in the brine. If these ratios are at least 1:1 then the salt and sugar amounts in the USDA recipe can also be changed to match those in the traditional recipe if desired; however, the texture may not be optimum. A small test batch is advisable.

## Summary

### SAFE IF:

**a.** Recipe is 1 part of vinegar to 1 part of water (or less). Vinegar used is 5% acidity—standard commercial product. If too sour, sugar is added, but vinegar is not reduced.

**b.** Boiling-water processing is used to prevent molding.

### TOP QUALITY IF:

**a.** Fruit or vegetable is fresh and young.

**b.** Pickling is done promptly after harvesting.

**c.** Spices and herbs are fresh.

## References

Boggess, T.S., Heaton, E.K. and Shewfelt, A.L. 1974. Processing sweet pickled peaches. Home Ec. Res.

Report No. 28. Washington DC: U.S. Govt. Printing Office.
Desrosier, N.W. and Desrosier, J. 1977. The technology of food preservation. 4th ed. Westport: AVI Publishing Co.
Fellers, P.J. and Pflug, I.J. 1965. Quality of fresh whole dill pickles as affected by storage temperature and time, process time, and cucumber variety. Food Tech. 116:416.
Fleming, H.P. 1981. Spoilage problems of pickles. Presented at natl. workshop for Ex. Food and Nutr. Spec. Chicago, Ill. October 1, 1981.
Monroe, R.J., Etchells, J.L., Pacilio, J.C., Borg, A.F., Wallace, D.H., Rogers, M.P., Turney, L.J. and Schoene, E.S. 1969. Influence of various acidities and pasteurizing temperatures on the keeping quality of fresh-pack dill pickles. Food Tech. 23:71.
USDA. 1988. Complete guide to home canning. Ag. Info. Bull. No. 539. Washington DC: U.S. Dept. of Agriculture.
USDA. 1965. Effect of household processing and storage on quality of pickled vegetables and fruits. Home Ec. Res. Report No. 28. Washington DC: U.S. Govt. Printing Office.

## CONSUMER QUESTIONS

**Q.** *What amount of vinegar is a pickle recipe supposed to contain?*
**A.** Pickling requires at least as much vinegar as other liquids, as a general rule of thumb. Exceptions include recipes with large amounts of sugar and/or salt so the osmotic pressure combines with the pH level to inhibit microbial growth. These exceptions need to be tested to ensure safety (i.e., consider recipe source).

| *Vinegar* | : | *Water*[a] | |
|---|---|---|---|
| 1 | : | 1 | OK |
| 2 | : | 1 | OK |
| 3 | : | 1 | OK |
| 1 | : | 2 | NOT SAFE |
| 1 | : | 3 | NOT SAFE |

[a]Since other liquids such as fruit or vegetable juices are primarily water, they should be counted on the water side.

If this is too sour for your taste, add a little sugar. Vinegar is necessary to preserve the product. *C. botulinum* and other bacteria that cause human illness cannot grow in such acid conditions. However if you change the amount of vinegar, they may be able to grow, multiply, and produce toxin in the vegetable or other pickled product.

**Q.** *Is the ratio of fresh produce to vinegar important?*
**A.** The quantity of fruits and most vegetables to the pickling solution can vary, as these commodities have little buffer capacity. The ratio of vegetable or fruit to vinegar will not change the pH. Cover the fruit or vegetable well to ensure even penetration. Meats, fish, and legumes contain large amounts of protein so have a greater buffer capacity. Changing these tested recipes is not recommended.

**Q.** *Can I use homemade vinegar in pickling?*
**A.** Commercially manufactured vinegars are quality controlled to be 4–6% acid. The label will usually state the percentage. Home prepared vinegars may not reach this acidity so should not be used for pickling. Save them for preparing salad dressings.

**Q.** *Can the amount of salt be reduced?*
**A.** This amount can be adjusted if the amount of vinegar in the recipe is *at least* as great as the amount of other liquids, since then salt is not necessary to preserve the product from bacterial growth. In such cases, the amount of salt can be decreased "to taste." However, the texture of the pickles *may* be altered when the amount of salt is either increased or decreased.

**Q.** *What type of salt is required for pickling?*
**A.** This may influence the appearance of the pickled product. Iodized salt will make pickles slightly darker, as it reacts with plant pigments to form dark compounds. Table salt, whether or not it contains iodine, does contain anti-caking ingredients. These sometimes make pickle liquid cloudy. Pickling salt does not contain iodine and may or may not contain anti-caking ingredients (check label and compare prices). Both pickling salt and table salt will preserve the product and impart the same flavors to the pickles.

**Q.** *How can I make unique pickled products?*
**A.** The amount of sugar *may* be adjustable to

taste. The same rules for adjusting the amount of salt apply. Herbs and spices can be experimented with to give new flavors. In some cases, a color change will result. A different fruit or vegetable may be pickled if the amount of vinegar is at least as great as the amount of other liquids, then this pickling brine recipe can be used to pickle any fruit or vegetable.

**Q.** *What about alum in pickles?*
**A.** Older pickle recipes often contained alum. The active ingredient in alum is aluminum. Aluminum is linked to toxic effects if it is consumed in a large amount over time. Most people do not eat large amounts of pickles; however, since there is the possibility of a health hazard, most professionals are not recommending using alum in foods. Alum was responsible for giving pickles a crisp, crunchy texture. Simply removing alum from a favorite recipe usually results in a softer pickle. The USDA has developed recipes without alum that result in crisp pickles. See if your local Cooperative Extension Service distributes these recipes.

**Q.** *What is the purpose of lime in pickles?*
**A.** An alternative to alum is the addition of lime. The calcium in the lime makes fruits and vegetables firmer. Commercially canned products often contain calcium salts to improve their texture. However, lime used in foods needs to be free of contaminants. Use only "food grade" lime. This can be purchased from most drug stores by asking the pharmacist.

**Q.** *Can I change the size of the fruit or vegetable?*
**A.** Since the acid in the pickling brine preserves the food, this acid needs to penetrate the entire piece. If the food mass is too large, this penetration is too slow. The maximum size is usually a small apple, or small pear, or small ear of corn. It takes approximately 3 weeks for a pickling solution to penetrate food throughout a jar, therefore, pickling 3 weeks in advance of eating is recommended.

**Q.** *Is processing in a boiling-water canner necessary?*
**A.** Sealing keeps yeasts and molds out which could ruin the quality and possibly the safety of your product. Sealing the jars in a waterbath will ensure a longer shelf life and requires such a short time (usually 5 to 20 min), that the product is not cooked to the point of texture and flavor changes. Processing times will vary with the product being pickled.

**Q.** *What is the effect of metals in pickling?*
**A.** Boiling the pickling solution in a glass or enamel pan is recommended to prevent metals from reacting with the acid in the vinegar to produce off-flavors and darkening. Local water supplies with large amounts of iron will cause discoloration of pickles. Lighter colored fruits and vegetables (onions, pears, white peaches) are a particular problem. The discoloration becomes more apparent with storage time.

**Q.** *Is it possible to decrease recipe amounts?*
**A.** Pickling recipes may be cut proportionately. Making only 1 pint of a new recipe to see if you like it may be a good idea.

**Q.** *Is refrigeration necessary after opening the jar?*
**A.** Refrigeration temperatures slow microorganism growth which could result in a quality spoilage of the pickled product. The pH should be too low for growth or toxin production of human pathogens. Refrigeration prolongs the shelf life.

**Q.** *What is the effect of acid on plant pigments?*
**A.** Chlorophyll (cucumbers, peas, beans) reacts with acid to give an olive green color. This may be somewhat camouflaged with green food color. Carotenes (winter squash, carrots), including lycopenes (tomato, watermelon), are not affected by acid. Anthocyanins (most red fruits) become redder with acid. Some anthocyanins exist in a predominately colorless form (leucoanthocyanins). Adding acid and heating will shift these to a predominately red form, giving the commodity a pink tinge, as sometimes occurs in pickled pears, cauliflower, and garlic.

**Q.** *Why did the dill in my pickled cucumbers*

*turn red?*
**A.** Anthocyanin pigments naturally present in the dill will turn red in acid conditions. A harmless yeast on the dill may also grow and cause it to appear red.

**Q.** *Why did the garlic in my pickles turn blue-green?*
**A.** This is the effect of copper on a light-colored food such as garlic. The exposure to copper could have come from the water supply or from copper cooking utensils.

**Q.** *Can I use salt substitutes when making pickles?*
**A.** Potassium chloride can be substituted for sodium chloride (table salt) in sweet pickles acceptably for many consumers; however, it tends to taste unacceptably bitter in dills.

# 5 Fermented Pickles

Fermentation is a preservation method that dates back to early history. When Ch'in Shih Huang Ti was constructing the Great Wall of China in the third century B.C., the laborers were given mixed fermented vegetables as part of their rations, Caesar's soldiers had pickles as a delicacy, and Cleopatra is thought to have consumed them as a treat. Today, an estimated 75% of U.S. households eat pickled products at least once a week. However due to their acidity, they are usually used in small amounts at meals or as an accent food in mixtures. Pickled foods are not significant in the average U.S. diet, but small batches are made in many households and consumers have questions about the fermentation process. Cucumbers, olives, and cabbage are the most frequently fermented foods made in U.S. households. Turnips are often fermented in Germany, and blends of fermented vegetables are most common throughout Asia.

Fermentation is the anaerobic (without oxygen) or partially anaerobic breakdown of sugars. Vegetables are not usually thought of as sugary foods, but they contain adequate sugars for fermentation. During fermentation, naturally present bacteria break apart sugars, chiefly glucose, to form acids (Fig. 5.1). In fruit fermentation by yeasts, such as in making vinegar, a 2-carbon alcohol is first formed, then the alcohol can be converted to acetic acid.

Bacteria may be either aerobic or anaerobic; some are able to tolerate either condition. Aerobic species live in the presence of relatively high oxygen concentrations, such as in air. Microorganisms growing on surfaces are aerobic. The anaerobic bacteria grow in the presence of relatively low oxygen concentrations such as under the surface of a liquid. Fermentation bacteria are anaerobic. Good recipes provide an anaerobic environment for the vegetables to favor the growth of fermentative bacteria.

Successful fermentations require only sugar (in the plant) and microorganisms (also on the plant); however, to speed the formation of liquid and encourage bacteria that produce desirable flavors, salt is usually added. Salt may be added either dry or combined with water in a preformed brine. Some recipes also add acids (vinegar) to rapidly lower the pH and discourage growth of undesirable bacteria. Fermenting vegetables in an acid brine is an old preservation method. It is speculated that preformed brines were first used because dry salting did not draw sufficient water from plant tissues and

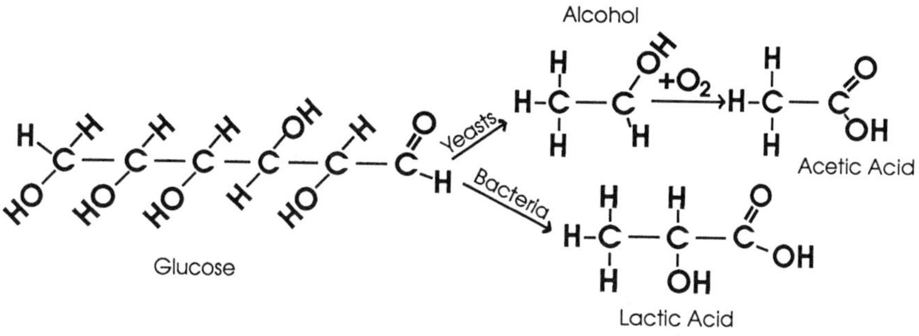


**5.1.** In fermentation, 6-carbon glucose (a simple sugar) is changed to 2- and 3-carbon acids.

a brine allowed sand and other impurities from the natural salt to settle in the bottom of the container resulting in a less gritty pickle.

Fermented foods have built-in protection from organisms that could cause foodborne illness. Preservation by fermentation depends upon the combined effects of acid, salt, carbon dioxide, anaerobic conditions (a low oxidation-reduction potential), enzyme activity, a rapid growth of certain microorganisms, and the presence of antimicrobial agents often produced by fermenting microorganisms. These factors act together to prevent growth of pathogenic bacteria (Hesseltine, 1983). Fermentation also reduces the growth of those bacteria, yeasts, and molds which would decrease the quality of pickles.

There are four main types of pickling processes:

**1.** Pickled products prepared directly from vegetables without undergoing fermentation. Quick (fresh) pickles with added vinegar fall into this category.

**2.** Pickled products prepared by dry salting such as sauerkraut.

**3.** Products fermented in a weak brine solution such as dill pickles.

**4.** Vegetables fermented in a high-salt brine such as salt stock pickles that are freshened (de-salted) before consumption.

Though pickling involves low human technology, the actual fermentation process is complex. In food systems, microorganisms ferment sugars by complete oxidation, partial oxidation, alcoholic fermentation, lactic acid fermentation, butyric fermentation, and other minor fermentative processes (Desrosier, 1970). Lactic acid fermentations in which the sugar in the food is converted to lactic acid and other end products (Fig 5.1) is of primary importance in pickled vegetables, and recipes are designed to promote it.

Fermentation is carried out by a sequential series of different species of microorganisms. They use different substrates, and thrive in different environmental conditions. As one group's products accumulate, the environment changes, becoming more optimum for another species, and the first group's population declines as the second group's increases. This process is then repeated (Fig. 5.2).

The initial fermenters are naturally present on the surface of vegetables, but in low numbers compared to aerobic bacteria. Anaerobic conditions and a proper salt concentration are important for the first fermenter, *Leuconostoc mesenteroides*, to outgrow the aerobic soil and water bacteria such as *Achromobacter*, *Enterobacter*, *Bacillus*, *Escherichia*, *Flavobacterium*, and *Pseudomonas*. A good fermentation recipe will specify the amount of salt, water, and vegetables so conditions that favor *Leuconostoc mesenteroides* are initially present.

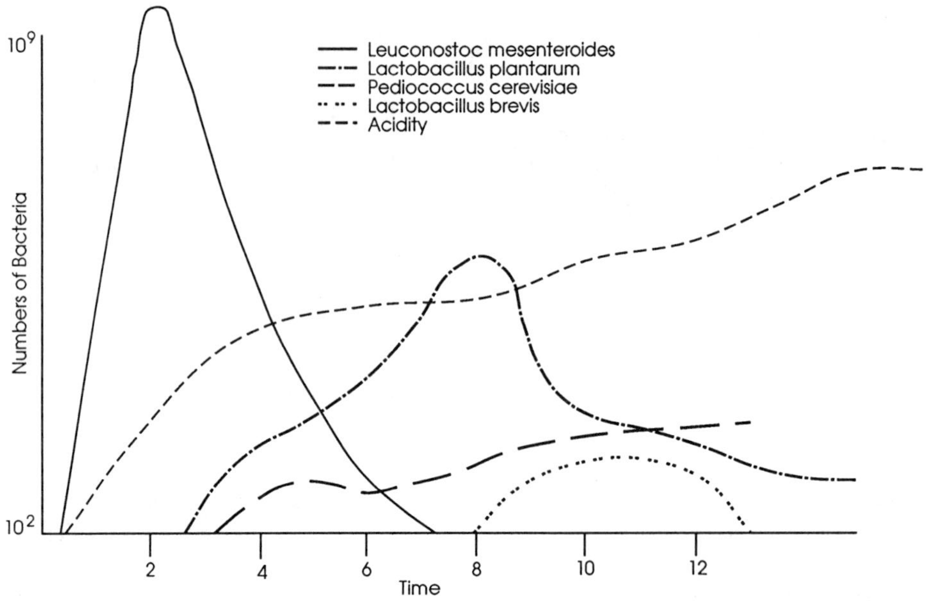


**5.2.** Sequential growth of bacterial species with acid production during fermentation of sauerkraut.

## Home Methods of Fermenting Pickles

The traditional pickle crock our mothers and grandmothers could purchase at the local store is now more frequently seen at antique auctions. These crocks were heavy (averaging 10 lbs and up) glazed stoneware containers that could hold a salt-acid brine without chemical reactions occurring on the walls. Since most of the other large kitchen containers of the time were metal, these crocks were used extensively until food-grade plastics were marketed. Today, fermenting is often done in 5-gallon plastic pails that once held mayonnaise, relish, or catsup for restaurants. They can be purchased from food service establishments for several dollars and are lightweight which makes draining them much easier than stoneware crocks. Plastics from nonfood usages may be unsafe and should not be used. A heavy dinner plate or glass casserole lid is usually placed on top of the vegetables to keep them under the brine. Smaller fermentations are done at the household level in quart or 2-quart jars. A brine-filled, pint-size double freezer bag or a weighted-down plastic grid works well in this situation to cover the top. A heavy, clean terry-cloth towel over the weight and container top reduces mold contamination.

Dill is added as a flavoring for some pickles and also contributes desirable bacteria. Few home gardens contain dill plants, so supermarkets carry the heads during pickling season. The dill is poorer quality when the buds have opened.

Most home fermentations are done in a cool dark part of the house such as a basement or pantry. Containers should be checked daily and yeast films removed as they form. It is not uncommon for a good fermentation to have off-odors during the early stages; these are not reasons for discard.

There are several home methods for determining if fermentation is complete. Usually good cured cucumbers are firm, crisp, a translucent olive green, and do not have white spots. Another test is to tap the sides of the fermentation container. If bubbles do not then rise to the surface, fermentation is complete. Some householders test the brine with litmus paper purchased from a drugstore. When a narrow range pH paper indicates a pH of 4.0 or below, the fermentation has progressed far enough.

It is recommended that canned homemade pickles be pasteurized in a boiling-water canner, when the quality of the fermented product is optimum, to inactivate spoilage enzymes. This also creates a vacuum seal so spoilage by mold is prevented. Processing has not always been recommended, so older recipes often do not have this end step. Products from these older recipes can be easily processed in a water canner too. A current United States Department of Agriculture (USDA) recipe for each specific product will state an accurate processing time. For some products, a lower temperature (185F, 85C) is recommended; this prevents softening.

## Sauerkraut Fermentation

Sauerkraut is fermented, shredded cabbage. It is usually successful at the household level because the recipes are uncomplicated and minor texture problems in small pieces of cabbage are not as apparent as texture problems in a whole fermented cucumber. Sequential growth of various microorganism populations is critical in making good sauerkraut (Fig. 5.2).

Cabbage has been a prominent food throughout much of recorded history. A 200 B.C. manuscript credits cabbage as being the most important vegetable the Romans cultivated. From 200 B.C. to 450 A.D. it was the main plant used for treatment of disease in the Roman Empire. Originally, cabbage leaves were acidified by the addition of sour wine or vinegar. In later years, salt was added. In 1607, Germans were fermenting cabbage pieces with salt, berries, and seasonings; very similar to our present day sauerkraut. In the late 1700s, cabbage and salt were the only ingredients used and cabbage was thought to possess healing properties. Today sauerkraut recipes are still simple: shredded cabbage and salt, and the consumption of vegetables from the cabbage family is considered a strong asset in the diet.

In sauerkraut fermentation *Leuconostoc mesenteroides* first establishes a good environment for continuing lactic acid fermentations. *Leuconostoc* cells are small compared to cells of other organisms competing in the environment (a pickle crock). Smaller cells have proportionately larger surface areas with which to take up nutrients from their surroundings so will multiply and ferment more rapidly. This gives *Leuconostoc mesenteroides* an edge.

*Leuconostoc mesenteroides* produces ethyl alcohol and carbon dioxide in addition to lactic acid. Carbon dioxide replaces the air and facilitates anaerobic conditions needed for fermentation. If the vegetables come into contact with air, yeasts and molds can grow on the exposed surfaces. Their enzymes cause softening, darkening, and off-flavors which are then dispersed throughout the kraut jar. This is the reason behind the recommendation of weighting vegetables down below the surface. Brine or water-filled plastic bags conform to the shape of the fermenting container and are effective in holding vegetables below the liquid. Some of the gas bubbles seen in these early stages are simply being released from air spaces inside the vegetable itself and some are the result of microorganism activity. Gas in the early stages is not a cause for concern. It is not believed that this first stage contributes many flavorful products.

Under optimum conditions, *Lactobacillus plantarum* is the next organism to predominate. It produces the largest amount of acid in the fermentation process and also removes a bitter-flavored compound (mannitol) produced by *Leuconostoc*. If the fermenting environment is not ideal, *Lactobacillus plantarum* does not outgrow other microorganisms. High temperature and high salt concentrations favor *Pediococcus halophilus* (formerly *P. cerevisiae*) as the next major organism; *P. halophilus* populations then increase and are major contributors to the acid production. *P. halophilus* produces kraut of inferior flavor because it is a homofermenter; it produces primarily only one acid (in this case lactic acid). *Leuconostoc* and *Lactoba-*

*cillus* convert glucose to approximately 50% lactic acid, 25% acetic acid and ethyl alcohol, and 25% carbon dioxide. Some heterofermenters also produce mannitol and dextran from sucrose. A variety of products gives a more complex flavor which distinguishes fermented pickles from acid-added pickles. The presence of complex flavor from heterofermenters is a major quality factor in sauerkraut.

The kraut can be canned at the end of the *L. plantarum* stage for an acceptable product. If the fermentation is allowed to continue and if enough sugar and mannitol are left as substrates, *Lactobacillus brevis* is the next organism to predominate. This organism is a gas former, and its products result in a sharp-acid flavor that most find less desirable. Gas production in the fermentation jar at the end of the process indicates probable quality spoilage (still safe to eat) due to *L. brevis*. Note: Some *L. brevis* multiplication during fermentation is desirable and this organism is present in small numbers in the early stages.

An optimum cabbage for kraut making contains at least 3.5% sugar within its tissue. As the fermentable carbohydrate in the cabbage is used, the level of acid formed ranges from 1.5–2.0%, expressed as lactic. As a general rule, one-half as much acid is produced as there is sugar in the cabbage. During fermentation the green chlorophyll changes to an olive-green pheophytin, and the cabbage tissue changes from opaque to translucent as air in the interstitial spaces of the cabbage tissue is replaced with liquid. These same changes in appearance occur when fresh cabbage is cooked.

## Household Sauerkraut Production

Microorganisms that can produce a top quality product are present on the vegetable tissue, but controlling the fermentation environment so they predominate in correct sequence is up to the food preparer. Salt, temperature, and preventing aerobic conditions are variables that householders can control. Salt concentrations can be controlled by accurately weighing or measuring both the cabbage and the salt. Fermentation temperature is adjusted by placing the crock in a cool place (air conditioned room, basement) if the weather is hot.

Salt added to the cabbage draws water and plant nutrients out of the vegetable's tissues. These nutrients are excellent substrates for the growth of lactic acid bacteria. Salt also raises the ionic (salt) concentration to a level where the enzymes that soften tissue cease to be active and the growth of organisms that are undesirable in the fermentation process such as the putrefactive genera are inhibited. Putrefaction is the anaerobic breakdown of protein. Plant cells contain enough protein for this reaction to be a problem. In putrefaction, carbon dioxide is produced as in fermentation reactions, but other gases such as hydrogen sulfide (rotten egg smell) are also released. Most North Americans find these odors reminiscent of sewage and offensive in a food. A putrid fermentation usually involves contamination along with the aforementioned conditions that encourage growth of the putrefactive bacteria. Household fermenting is not a sterile process, but care should be taken to use clean utensils, place the bag of brine on a clean surface when it is removed, and to work quickly while the top is exposed.

A proper salt concentration not only enhances progression of the various lactic acid bacterial populations, but also provides a pleasing acid-salt balance in the flavor of the finished product. Commercially, salt is most commonly added to shredded cabbage at 2.25% by weight. Household recipes have varying amounts of salt, but usually are around 2.5%. Even distribution of the salt is important for the fermentation to proceed well in all areas of the container. Householders mix the salt and shredded cabbage well in a large bowl before packing into jars. A higher salt concentration of 3.5% results in 90% inhibition of *Leuconostoc mesenteroides* and *Lactobacillus brevis*, (organisms which produce desirable complex flavors) and the final product (produced by homofermenters which then predominate) usually has a harsh, unacceptable flavor.

The fermentation temperature has a great effect upon microbial growth and subsequently upon the finished product's quality. Top quality krauts are fermented at 65F (18C) or lower,

45.5F (7.5C) minimum, and have the highest quality flavor, color, and ascorbic acid content from the multiplication of the heterofermentative lactic acid bacteria (Pederson, 1979). Many householders have a basement or garage this temperature in the autumn when kraut is usually made. The total fermentation time is about 20 days at this temperature. At temperatures below 45.5F (7.5C), the *Leuconostoc mesenteroides* grows very slowly, attaining an acidity of 0.8% in about 1 month. Lactobacilli and pediococci grow even more slowly and fermentation may not be complete for 6 months. At 90F (32C) the fermentation is complete in 8–10 days with most of the acid produced by the homofermentative *Lactobacillus plantarum* and *P. halophilus*. A less complex flavor results.

### Sauerkraut Problems

Top quality sauerkraut is not produced in every batch. Usually minor quality defects are the problem, but occasionally, the entire lot must be discarded.

Texture defects are common. They can come from several different sources. Enzymatic breakdown of plant tissue is the main cause; the result can range from soft to slimy kraut. The enzymes present in cabbage can be controlled by initially adding 2.5% salt to the shredded cabbage, which results in quick acid production. This enzyme activity is inhibited by both salt and acid. Enzymes from molds and yeasts can be eliminated from the kraut if the cabbage is kept below the surface of the liquid. Molds and yeasts are aerobic and can grow on floating cabbage, but not cabbage in anaerobic conditions. Processing finished kraut in a boiling-water canner eliminates the possibility of continued enzyme activity because heat destroys the enzymes.

Poor flavor kraut lacks the proper combination of acids and other products of fermentation. Certain organisms need to be present in a specific sequence for the finished fermentation to have optimum flavor. These microbial populations can be controlled by selecting fresh, sweet cabbage, adding 2.5% salt, by storing the fermentation container at temperatures below 80F (27C), and by stopping the fermentation when the quality is optimum. White scum is usually due to yeasts. Their presence is not a reason for discard of the kraut, but the scum should be removed daily.

Pink sauerkraut is not common, but when it occurs, householders are usually concerned about the safety of consuming that batch. Cabbage contains a plant pigment that can exist in a colorless form in slightly acid conditions, a blue form under near neutral conditions, or a red form in acid conditions. In the colorless form this pigment is called leucoanthocyanin, and in the colored forms it is referred to as the anthocyanin pigment. Occasionally, as acid in the fermenting kraut increases, the cabbage will take on a light red (pink) hue. The pigment is harmless to consume, as is pink sauerkraut. Pink kraut is more likely when the season has been dry. Pink kraut can also result from yeast growth, especially if the temperature is high and the salt was unevenly distributed. This also is not harmful.

Sauerkraut may darken at the surface due to oxidation. This darkening may be in spots, or an overall brown to black color. Enzymes of the plant and enzymes from yeasts and molds growing on the surface can oxidize kraut. Uneven salting and a high fermentation temperature increase the rate of color change.

## ■ Fermented Cucumbers

North Americans consume about 27 billion cucumber pickles a year. Forty percent of these were fermented types in the mid-1980s (McNish, 1986), but with fermented cucumbers becoming more expensive on grocery store shelves, many householders now must make their own top quality pickles.

Cucumbers progress through a lactic acid bacterial fermentation similar to sauerkraut but unlike sauerkraut, spices and vinegar are often added in fermented cucumber recipes. During fermentation, the cucumber turns from light green to opaque white to a translucent olive green as air between cells is replaced by water, and chlorophyll is converted to pheophytin—an olive-green pigment—with the decrease in pH.

Texture changes also occur. When the cucumbers are first placed in the brine, they lose water—across an osmotic concentration gradient—because the brine contains more salt than the vegetable tissue; the salt pulls water out. The cucumbers initially lose weight and become tough and rubbery. As the fermentation progresses, the cucumbers absorb brine which restores lost weight (in some cases above the original weight) and they become crisp. Changes in the plant pectins also occur during fermentation which make the cucumbers more permeable to sugars, spices, and vinegar. As these substances move into the plant tissue, flavor increases. Fermented pickles contain a variety of acids, but lactic is the predominant one, accounting for 90% of total acid (USDA, 1965). Processed fermented dill cucumbers are on the average 0.94% acid in their flesh and 0.91% acid in their brine. After 9 months storage, the flesh averages 0.84% acid and the brine 0.82% (Desrosier, 1970).

Simple fermentation methods used for centuries are still practiced at the household and the commercial levels. Householders making pickles for the first time are often surprised at the lack of human intervention needed for the fermentation process. The legend of a clumsy Mesopotamian dropping a fresh cucumber into a vessel of salt water about 2,100 B.C. and removing it weeks later to find the world's first pickle (McNish, 1986) may be an accurate representation of the origin of the pickle industry. The basic process has not changed greatly. Commercially, pickles are fermented in large vats. Traditionally these vats were made of wood and open on top. Fresh, unwashed cucumbers with natural contamination from the environment were used. Molds were controlled with ultraviolet rays from the sun. The open vats allowed insects and surface dirt to enter also.

In the last 10 years, the trend has been toward using washed cucumbers, lactic acid starter cultures, and clean fiberglass vats with small openings on top. These clean-fermented pickles have a slightly different flavor, but consumers first accepted blends with the traditional, and now readily purchase the clean-fermentation types. In household recipes it is usually recommended that cucumbers be washed free of visible dirt but not thoroughly cleansed; some starter bacteria need to remain on the cucumber so they can be introduced into the fermentation container.

Removing gas from the commercial vats is a problem. Vats have exploded from gas accumulation. The small quantities fermented at the household level do not build up such pressure. Householders concerned about gas in the early stages of fermenting pickles need to realize that it indicates desirable organisms are fermenting; it is not cause for discard.

Cucumbers are often started in a low-salt brine so as not to draw water out too fast (osmotically shock the tissues) and then more salt is added each day; gradually at first and then in larger amounts on subsequent days when shock is much less likely. Examples of this type of pickle are the 14-day, 7-day, and salt stock. Fermented cucumbers are not considered to be low-salt foods from a nutritional standpoint, but some fermentations contain much more salt than others. The methods that involve 3–5% salt brines are called low-salt pickles, the recipes with 10% or higher concentrations are referred to as high-salt pickles. A low-salt brine for fermented dill cucumbers has recently been found to result in more rapid acid formation and ultimately a better quality pickle than achieved with higher salt concentrations. When cucumbers are placed in a 5% (3/4 c/gal water) brine, they absorb salt until an equilibrium is reached; usually at this point, there is about 3% salt left in the brine. In approximately 2 weeks depending upon the fermentation temperature, the acidity with low-salt methods reaches 0.6–1.0% as lactic with a pH of 3.4–3.6. The higher salt brines (10% and above) not only have a slower lowering of the pH, but have increased gas evolution and more bloaters. Bloaters are hollow-center pickles that float. Very high salt concentrations of 10 to 13% (40°–50° salometer brines) inhibit lactic acid bacteria. Fermentation is greatly slowed if it occurs at all.

Cucumber fermentations usually involve brine strength adjustments throughout the process. The brine strength can be tested every several days with a salometer, (see Appendix for use and purchase outlets). In making salt stock pickles, the brine strength should not fall below 40° salometer and should increase 3°–4° each week until fermentation is complete (5–6 wk), at which time the brine is 15% (60° salometer) (Dunn, 1977). Some old recipe books recommend floating an in-the-shell raw egg on the brine to test it; floating indicates a brine strength of 10% or greater, which is not a measurement needed for most recipes today. Householders may simply adjust the brine by following a recipe and adding specified amounts of salt at the recommended time increments. Recipes from a reliable source usually produce good pickles this way. It is important to measure the volume carefully and measure the salt accurately. More salt can be added to the top of the weight on top of the container if needed. The added salt will then gradually dissolve into the brine.

Genuine dill pickles are the most common commercial and household fermented type. They are completely fermented in low-salt brines—5% or lower. To make genuine dills, fresh cucumbers are placed in the brine with dill and spices. Some recipes add vinegar to the brine, but this may inhibit some of the desirable fermentors. Spices and herbs have not been found to inhibit microbial growth.

The first few days of fermentation, the brine is cloudy from bacterial growth and gas liberation. The bacteria settle on the cucumber surface and the bottom of the container. At 5%-salt concentrations and lower temperatures, *Leuconostoc mesenteroides* is the first fermenter. At higher salt concentrations or higher temperatures, the fermentation is usually initiated by *Streptococcus faecalis*, *Pediococcus halophilus* (formerly *P. cerevisiae)*, *Lactobacillus plantarum*, and sometimes *Lactobacillus brevis* is also present (Pederson, 1979).

After 2–6 weeks fermentation at 72F (22C), 0.6–1.0% lactic acid will accumulate; pH is 3.4–3.6. The characteristic genuine dill flavor is due to fermentation products brought about by biochemical changes from the microorganisms in addition to flavor contributions from the dill and spices. The fermented pickle flavor has been described as a distinctively mildly acid, dill-spicy, salty taste. This complex flavor is highly regarded among pickle connoisseurs; however, it is not available at many retail markets due to the difficulty of mass producing it cost-effectively. Though fermentations usually proceed well, some lots will spoil and need to be discarded. Acid-added (quick) pickles are not so risky to produce.

The brine is often cloudy when fermentation is completed but it contains the complex flavors of fermentation that cannot be duplicated artificially in a fresh, clear brine that pickles are often finally packed in. Filtering the fermentation brine before packing gives it a more acceptable appearance for some consumers. Either the fermentation brine or a freshly made brine is safe to use.

Pickles should be packed tightly into jars for pasteurizing to prevent floating above the brine. At the commercial level, this is done by a "pickle pusher" who uses a rubber mallet to hammer in the last pickles. This tight packing works since they are preserved by their fermentation and not solely by heat treatment as are other canned vegetables. The heat treatment is primarily to establish sealed anaerobic conditions and inactivate softening enzymes. Convection currents around the pickles are not important for safety as they are in low-acid canned foods which need to be packed loosely.

Though the lower salt brines result in better quality pickles and are helpful to those wishing to decrease some salt in their diets, pickles made by this method are not "preserved" for long-term storage; they require additional processing. Cucumbers that are fermented only in low-salt brines need to be stored at refrigerator temperatures or pasteurized to inactivate tissue-softening enzymes since the salt concentration is not high enough to inhibit them. In an acetic acid medium, such as brine poured over packed fermented pickles, 90% of *Leuconostoc mesenteroides*, *Lactobacillus plantarium*, and *Lactobacillus brevis* cultures are killed by 20–30 seconds of exposure to 150F (66C). To reduce the microorganism level to $1 \times 10^9$ cells or less per container (a standard level) takes 3.3–5 minutes of this heat exposure. USDA processing times of 30 minutes at 180F (82C) or 15 minutes at 212F (100C) for quarts of fermented dills at 0–1,000 feet (USDA, 1988) take into account heat-transfer efficiency and provide an extra margin of heat treatment. Without pasteurizing, the fermentation process continues and growth of aerobic organisms is a possible problem in unsealed jars. Some aerobic organisms can potentially make fermented pickles unsafe as they metabolize the acids. The higher salt brines—10% and above—do not need further processing for preservation, but a sealed jar may increase shelf life by eliminating mold growth. Vacuum-sealed jars do not require monitoring during storage.

Some USDA recipes (fermented dills, sweet gherkins, 14-day pickles) may be optionally processed in temperatures of 180–185F (82–85C) for 30 minutes instead of in boiling water (212F, 100C). The lower processing temperatures do prevent spoilage in the USDA specified recipes and the pickles are not as heat softened as with traditional processing. However, determining the water temperature is more involved than when a boiling temperature is used. These simmering temperatures are not easily visually determined, so a thermometer (candy or jelly) must be placed in the water without touching the sides or bottom of the canner periodically during the 30 minutes processing. It is recommended that sauerkraut and some of the USDA quick pickle products be processed at 212F (100C) instead of the lower temperature to avoid possible spoilage.

## Problems of Fermented Cucumbers

Householders fermenting cucumbers for the first time are often concerned about the pickles appearing spoiled. When reliable recipes are used with good cucumbers, fermentations rarely become unsafe or of unacceptable quality. Commercial fermentations also become unsightly during normal production. Bick's Pickles, near Toronto, Ontario, ferments pickles the traditional way. When a commuter train track was constructed beside their wooden pickle vats, the company started receiving complaints about the foaming mass of grey scum, wood splinters, dead insects, and bleached cucumbers floating in the vats. Bick's Pickles, unable to convince the nonmicrobiologist consumers that nothing was wrong with their brine, constructed a 20-foot-high fence between the train and the vats (McNish, 1986). Other pickle producers have had similar image problems but on a smaller scale as their vats were not along a main arterial road. However, many an employee claims never to have eaten pickles again after a season working as a vat skimmer and viewing the fermentation process.

Cucumbers in low-concentration brines may soften due to enzymes which can come from the vegetable itself, from yeasts, or from molds, because the salt and acid concentrations are not high enough to inhibit these microorganisms or the naturally present plant enzymes. Molds may grow on the fermentation surface or be present on remnants of the cucumber blossom. Careful removal of all blossoms before fermentation is an important step to decrease the amount of molds. To avoid the softening problem, some dill recipes initially cover the fresh cucumbers with a high-salt brine of 9.5–10.5% salt (36°–40° salometer) (Pederson, 1979). These finished pickles have a lactic acid content of 0.7–1.0% and between 4.5–5.5% salt. The 5% salt brine dills average 3% salt in the finished product. However, higher salt content in the finished product tends to overpower some of the more subtle fermented dill flavors. Fermenting at 4–8% salt and then holding at 10–12% salt concentrations is recommended for top textural quality and acceptable salt flavor for those not wishing to process in a boiling-water canner (Fleming, 1981).

Householders wanting to prepare low-salt pickled vegetables will have greater success using quick pickle recipes in which added acid and processing preserves the vegetable tissue. It is possible to produce a batch of acceptable flavor and safe to consume no-salt fermented cucumbers or sauerkraut, but it is more common that batches without salt will need to be discarded before fermentation is complete due to undesirable fermentation.

Commercially brined and fermented pickles have never been a reported source of *Clostridium botulinum* poisoning (Hesseltine, 1983). However, other microorganisms are responsible for numerous quality spoilage problems in fermented cucumbers. Yeasts are one group of troublesome organisms.

A white film on the surface of the liquid during fermentation indicates the presence of yeasts which require air to grow. Yeast activity can be decreased by covering the fermentation container with a brine-filled bag to enhance anaerobic conditions. The presence of yeasts is not reason enough to discard the entire batch, but the film should be removed daily, and the slime rinsed off the brine-filled bag to reduce recontamination. The main concern with yeast contamination is enzymes produced by yeasts. These enzymes soften plant tissue causing soft, slippery, and/or bloated pickles. Yeast enzymes also break down proteins and lipids which results in undesirable flavor changes. If yeasts were allowed to grow and multiply during fermentation, safety spoilage may occur. Yeasts use lactic acid for their metabolic processes—lactic is the main acid in fermented foods—so they can potentially decrease the acidity of the brine, making the growth of human pathogens possible.

The presence of molds is not a positive sign, but they float and can be skimmed off along with the slime from yeasts. If allowed to flourish, mold enzymes can soften pickles, and it is possible for molds to use acids and increase the pH of a food.

Texture is an extremely important characteristic in pickle evaluation. Commercial pickle manufacturers frequently introduce new ad campaigns, but the old theme of a crisp, crunchy product is still highlighted. Householders also strive for crisp pickles. Excessive softening of cucumbers during fermentation may be caused by solubilization of pectin substances (Sistrunk and Kozup, 1982). Pectic enzymes in the cucumber such as protopectinase, pectin methylesterase (also known as pectase), and polygalacturonase contribute to the softening of plants during ripening and aging (Fig. 5.3). These enzymes are also manufactured by yeasts. Protopectinase converts the insoluble, hard textured protopectin into colloidal pectin or water soluble pectinic acids; the net effect is softer tissue. Pectin methylesterase cleaves the methyl esters from pectin to produce poly-D-galacturonic acid, then polygalacturonase is able to break the molecule down further into

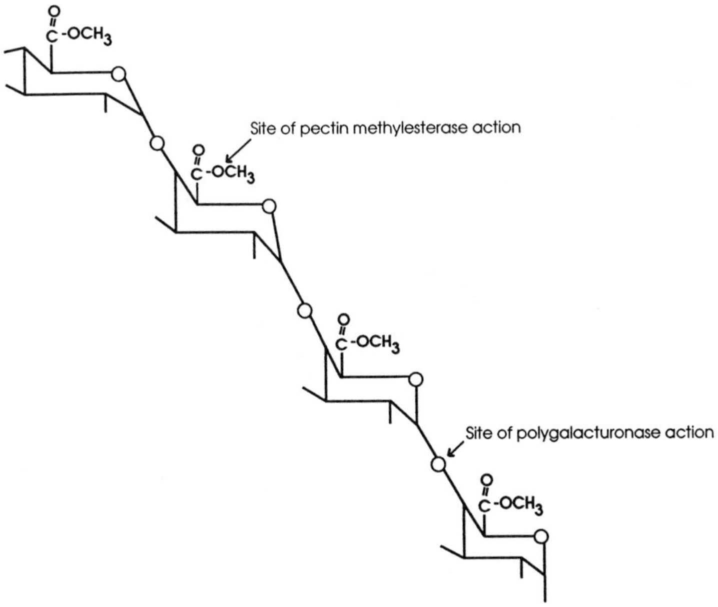


**5.3.** Enzymes that catalyze changes in the pectic substances that give plant cells structure.

single D-galacturonic acid units (Fennema, 1985); all contribute to the softening of the pickle because shorter chains of pectin are not as firm as the original long molecule. Heat treatments such as blanching raw cucumbers and pasteurization temperatures (180F, 82C) after fermentation can inactivate these enzyme systems resulting in a firmer end product. Pickle softening may also result from too high or low concentrations of salt or acid, the storage time, and storage temperature above 86F (30C) (Fleming et al., 1978).

The addition of calcium salts has a firming action on plant tissue due to its bonding between pectin molecules (Fig. 5.4). As calcium bridges are formed between pectin molecules, the structure of the molecule becomes more rigid. This procedure is commonly used for commercially pickled cucumbers and has at times been a popular addition to home processed pickles. In the 1970s and early 1980s the calcium, usually as lime, was not included in the ingredients of published household recipes. This trend may have reflected a growing consumer desire for "additive-free" food or perhaps a fear on the part of recipe authors that householders would not use food-grade lime. Calcium appeared again in the 1988 USDA pickle recipes. Calcium is currently often added to commercially canned tomatoes, potatoes, green beans, peas, and other plants for an improved firmer texture in these processed products. Alum has also been used in the past but is no longer recommended because of off-flavor and concern about human intake of large amounts.

As with other preservation methods, starting with top quality produce is important to quality of the finished product. Overripe cucumbers are softer than those of optimum ripeness. Their use in pickling results in a softer finished product too.

Large cucumbers and cucumbers packed in high-concentration salt brines are prone to bloat. Bloaters are cucumbers with hollow interior sections caused by gaseous fermenta-

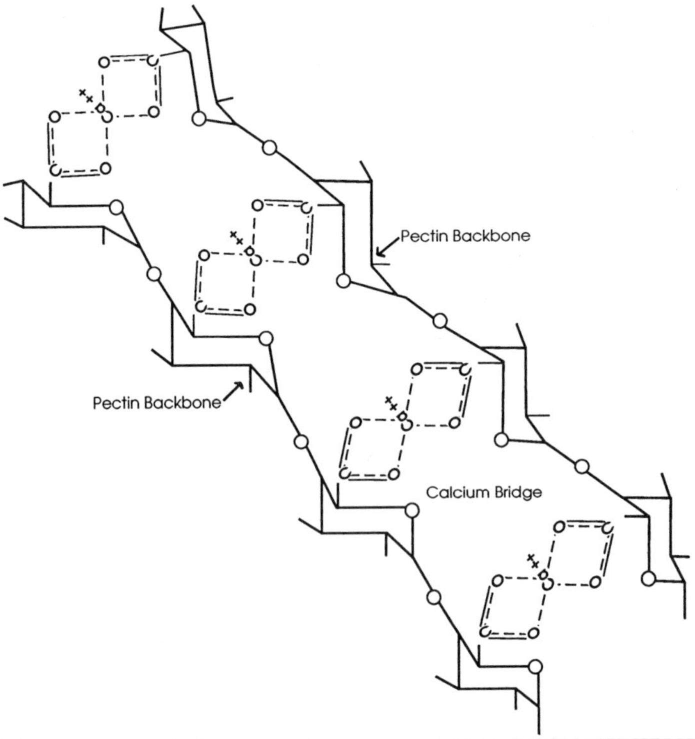


**5.4.** Calcium bridge to provide structure by linking pectin molecules in plant materials.

tion. Piercing the cucumbers with a fork helps the gas to escape—in addition to aiding brine penetration. Commercially, nitrogen is pumped through the vats to decrease bloating.

## Commercial Pickled Products

### Dill Pickles

The primary herb flavoring is dill. There are three types of fermented dills. For all, the brine usually has added vinegar at the beginning. Genuine dills go through a long natural fermentation. Overnight dills, also called fresh fermented, proceed through fermentation for 1–2 days and then are refrigerated for continued slow fermentation. They retain some characteristics of fresh cucumbers and also have a little of the fermented taste. Processed dill pickles were originally salt stock before freshening and finishing in a dill solution. In addition to these main types of dills, kosher dills are popular. They contain garlic or onion, and often red pepper, bay leaf, and/or mustard seed. They can be either fermented or quick types.

### Sour Pickles

These are salt stock cucumbers which have been finished in a spiced vinegar and water solution. Whole sour cucumbers are popular marketed either by themselves or with other sour vegetables. Chow-chow usually refers to chopped, sour mixed pickles (often cabbage) flavored with mustard, turmeric, garlic, cinnamon, cloves, ginger, nutmeg, cayenne, and pepper. Piccalilli can be sour pickles chopped into relish size. Sour relishes usually contain a mixture of chopped sour vegetables.

### Sweet Pickles

Homemade sweet pickles are usually not fermented, but some commercial types are. These are pickles in which the vinegar has been drained, and then they are finished in sweet, spicy syrup (or liquid). The liquid is added a little at a time until the desired sweetness is reached. They are then aged. Plain sweet pickles are marketed by the shapes they are cut into; sliced, chips, chunks, wafers. Candied pickles are extra sweet. They are usually sliced into chips. Sweet dill pickles are sweetened genuine or processed dills. Mixed sweet pickles are sweet pickled cucumbers combined with other sweet pickled vegetables such as cauliflower, onions, sweet peppers, and/or green tomatoes. Sweet piccalilli and sweet relishes are finely chopped sweet pickled cucumbers alone or in combination with other sweet pickled vegetables.

### Quick Pickles

As described in Chapter 4, fresh pack pickles are found in the same grocery section. These are made from fresh fruits and vegetables. Vinegar, seasonings, and sometimes sugar are added. The product is then pasteurized. These are preserved by the low pH and vacuum pack produced by boiling-water processing. The proportion of vinegar to water should not be less than 1:1. Natural fermentation with its complex flavors does not occur. The most common type in the fresh cucumber category is the bread and butter style in which cucumbers are cut crosswise and canned in a lightly spiced, sweet vinegar. Polish-style dills are usually quick dill flavored pickles. The flavor is similar to that of overnight dills but quite different from the wholly fermented genuine dills. There are many types of quick vegetable relishes and other pickled vegetables and fruits. Traditionally, the most variety was obtained through household recipes, but gourmet sections of some supermarkets carry a large selection.

## Summary

Fermented pickles and sauerkraut from *Raw Vegetables*

**SAFE IF:**

No spoilage

**TOP QUALITY IF:**

**a.** Top quality vegetables are used.

**b.** Recommendations for salt and added vinegar are followed.

**c.** Fermentation is done at cool room temperatures.

**d.** Fermented product is canned after fermentation is complete and processed with short boiling-water processing to halt fermentation and give airtight seal.

## References

Desrosier, N.W. and Desrosier, J. 1977. The technology of food preservation. 4th ed. Westport: Avi Publishing Company.

Dunn, C.M. 1977. Quality in home made pickles and relishes. No. B2267, Madison: Coop. Ext. Programs, Univ. Wisc. Extension.

Fennema, O.R., ed. 1985. Food chemistry. New York: Marcel Dekker.

Fleming, H.P. 1981. Spoilage problems of pickles. Presented at the Nat. Workshop for Ext. Fd. and Nutr. specialists. Oct. 1, 1981, Chicago, Ill.

Fleming, H.P., Thompson, R.L., Bell, T.A. and Hentz, L.H. 1978. Controlled fermentation of sliced cucumbers. J. Food Sci. 43:888.

Hesseltine, C.W. 1983. The future of fermented foods. Nutr Rev. 41:293.

James, C. and Buescher, R. 1983. Preference for commercially processed dill pickles in relation to sodium chloride, acid, and texture. J. Food Sci. 48:198.

McNish, J. 1986. They make pickles the old-fashioned way, and it's disgusting. Wall Street J. April 15, 1986.

Pederson, C.S. 1979. Microbiology of food fermentations. 2d ed. Westport: Avi Publishing Company.

Sistrunk, W.A. and Kozup, J. 1982. Influence of processing methodology on quality of cucumber pickles. J. Food Sci. 47:949.

USDA, 1988. Complete guide to home canning. Ag. Inf. Bull. 539. Washington DC: U.S. Dept. of Agriculture.

USDA, 1965. Effect of household processing and storage on quality of pickled vegetables and fruits. Home Econ. Res. Report No. 28. Washington DC: U.S. Govt. Printing Office.

## CONSUMER QUESTIONS

**Q.** *Can salt be omitted from sauerkraut recipes?*

**A.** Salt permits the naturally present lactic acid bacteria to outgrow the other types that it inhibits, which increases the probability of a fermentation rather than spoilage. It also inhibits enzymes which can make the cabbage soft. Attempts at leaving salt out should be made only by people prepared to discard many batches, and able to determine when to discard.

**Q.** *Can salt be left out of brined cucumber pickles?*

**A.** See above. Salt is needed in specific concentrations to encourage growth of desirable fermenters and to inhibit softening enzymes. Two percent brines are more likely to result in spoilage. Five percent salt brines (3/4 c/gal water) result in top quality pickles instead of the 10% brines found in some recipes.

**Q.** *Is a white filmy yeast growth on the surface of fermenting foods a problem?*

**A.** Yeasts require air to grow. Covering the fermentation container with a brine-filled bag to enhance anaerobic conditions decreases the occurrence of this. Yeasts use lactic acid for their metabolic processes—lactic is the main acid in fermented foods—so they can potentially decrease the acidity of the brine making the growth of human pathogens possible. Yeasts also have enzymes that soften pickle tissue. The film should be removed daily.

**Q.** *What causes soft or slippery pickles?*

**A.** Most likely due to enzyme activity. Yeasts (see above) or molds may be involved. The softening enzymes could come from the cucumber itself. Too weak a brine will allow these enzymes to remain active. Leaving remnants of the cucumber blossom attached will contribute a much greater load of enzymes into the jar, as the enzymes are particularly concentrated there. Yeasts manufacture enzymes which can cause softening. Surface yeasts grow as scum on the top of fermenting vegetables and on pickles not covered by brine. Their enzymes can hydrolyze pectic materials in the cucumber which softens it. Remove surface scum daily to prevent these effects.

**Q.** *How can shriveled pickles be prevented?*
**A.** Water loss (wilting) between the time the cucumbers are harvested and put into brine is the most common cause. Too high salt or too high sugar concentration in brines initially will draw excessive water out of the cucumber and cause shriveling. Try a 5% salt brine (3/4 c salt per gal of water).

**Q.** *Can hollow pickles be prevented?*
**A.** They may have grown that way on the vine, especially if not receiving enough water. Especially dry summers result in a lot of pickle-quality complaints. When pickling whole cucumbers, it is a good idea to slice samples to check the interior. If using cucumbers from a home garden or U-pick field, harvest as close to the fermenting time as possible. Holding for over 24 hours increases the chance of hollow pickles. Hollow sections could also be the result of bloating. This may be caused by yeast growth, fermenting at too high a temperature, or packing large cucumbers in high-salt brines.

# 6 Curing

■ New Year's hams, Fourth of July hot dogs, Octoberfest sausages, Sunday brunch bacon, and sack lunch bologna; cured meats have an important place in the American diet today, but in past centuries they were often the only meats available for entire communities. Curing is a low-cost, low-technology method that has been practiced by many civilizations wherever salt was available. Salt was used for curing fish in Asia as early as 3,500 B.C., salted meats made into sausage were mentioned in Homer's *Odyssey* around 800 B.C., and by 5 B.C. salted meat slabs and strips were a staple in the Mediterranean area. In the early days of curing, salt mined from naturally occurring deposits was rubbed on the surface of meats. This salt often was a mixture of sodium chloride (table salt) and sodium nitrate (saltpeter). The nitrate salt produced a distinctive flavor and pink color that did not deteriorate with cooking and was enjoyed by people then as now. Though nitrated meats may have been produced accidently at first, the salt deposits with nitrate were quickly favored.

The basic reactions within cured foods have always been the same and household curing procedures have changed little; however, commercial curing has undergone radical transformations. The commercial hams, bologna, and sausages displayed in retail stores now are so different from the products of only 15 years ago that they are a source of many consumer questions. Since there is so much confusion about them and holiday times find food professionals inundated with questions, commercially cured foods are discussed in this chapter along with household curing methods.

The United States has a long history of curing meat with salt and smoke treatments. Native Americans boiled sea water for salt and also mined it from the earth. They commonly used it to preserve buffalo meat and fish for times when game was scarce and for eating away from home. The European colonists were familiar with salt preservation and gave great value to America's salt deposits. Several large battles were fought on U.S. soil during the American Revolution and the Civil War over control of salt mines (Fort Ticonderoga and Saltville, Virginia). In colonial New England a brine-filled pork barrel was in every reputable cook's cellar, and early Southern plantations depended upon their smokehouse and ice house for a stable food supply. Refrigeration has replaced the ice house, but smokehouses are still common in the rural South. Curing continues at the household level with the original technology, but since the 1970s the use of nitrate salts and the overall salt concentration has significantly decreased. In today's household, meats that have been lightly cured for flavor variety and preserved by refrigeration are quite a different product from those cured a few decades ago as a preservation method and require different handling.

## Curing Procedures

There are several methods for curing meats. Dry cures, pickle cures, and combinations of the two are used by householders. With dry cures, the salty seasoning mixture is rubbed on the surface and the salts draw moisture from the tissues forming a brine. Penetration of the brine into the meat is by diffusion. In conventional pickle curing, the meat is submerged in a preformed brine and then diffusion takes place. If several hams are being cured in the same container by either method, they need to be repacked during the diffusion period to assure uniform penetration.

Dry curing has the advantage of being low technology and usually there is little spoilage. Spoilage around the ham bone is the most common type of bacterial problem. The exact set of conditions which result in this problem is not known; however, it has been suggested that slow salt penetration is a factor. Dry curing is used for commercially cured country hams which are premium priced when they are available. Properly cured country hams do not need refrigeration. Householders most often dry cure in traditional barrels but commercially, watertight shelves for efficient use of space and metal pressure boxes are popular. The pressure boxes force the meat juices out faster than the salt alone would draw them out. They are commonly used for dry cure bacon. Dry curing usually takes 2–2.5 days per pound for hams and shoulders, and 7 days per inch for bellies. An average bacon slab requires 10–14 days. These meats tend to have higher salt content than the pickle cured. Dry cured hams are too salty to be cooked directly after aging; they must first be freshened. A common way of freshening is a 24–36 hour soak, followed by scrubbing the outside of the ham with a vegetable brush, and simmering several hours in a large kettle of water, before baking in the oven.

Pickle curing has the advantage of producing a less salty food which many consumers prefer now and it requires less labor. The pickle consists of a brine of water, salt, sodium nitrite, perhaps other seasonings, and sugar. (The word brine and pickle are interchangeable in curing.) Sweet pickle cures contain more sugar. The brine or pickle is added to the meat by immersing or injecting. Conventional pickle curing involves immersing the meat in brine. It can be used for all cuts and is applicable at the household level. Immersing is the chief method used today for commercial corned beef. The curing time for immersed pickling is about the same as for dry cures: 2–2.5 days per pound for hams but time varies with the brine strength. Curing solutions are more concentrated than the final salt levels in the meat. The density of commercial brines is usually 46% salt (70° salometer) at 40F (4C). Household pickle cures most often are 15%–24% salt, but wider ranges in recipes exist. Substituting a 50:50 mixture of potassium and sodium chloride for table salt produces an acceptable ham for some consumers who must follow decreased sodium diets; however, higher levels of potassium chloride tend to be too bitter.

In the 1940s injecting brine into meats became popular. Artery pumping and stitch pumping are the two main types of injection. Both are used extensively by the meat industry and can be done at the household level. When the pickle is injected, a long holding time for diffusion to take place is not necessary. This makes injection a less expensive way to cure and very attractive to commercial curing operations.

Artery pumping forces brine into the meat through the slaughtered animal's still intact circulatory system. The arteries have strong walls to withstand pressure from the heart's pumping, consequently they are used instead of veins. The historical story contends this process originated when a New Zealand undertaker experimented with his embalming equipment. Artery pumping is currently used for commercially curing some hams and occasionally picnic cuts. This method is labor intensive so expensive cuts are used to make it cost effective. Householders use artery pumping for hams. The pumping needle is inserted in the ham's femoral artery and the brine is added until there is an 8 to 10% weight increase. Some processors also put a pump in the branches of this artery which may result in a more uniform cure. Up until the 1950s pumped hams were held in brine 2 weeks before smoking for diffusion of the cure. Currently they are held from 1 hour to 3 days depending upon the processor. Modern pumping results in a mildly cured product which requires refrigeration for preservation.

Some artery-pumped hams are rubbed with dry cure and then held for a short time; they are classified as combination-cured. Combination hams have many of the desirable characteristics of each method, but do not require the long curing times that the solely dry cured products need.

Stitch pumping forces the brine directly into the muscle portion of meat through a needle(s) with several openings. In commercial hams, this is done with an automated assembly line where several needles puncture the muscle and inject simultaneously. Generally, 10% by weight of a 42% salt (65° salometer) pickle containing 150 ppm nitrite plus alkaline phosphate is injected. Stitch pumping is used for less expensive cuts such as picnics, shoulders, and bellies. Approximately 80% of the bacon on the market is stitch pumped. It has greater shrinkage when cooked than the dry cured; but when multiple needles are used, the brine distribution is excellent and it is a very cost-effective method for industry.

## Curing Ingredients

Various cures (curing salt mixtures) are available for household curing but all contain salt, sweetener, and nitrite. Optional ingredients include spices, baking soda, phosphates, sodium erythorbate, sodium ascorbate, hydrolyzed vegetable proteins, and monosodium glutamate.

### Nitrate and Nitrite

Nitrate additions contributed flavor, the distinctive red color of cured meats, retarded fat rancidity, and had a preservative effect against growth of *Clostridium botulinum*. In the early 1900s, it was discovered that these effects occurred only after nitrate ($NO_3$) was converted to nitrite ($NO_2$) in the food; nitrite was the active compound. Therefore, in 1925 the USDA approved additions of sodium nitrite with nitrate. Due to possible carcinogenicity, the use of both of these compounds is now strictly controlled. Federal meat inspection regulations of 1978 made nitrate additions illegal for all commercially cured meats except some fermented sausages where botulism poisoning is more of a risk. Nitrate is still readily available to consumers and some householders continue to use it for curing. Nitrate is not considered more toxic than nitrite; it was simply deemed an unnecessary additive. Additions of nitrate had been used only to provide a reserve that could be converted to nitrite in case the added nitrite became depleted during curing (Fig. 6.1). Since wide availability of refrigeration has made *C. botulinum* growth in cured foods uncommon, it was decided that this margin of safety against bacterial spoilage was no longer needed. Other compounds that inhibit *C. botulinum* growth such as ascorbate, erythorbate, and lactic acid cultures are now of interest in curing mixtures. A 1986 regulation specifies the amounts of these required to increase safety after nitrate was removed and nitrite levels limited in bacon. Other cured meats do not have such a specification. No minimum level of nitrite in cured meats is established. Since the functions of nitrite include inhibiting the growth of food poisoning and food spoilage microorganisms and retard-

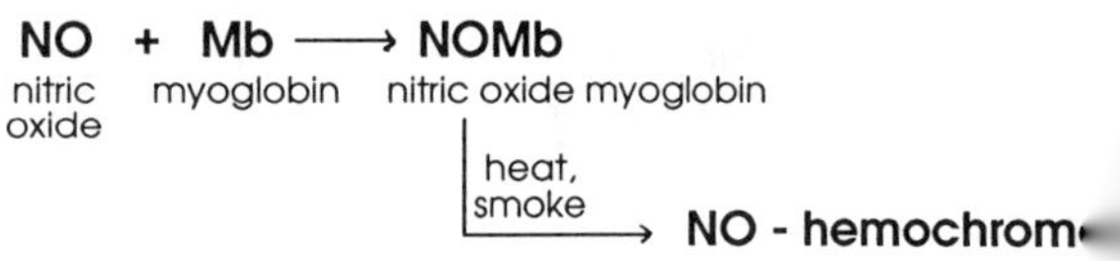


**6.1.** Reactions leading to cured-meat color.

ing rancidity, lower levels of nitrite in cured products have been accompanied by labels stressing refrigerated storage.

## Salt

Salt preserves by dehydrating the meat and altering the osmotic pressure which inhibits bacterial growth. See Chapter 3 for the exact $a_w$-preservation mechanism). Sodium chloride (table salt) is the only ingredient necessary for curing by definition but salted meats with the high concentration of salt required for preservation are uncommon now. Salt used alone gives a harsh, dry, very salty product and causes the lean portions to turn dark. Fat backs are often only salted, then added to foods such as pork and beans for flavor. Salted clear plates, jowls, and heavy bellies are also used as seasoning meats. Though today's cured meats contain lower salt levels than in the past, they still have a fairly high final salt content: hams about 3% and bacon 2%. Country hams containing sodium nitrite must have an internal salt content of at least 4% (CFR, 1990).

## Sugar

When sugar and corn syrup solids are added to cures, they soften the harsh flavor of salt and retain some moisture in the meat for palatability. This sugar-bound water is not available to microorganisms. Usually sucrose (table sugar) is the sugar used at the household level. Corn syrup and corn syrup solids—a dehydration product of corn syrup—are less expensive than sucrose which makes them desirable to the food industry. These sugars contribute flavor and are the main energy source for nitrate-reducing bacteria in the curing solution. These bacteria were especially important for safety when only nitrate was added to cures and the meat was stored unrefrigerated since they converted nitrate to the essential nitrite, which inhibits *C. botulinum*. Sugar also helps to lower the $a_w$, though the concentrations of sugar in curing are not high enough to significantly reduce the $a_w$ and reduce bacterial growth. Sugars participate in the Maillard browning reaction which results in a pleasing color and flavor; however, this can be a problem in some bacon cures where the result is bacon that overbrowns when fried.

Commercial operations most frequently use 20–30 pounds of sugar per 100 gallons of brine even though consumers have shown greatest preference for 16 pounds per 100 gallons of brine (2% sugar) in taste tests of cured hams (Pearson and Tauber, 1984). The amount of added corn syrup solids is limited to 50 pounds per 100 gallons of brine under federal regulations. Some householders take great pride in their sugar-cured hams and since these recipes are not used by the food industry, they truly have a unique product.

## Ascorbates

Sodium ascorbate (vitamin C) and sodium erythorbate (similar to vitamin C) are individual compounds in the larger category of ascorbates. Both are reducing agents which inhibit the formation of nitrosamines in cured meats (Fiddler et al., 1973). Sodium ascorbate or sodium erythorbate can be added at the maximum level of 0.875 ounce per 100 pounds sausage emulsion or 87.5 ounces per 100 gallons of pickle solution. In bacon, ascorbates are added at 500 ppm levels which significantly reduces or eliminates nitrosamine formation. The exact mechanism is unknown.

Ascorbates are added to cures for additional advantages. They reduce metmyoglobin to myoglobin, thereby reducing the processing time of frankfurters by one-third through elimination of the waiting time for nitrite to form a stable pink pigment. With the addition of ascorbate, cured emulsions such as sausage, bologna, and hot dogs can be transported directly to the smokehouse after stuffing.

When exposed to light, sliced cured meats are especially prone to color fading. Retail meat displays often have the package turned upside down to shield the meat and a picture of a ham slice is placed on top to attract consumers. The antioxidant properties of ascorbates retard this fading and also retard rancidity which can be a significant spoilage problem. The mechanism for both reactions probably involves prevention of heme-catalyzed lipid oxidation. When the ascorbate is depleted, the heme products are then degraded and apparently catalyze lipid oxidation. Ascorbates can increase shelf life, but not prolong it indefinitely.

Vitamin E (tocopherol) is also sometimes added to curing mixtures as a reducing agent. Its name on the label is familiar to consumers and results in fewer "additive" questions.

## Sorbate

Potassium sorbate inhibits mold growth. It is allowed at a 2.5% level in dipping solutions for stuffed dry sausages to retard molds on the casings. When sorbate is used in combination with low levels of nitrite, it inhibits *C. botulinum* (Sofos et al., 1979 and Sofos and Busta, 1981). Research suggests that addition of potassium sorbate at a 0.26% level in cured bacon would be desirable to permit reduction of the levels of added nitrite. Possible adverse reactions to sorbate when these products are eaten has prevented its addition.

## Phosphates

Phosphates are added to increase the water binding properties of cured meats. This is a cost-effective way of increasing the finished yield for commercial manufacturers since water is significantly less expensive than meat, and consumers prefer the moist, juicy product which results with phosphates in the cure. Phosphates improve water retention by raising the pH, which causes the muscle proteins to unfold, thereby making available more sites for water binding. When phosphates are added to pumping brines, finished yields of hams can reach over 100%. Phosphates also combine with trace metal ions which retards rancidity.

Intact hams and shoulders may be cured with the addition of sodium tripolyphosphate, sodium hexametaphosphate, sodium acid pyrophosphate, sodium pyrophosphate or disodium phosphate. Sodium acid pyrophosphate is the only phosphate permitted in sausages. The maximum level of residual phosphates allowed is 0.5% in the finished food.

There is some concern about the amount of phosphates in the diet of people who consume large amounts of commercially processed foods. Since phosphates are classified as a safe substance they are a common carrier for food additives. Since many people already have inadequate calcium intakes, phosphate food additives may compound this nutritional problem. The phosphates in cured meats alone do not have a significant nutritional impact in the average North American diet.

## Cured Color

There are several biological compounds that contribute to the color of fresh meat, but myoglobin is probably the most important since it is the most abundant. Hemoglobin is still present in bled meat but its reactions with nitrite are similar to those with myoglobin so only myoglobin will be discussed here. When combined with oxygen, the myoglobin pigment is in a bright red form (oxymyoglobin). This is the color of raw steaks and ground beef in the market. Under acidic and reducing conditions in the presence of nitrite, the myoglobin is converted to nitric oxide myoglobin—the red color of uncooked cured products. Acid for this reaction may be produced by the muscle tissue itself and bacteria (especially micrococci) produce the reduced conditions. These reduced conditions were also necessary for reducing nitrate to nitrite, when nitrate was added as the source of nitrite.

Figure 6.1 shows a possible pathway for cured meat color. Some other theories involve nitric oxide metmyoglobin as an intermediate step. When cured products are heated, the proteins denature and nitrosyl hemochrome is formed, which is the pigment responsible for the familiar pink color of cooked cured meats.

## Nitrosamines

Nitrite changes to nitrous acid which reacts with secondary amines, forming nitrosamines (Fig. 6.2). This reaction can take place in cured foods or within the human stomach after ingestion of foods containing nitrite. Nitrosamines have been shown to be carcinogenic compounds in animal studies. This has prompted questions on the relationship of

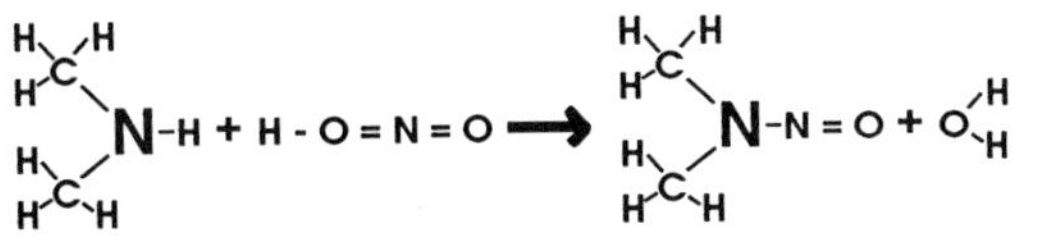


**6.2.** Nitrosamine formation in cured meat.

human cancers and consumption of cured meats.

Nitrite can form nitrosamines in dry curing premixes, so in 1973 the FDA required that the nitrite be packaged separately from the other ingredients unless a chemical buffer to keep the nitrite separate is used. Nitrosamines have not been found in these modern mixtures. As with commercially cured meats, nitrite has not been eliminated from home curing mixes based on the argument that it prevents the outgrowth of food poisoning organisms, particularly *Clostridium botulinum* in cured meats that are held at room temperature. It is considered that the risk of botulism poisoning is greater than risk of cancer from the low—several ppb for most—levels of nitrosamines which may be found in these products. Some studies suggest nitrite is carcinogenic in its original form; however, after reviewing the research, the National Academy of Sciences decided more studies were needed before removing this compound since it had a proven beneficial effect on food safety (Natl. Academy Sci., 1981).

Bacon poses a special problem because nitrosamines can easily form in bacon as it is normally prepared for consumption. The formation of nitrosamines through a reaction between nitrite and secondary amines is accelerated by high heat such as frying. The amount of nitrosamines formed during frying is reduced greatly by keeping initial levels of added sodium nitrite below 120 ppm and by the presence of α-tocopherol (vitamin E) coated salts, potassium sorbate, and sodium ascorbate. These additives are commonly in commercial bacon; unfortunately their imposing names have caused consumer concerns about chemicals in the food supply. Frying the bacon at lower temperatures for shorter periods of time (thoroughly heated vs crispy) also reduces the nitrosamine formation (Gray et al., 1982). Nitrosamine formation is very unlikely in microwaved bacon which does not reach high temperatures. Discarding the bacon fat instead of cooking with it is now recommended since the fat contains 2 times the nitrosamines of the bacon strips. Nitrosamines are also volatile, thus one should avoid breathing the air directly over the frying pan. Concern about nitrosamines in the diet has prompted the following federal regulations for commercial curing in the United States: maximum levels of nitrite allowed are 2 pounds per 100 gallons pickling solution that is pumped into the meat at a 10% level, 1 ounce per 100 pounds meat for dry curing processes, and 1/4 ounce per 100 pounds chopped meat. The maximum amount of sodium nitrite that can be added to bacon is 120 ppm nitrite with 550 ppm sodium ascorbate. This level of ascorbate in combination with α-tocopherol-coated salt significantly reduces N-nitrosamine formation when bacon is fried. The level of nitrite cannot be over 200 ppm in the finished food product—bacon, ham, sausages. No sodium nitrite nor sodium nitrate is allowed in infant, junior, or toddler baby meats.

Nitrates naturally occur in some vegetables. Some nutritionists speculate that 72% of nitrites and nitrates in the adult American diet come from vegetables. Beets, celery, lettuce, spinach, radishes, and rhubarb each contain approximately 200 mg nitrate per quarter pound of vegetable (2,000 ppm). Vegetables such as cabbage, kale, mustard, and turnips are also high in nitrate. Most adult Americans ingest 100 mg nitrate, 1 mg nitrite, and 1 g preformed nitrosamines daily. Vegetarians ingest an average of 268 mg nitrate daily. The nitrate content of vegetables increases when they are grown in nitrate-fertilized (both organic and synthetic fertilizers) soils (ACSH, 1989).

Perhaps it places the threat of cured meats into perspective to note that human saliva contains nitrite. Some researchers feel it is a larger contributor of nitrites to the stomach than foods are for the average person living in the United States. An estimated 9% of nitrites in the diet usually comes from cured meats and

19% from human saliva (ACSH, 1989).

## Smoking

Smoking is an optional treatment after curing which householders use now primarily for flavoring. The term smoked does not always indicate the ham has been cooked. Over 300 different compounds have been isolated from wood smoke, but not all of these are known to be present in smoked meat. Some of these compounds cause desirable changes in the food and some raise health concerns. Smoke compounds added to meat by hanging it over burning wood do not penetrate the meat well and their action is primarily on the surface; however, smoke compounds injected to the center portions for flavor will have an effect throughout the product.

The antioxidant properties of smoke have been beneficial to householders preserving their own meats. Smoking protects the meat from surface oxidation and thus extends the shelf life. Wood smoke that is produced by smoldering contains the largest amounts of phenols with high boiling points. These compounds have the greatest antioxidant properties. Most commercial processors use smoldering sawdust to produce smoke economically, but woods such as hickory, red oak, apple, and mesquite are favorite choices for household smokers.

Phenolics, carbonyls, and oxides are smoke compounds which penetrate the surface of smoked meats. Smoked meat flavor and aroma are primarily due to these compounds. Some of the phenolic hydrocarbons in smoke and smoked foods have been found to be carcinogenic in laboratory animals, and implicated in human cancer cases of Baltic Sea and Icelandic peoples who consume large amounts of smoked fish. Due to these concerns, phenolic hydrocarbons that do not contribute to flavor, color, or aroma have been removed from many liquid smoke preparations. The Maillard browning reaction during smoking is in part responsible for the color of smoked foods, but compounds in the smoke itself also contribute to color changes (Ruiter, 1979). Nitrous oxide in smoke can react with secondary amines in the meat to form nitrosamines but this reaction is not favored under the slightly acidic pH of the meat (Doerr et al., 1966); thus it is not a great concern for consumers. Organic acids in the smoke are responsible for surface coagulation of proteins, which is very important for skinless frankfurters and sausages so that they maintain their shape and also for the peelability of those with inedible casings.

The threat of trichinosis has determined cooking endpoints for pork products for decades. This cooking is most often done at the same time as smoking. Whole pieces of pork are required by federal regulations to reach an internal temperature of 137F (58C) to kill any trichinae present; most commercial operations heat to 140F (60C) for a margin of safety. Federal classifications label "smoked hams" those that have reached an internal temperature of 140–147F (60–64C), and "fully cooked" (also called cooked, ready-to-eat, and ready-to-serve) as hams which have reached temperatures of at least 148F (64C). Surface microbial growth is retarded in smoked meats due to the effects of drying, heating, and compounds (acetic acid, formaldehyde, creosote, phenols) in the smoke.

Smoking does have an effect on nutrients, but this is not significant in the average U.S. diet. Heat-labile thiamin is often destroyed and the nutritive value of proteins may be decreased when carbonyl compounds in the smoke react with basic amino acids causing a loss of available lysine.

## Hams

Several types of hams can be cured at the household level. They are distinguished by pork cut and cure.

### Country Hams

American dry cure hams are primarily produced in Virginia, in Suffolk, Smithfield, Richmond, Southampton, and Jamestown, where this uniquely flavored smoked ham is thought to have originated. Visitors from England during colonial times were surprised at such an exceptional flavor from a cure process similar to theirs. To this day, the exact cure seasonings used in many Virginia hams are closely guarded family secrets, and households take great pride in the quality of their hams. Country hams are not cooked but their dry curing process eliminates any trichinae threat (Pearson and Tauber, 1984). In dry curing, the salts are applied initially and then again 5–14 days later. The cure slowly penetrates from the outer edges to the interior. The total curing time for average size hams is 30–40 days after which they are held 20 days at 36–40F (2–4C) to allow salt concentrations to become uniform throughout the ham. This process is called salt equalization. At the end of curing, the innermost portions are not as salty as the outer so salt equalization is important for both flavor and preservation.

Bacterial growth is not a concern in properly cured, salt equalized country hams. After salt equalization country hams are aged 6–12 months. Following aging, they may be smoked 1 1/2–2 days. A 35% shrink in weight is common for these long-cure uncooked hams. Country hams have a final moisture content of 50–60% and are 4–5.5% salt. Their low $a_w$ ($<0.92$) allows safe storage without refrigeration (CFR, 1990). Mold growth may occur during salt equalization or aging of country hams. Sometimes the mold is removed with a cloth soaked in vegetable oil or vinegar before the commercially cured ham reaches consumers. Some consumers simply expect a moldy outside along with the yellow fat and dark red, very firm lean characteristic of this product.

When nitrates were still used in dry cures, the chemical changes that took place during curing were slightly different. In the old-time scenario, after the cure was rubbed on the outside of the meat, nitrate was gradually reduced to nitrite. This slower availability provided better color development than nitrite alone can achieve. Long-time household curers may complain about this color difference in nitrite-only hams. Certain ethnic groups in the United States dry cure specialty products at home, and small cured-meat businesses usually offer a great variety of products not seen in supermarkets. Consumer questions regarding these unfamiliar foods are common. Such uncooked hams include Westphalian (center muscle of shoulder) from Germany, prosciutto from Italy, and Scotch hams. These all undergo a 2-week dry cure followed by a 2-week pickle, a 1-month ripening period, and 7 days in the smokehouse. Westphalian hams possess a distinctive flavor due to smoke from juniper berries and twigs added to a beechwood fire. Some Scotch hams are produced in the United States. These unique dry cure hams have the skin, visible fat, and bone removed. The meat is then rolled and either tied or put into a casing. Prosciutto, originally made in Italian households, is now manufactured commonly in the United States through a dry cure process similar to American country hams, but first the aitchbone is removed to allow the meat to flatten. After 45 days of dry cure (salt, sugar, allspice, pepper, nutmeg, mustard, coriander, nitrite), weights are placed on top or pressure molds are used to flatten to 2 inches. After flattening, prosciutto hams are scrubbed with water and smoked several days, then rubbed with white and black pepper until thoroughly coated and aged 30 days.

### Sectioned and Formed Hams

This type of retail ham is the one most commonly served at home but it is not made there because special additives and tumbling equipment are required. These are also the most popular hams among food service establishments. Ten- and 15-pound restaurant-pack, boneless buffet hams are marketed as hams in which the quality is uniform throughout and each slice is the same diameter. Such popularity generates many consumer questions.

Sectioned and formed hams are made up of intact muscles and large sections of muscles bound together with particles of meat to form a single, solid piece. The binding substance may be nonmeat additives, meat emulsions, or myofibrillar proteins. These compounds form a protein framework on the surface of the pieces. Tumbling and massaging the meat is necessary to form the bonding matrix from the meat's own myofibrillar proteins. Polyphosphates are added before tumbling to increase the strength of the bonds and increase finished yields by water binding. Addition of salt helps extract the myofibrillar proteins for binding and increase the water holding capacity. Generally salt is added at the 2–2.5% level. Nitrite is added at 156 ppm level. The cure is absorbed readily by the pieces as they are tumbled which gives a product with more uniform color and flavor than achieved by conventional methods.

Tumbling has the added advantage of making the pieces soft for shaping. Shaping is usually done by forcing the products into casings or metal containers. This molded meat is then heated, and sometimes smoked to coagulate proteins and make a stable bond. An internal temperature of 135–155F (57–68C) results in a good bond and the meat is often held at this temperature to destroy microorganisms, which increases shelf life (Cross et al., 1971).

### Shoulders

Cured shoulders (picnic cut) are marketed as picnic hams. Shoulders are sometimes cured at the household level as their smaller size is appropriate for family consumption. Commercially, picnics are usually stitch pumped mechanically so are economically priced. Federal curing regulations are the same for butt hams and picnics.

Boston butts (cottage rolls) are boneless pork shoulder cuts. They are usually stitch pumped followed by immersion in a pickle and then smoked to an internal temperature of 142–152F (61–67C). Boston butts are often sliced and fried as a leaner alternative to bacon.

## Sausages

Sausage comes from the Latin word *salous* which means minced meat preserved by salt. Sausages may have been used by the Chinese as early as 1,500 B.C., and are mentioned in written works by 8 B.C. as being a common food. These ancient foods were low-moisture meat cylinders that had undergone fermentation. The interior of this type of sausage (and modern sausages in casings) is anaerobic. Both salt and the low pH of fermented sausages inhibit microorganisms; but in the past it is likely that these critical ratios were not always achieved, as the word botulism is derived from the German word for fermented sausage. Today, only a few types of sausages are fermented and most contain significantly less salt than the original products so must be preserved by refrigeration or freezing. The first sausages used intestines and stomachs of slaughtered animals for the casing. This resulted in a meat mixture with cylindrical shape; a tradition kept today with the synthetic casings.

The broad term *sausage* covers more variety than the ground meat on top of a pizza or the links commonly served beside eggs and toast at breakfast. Luncheon meat and frankfurters are also technically sausages. Sausages can be emulsions of finely chopped meat or mixtures of coarsely ground meats; they can be fresh or cured, unsmoked or smoked in a variety of ways, uncooked or ready-to-eat, and some may have water added or be fermented products. Traditionally sausages have not been low-fat foods and today's fresh pork sausage is only 65% lean meat.

Dry and semidry sausages are fermented by either naturally present or inoculated lactic acid bacteria. *Pediococcus cerevisiae* and *Lactobacillus plantarum* are the types most often inoculated. The fermentation takes place before stuffing. The lactic acid produced gives a distinctive flavor and has a preservative effect by lowering the pH. Summer sausage is fermented at 40F (4C) for 12–72 hours. The end pH is 4.8–5.0 (0.75–1.0% total acidity as lactic acid). After fermentation Thuringer and summer sausage are semidried to approximately 75% of their original weight. Hard salami, pepperoni, and cervelats are dried to about 65% of their original weight. The drying may take 10–100 days, depending on the product.

Like ham cures, sausage seasoning mixes can be family heirlooms. Most sausages available in the United States today are of European or Asian origin with the notable exceptions of Lebanon bologna originating in Pennsylvania's Lebanon County and scrapple which may have originated in Chester County, Pennsylvania, but was common on farms throughout the Midwest wherever a woman of German descent cooked.

Scrapple is still made today in U.S. households.

### Casings

Sausage is often a meat emulsion of too thin consistency to undergo processing without being contained. Casings are used throughout the world as seasoned, ground meat containers. The casing determines the size and shape of the sausage. It may serve as the processing mold, container for handling and shipping, and as a merchandizing unit for display. It must be not only strong to form the link (with string or metal clamps) but also flexible for meat contraction and expansion during processing and storage. At one time, sausage production was limited by availability of animal intestines, but now cellulose and collagen casings are popular. A variety is available for both home and commercial use.

Animal casings are edible but expensive so only top quality products warrant their use. The gastrointestinal tract of sheep, hogs, and cattle from gullet to anus is used for casings. Pieces of the tract are graded by diameter since this differs by location. Bladders are used for special sausages such as mortadella which is a dry sausage with a beef bladder casing. The animal parts are washed and treated with chemicals to remove soluble components before being sold as casings.

Preflushed casings are available packed in a brine solution. They need to be flushed again to remove any impurities and the salty brine, and to prevent color problems. Pretubed casings have a plastic tube on the end which makes them easy to transfer to the stuffing horn.

Regenerated collagen casings are more uniform in size than animal gastrointestinal tract casings, but have many of the same desirable characteristics. The collagen usually comes from beef hide. After production it is pushed through a die to form a tube. Small collagen casings are edible and frequently are used for fresh pork sausage links. The material used for the large-diameter casings is often treated with aldehydes to cross-link the collagen for increased strength. These cross-linked casings are too tough to eat.

Cotton cloth bags have been used to make cellulosic casings since the U.S. Civil War when both sides ran out of animal gut casings. The bags (1, 2, or 5 lb) are stuffed tightly with ground pork until the fat seeps through the cloth. They are then smoked which, in addition to flavor, seals the fat into an impermeable coating which keeps the inside moist. The cellulose may also come from linters—a fuzz removed from cottonseed after it has gone through the cotton gin. Linters are treated and extruded to form a tube. These are widely used by the commercial food industry and may be pigmented to enhance appearance.

### Sausage Ingredients

Spices provide flavor and in some cases bacteriostatic and antioxidant (ground mustard seed) properties. Black pepper is the most popular single spice, but most existing spices are used in one type of sausage or another.

Ice is often added in the mixing stage to prevent mechanical overheating and to dissolve the curing salts. Ice also adds fluid important for emulsifying the components and thins the consistency so the mixture will flow into the casings without tearing them. The final water content greatly affects the texture and tenderness of the sausage.

Some fermented sausages contain glucono-delta-lactone (GDL). GDL decreases the pH which accelerates the development of cured meat color. It can be added to dry and semidry sausages at a 1% level where it decreases the pH by 0.5. This rapid drop in pH also decreases the growth of microorganisms until fermentation takes place. GDL imparts a distinctive "biting acid" taste to the sausage which some find objectionable and others consider to be part of the flavor of a fermented sausage.

Commercially, acids are commonly sprayed on the surface of frankfurters and small sausages to coagulate surface proteins. This improves the surface consistency, surface color, and increases peelability. Acids can be incorporated into the cellulose casing material.

Originally made from chimney tar extracts, liquid smoke was first used in sausages 150 years ago. Liquid smoke yields a good color and a smoky flavor without the use of smokehouses which encounter concerns with environmental air quality. At the household level, it may be added directly to the meat in the emulsion-forming stage or basted on during cooking.

A variety of binders and extenders are used by commercial manufacturers and also by householders. Soy protein, milk protein, yeast protein, flours, and starch are common. They decrease meat costs, improve yield, improve slicing, may improve flavor, improve emulsion stability, and increase fat and water binding. With the exception of the cost advantages, these functional properties are not better than those of lean meat. Commercially, binders and extenders legally can make up 3.5% of the sausage, and the amount of isolated soy protein is restricted to a maximum 2% level. At higher levels (>3.5%), the product is labeled as imitation sausage.

Flavor enhancers such as MSG (monosodium glutamate), IMP (inosine monophosphate), and GMP (guanosine monophosphate) are allowed by the USDA in sausages, as are antioxidants such as BHA (butylated hydroxyanisole), BHT (butylated hydroxytoluene), catechin, quercetin, DMP (2,6-dimethyloxyphenol acid), phosphoric acid, and nitrite. MSG is readily available to householders—Accent is a common brand—and other additives may be present in purchased mixes for sausage making.

## Specialty Products

Corned beef was traditionally prepared from beef brisket but now the round is sometimes used for a leaner product both commercially and at the household level. Corning means preservation of meat by sprinkling with grains (or in Latin, corns) of salt. The same pumping and pickling processes of hams are used today for corned beef, but the seasonings traditionally include garlic, allspice, bay leaf, celery, and onions. Briskets can be pumped to 120% of green (raw meat) weight and then immersed in the pickle for a few days or only immersed for 2 weeks. The ready-to-eat types are then cooked in water or steamed to 152–160F (67–71C). Some corned beefs on the eastern coast do not contain nitrite.

Dried beef is mildly seasoned and often prepared from the round. It is dry or pickle cured (salt, sugar, nitrite) before drying and then may be smoked to an internal temperature of 90–100F (32–38C). The end moisture content is only 25–35% of the original meat. It may be sold uncooked.

Canadian bacon is made from the sirloin (strip) muscle of the pig which contains very little fat. The muscles are stitch-pumped, then placed in a pickle for several days. They are then rinsed with water, placed in casings and smoked and cooked to an internal temperature of 150–155F (65–68C).

Wiltshire side bacon is common throughout Western Europe. The "side" in this case is the shoulder, loin, belly, and ham; all removed from the animal in one piece. When cured and sold separately, the hind legs are called gammons and shoulders are fore ends. Wiltshire sides are pump cured, covered by a pickle for 7–10 days, removed from the pickle for 2–14 days of refrigerated storage, and then may be sold as is or smoked. Smoked pork loins are typically stitch pumped and pickle cured 3–5 days and then smoked and cooked to an internal temperature of 142–152F (61–67C). Windsor chops are slices cut from these smoked loins.

Pickled pigs feet can be made with a long, cold or a short, hot cure. During the long process, the feet are in a brine 2 weeks, then skinned, detoed, and simmered 3–4 hours until tender. After cooking they are chilled, split, and some bones are removed. The feet are then packed in a 40–75 grain vinegar with herbs and spices. In the short cure process, the cleaned feet are soaked in a cool pickle for several hours and then simmered in it 3–4 hours before being packed into jars with vinegar.

## Cured-Meat Spoilage

Questions on the consumability of household refrigerated hams are very common. If the refrigerator was at least 40F (4C), and the ham appears (odor, color, texture) to be unspoiled, it is safe to eat from a microbiological standpoint. Since cured-meat products are part fat, rancidity causes spoilage, even at refrigerated temperatures. Nitrite inhibits oxidative rancidity of unsaturated fats, but as the level of this additive decreases, rancidity becomes a more common reason for discard. Antioxidant additives are available to commercial operations to address this problem, but householders are finding the low-nitrite cured meats have shorter shelf life.

Bacterial spoilage can occur under refrigeration, though at a slower rate than at room temperature ($Q_{10}$—see Chapter 8). Anaerobic bacteria can ferment carbohydrates in sausage products to produce lactic acid and sometimes $CO_2$ gas. The gas can cause the sausage or the vacuum package to become swollen. Presence of gas is reason for prudent discard, though foodborne illness bacteria are probably not the culprits if the product has been refrigerated.

Most cured meats which are produced today do not contain high enough salt levels for preservation at room temperature. A more common problem than botulism from cured meats is the scenario in which sliced ham may become contaminated with *Staphylococcus aureus* from a foodhandler's hands. If the meat is then held at a temperature permitting multiplication (50–113F, 10–45C), enterotoxin may be produced which can cause food poisoning. Consumers are more familiar with salmonellosis from poultry than with *S. aureus* poisoning and may feel the vomiting and diarrhea from turkey ham within 2–6 hours after consumption is due to a "bad bird." However *S. aureus* is a classic problem in salted products (see foodborne illness in Chapter 2), and should be considered first.

Consumption of moldy country hams is controversial; however, mycotoxins have not been isolated from such hams. However, scientists generally advise moderation, pointing to a lack of cause-and-effect data and the century-old practice of moldy country hams being enjoyed by householders.

Bone sour is a spoilage problem primarily for the dry cure hams when the salt concentrations are not high enough soon enough in the center portions. Storage at warm (unrefrigerated) temperatures accelerates bacteria growth which makes bone sour more likely. Householders can easily test for bone sour before cutting the ham by inserting a skewer into the center portions near the bone and then noting the odor when it is withdrawn. Off-odors indicate bacterial spoilage and are reason to discard.

A rainbow sheen across the surface of sliced ham sometimes causes the householder to be concerned about the product safety. This iridescence is due to light being defracted by the thin layer of fat spread across the meat in slicing. It is not harmful.

Greening is a spoilage problem for cured meats. In bologna, it may be due to *Lactobacillus viridescens* or another greening bacteria entering during the emulsion forming process and not being destroyed during the cooking. Heating to 155F (68C) will kill these bacteria unless they are present in very large numbers. These bacteria produce hydrogen peroxide which oxidizes the myoglobin to a green-colored form (Niven et al., 1954 and 1957). Greening can also be due to chemical or metallic reactions. Nonmicrobial discoloration may also be the result of an oxidation of the cured-meat pigments. This process is accelerated by light (therefore packages of sliced ham are turned upside down), oxygen (vacuum packing decreases available oxygen), and higher storage temperatures (storage $<28$F ($-2$C) is ideal). If at least 70% of the meat pigments are cured, the meat is more resistant to color degradation.

## Nutrients

Cured meats and meat emulsions are good sources of high-quality protein. The protein content of cured muscle or organ meats is similar to that of the fresh. Meat mixtures—may be labeled as containing meat by-products—that contain large amounts of connective tissue proteins are low in histidine and tryptophan; they have a lower biological value than those made of lean meat proteins, but this is important only for very low protein intakes. Most of the raw meat used for processed, cured meats (bologna, sausages) is from trimmings, and very lean, tough cuts from the choice and select grade of animals. The carbohydrate content of meat and meat products is negligible even if sugar has been added during processing. The percent of calories from fat varies with the cured product, but sausages are usually high-fat foods.

Cured meats and meat emulsions are good sources of most B-complex vitamins, phosphorus, iron, copper, zinc, sodium, potassium, and magnesium; but usually low in vitamins A, D, E, K, and C. Liver is generally higher in vitamins and cholesterol than muscle tissue and this is reflected in the nutrient content of liverwurst and liver sausage.

Large amounts of the heat-labile thiamin and some vitamin $B_6$ are usually destroyed during each of the curing, smoking, cooking, and canning steps (Kylen et al., 1964), and dry cured hams contain significantly less thiamin than hams which are pickle cured. Mildly cured and smoked products lose 15% of the thiamin, and cooked processed meats typically lose 25% of the thiamin. Vitamin $B_6$ losses are approximately half those of thiamin. Riboflavin and niacin are stable during the heating process but may be lost in drip. Since cooked processed meats have lost some of their moisture and fat, their vitamin content is slightly concentrated. This is shown as higher values in food composition tables than for their raw counterparts.

## Summary

### SAFE IF:

**a.** Directions and proportion of salt and nitrite to be added are followed carefully for specific meat products.

**b.** Curing is combined with refrigeration or freezing.

**c.** Meat is adequately cooked before serving.

### TOP QUALITY IF:

**a.** Meat is lean and fresh.

**b.** Curing is completed and product is stored at refrigerator temperature.

## References

American Council on Science and Health. 1989. Natural carcinogens in American food. New York: ACSH.

Brady, D.E., Smith, F.H., Tucker, L.N. and Blumer, T.N. 1949. Characteristics of country style hams as related to sugar content of curing mixture. Food Res. 14:303.

Carpenter, Z.L., Kauffman, R.G., Bray, R.W. and Weckel, K.G. 1963. Factors influencing quality in pork. B. Commercially cured bacon. J. Food Sci. 28:578.

Code of Federal Regulations, Office of Federal Regulations, revised April 1, 1990.

Cross, H.R., Smith, G.C. and Carpenter, Z.L. 1971. Effect of quality attributes upon processing and palatability characteristics of commercially cured hams. J. Food Sci. 36:982.

Daun H. 1979. Interaction of wood smoke components and foods. Food Technol. 33 (5):66.

Doerr, R.C., Wasserman, A.E. and Fiddler, W. 1966. Composition of hickory sawdust smoke: Low-boiling constituents. J. Agric. Food Chem. 14:662.

Fiddler, W., J.W. Pensabene, E.G. Piotrowski, R.C. Doerr and W.E. Wasserman. 1973. Use of sodium ascorbate or erythorbate to inhibit formation of N-nitrosodimethylamine in frankfurters. J. Food Sci. 38:1084.

Fields, M.D. and Dunker, C.F. 1952. Quality and nutritive properties of different types of commercially cured hams. I. Curing methods and chemical composition. Food Technol. 6:329.

Gray, J.I., Reddy, S.K., Price, J.F., Mandagere, A. and Wilkens, W.F. 1982. Inhibition of N-nitrosamines in bacon. Food Tech. 36(6):39.

Kylen, A.M., McGrath, B.H., Hallmark, E.L. and Van-Duyne, F.O. 1964. Microwave and conventional cooking of meat. Thiamine retention and palatability. J. Am. Dietet. Asso. 45:139.

National Academy of Sciences. 1981. The health effects of nitrate, nitrite, and N-nitro compounds. Washington DC: National Academy Press.

Niven, C.F. and Evans, J.B. 1957. *Lactobacillus viridescens* NOV. Spec., a heterofermentative species that produces a green discoloration of cured meat pigments. J. Bacteriol. 73:758.

Niven, C.F., Buettner, L.G. and Evans, J.B. 1954. Thermal tolerance studies on the heterofermentative lactobacilli that cause greening of cured meat products. Appl. Microbiol. 2:26.

Pearson, A.M. and Tauber, F.W. 1984. Processed meats. 2d ed. Westport: AVI Publ.

Rice, E.E. 1971. *In:* The science of meat and meat products, ed. J.F. Price and B.S. Schweigert. San Francisco: W.H. Freeman and Co.

Ruiter, A. 1979. Color of smoked foods. Food Technol. 33 (5):54.

Saffle R.L. 1968. Meat emulsions. Adv. Food Res. 16: 105.

Sofos, J.N. and Busta, F.F. 1981. Antimicrobial activity of sorbate. J. Food Prot. 44:614.

Sofos, J.N., Busta, F.F. and Allen, C.E. 1979. Botulism control by nitrite and sorbate in cured meat: A review. J. Food Prot. 42:739.

Tompkin, R.B., Christensen, L.N., Shapris, A.B. and Bolin, H. 1974. Effect of potassium sorbate on salmonellae, *Staphylococcus aureus, Clostridium perfringens,* and *Clostridium botulinum* in cooked, uncured sausage. Appl. Micro. 28:262.

Wistreich, H.E., Morse, R.E. and Kenyon, L.J. 1960. Curing of ham: A study of sodium chloride accumulation. II. Combined effects of time, solution concentration and solution volume. Food Tech. 14:549.

## CONSUMER QUESTIONS

**Q.** *Why does ham spoil inside near bone?*
**A.** This is due to bacterial growth. The chance of bone sour can be reduced by making sure there is plenty of curing salt mixture on the meat and the entire piece is immersed in the brine. Keeping the ham refrigerated during the entire cure and salt equalization period is a good preventive measure as the growth occurs before the salt penetrates these inner portions of the ham. Sometimes this occurs even in commercial operations with good temperature control.

**Q.** *Why may there be air pockets in sausages?*
**A.** Sometimes this happens even to experienced stuffers. For the occasional air pocket at the end of the link, simply poke it with a needle to let the air out and press the casing back next to the meat. If there are many air pockets in the link, it should be eaten soon for best quality or the meat removed and stuffed into another casing. Sometimes meat mixtures which are not uniform and contain large chunks do not stuff well and are better molded by hand and oven-baked.

**Q.** *Why do casings seem to tear easily?*
**A.** Animal casings are fragile, as are some of the other edible types. They take practice to use and this learning by error can be expensive. Try using a stronger casing and forcing the meat into it more gently and slowly.

**Q.** *Should I not buy moldy country ham?*
**A.** These hams commonly do have mold on the outside. Some commercial retail outlets remove this mold with a vinegar or vegetable oil wipe before displaying the ham for sale. Moldy interiors should be discarded; however, mycotoxins have not been found in country ham with the traditional mold.

**Q.** *Are nitrates and nitrites a problem in the diet?*
**A.** These compounds may form nitrosamines in the meat or after ingestion. Nitrosamines have been linked to human cancers. Nitrates occur naturally in root vegetables and there is no evidence that consumption of root vegetables is linked to higher cancer rates; however, some theorize that eating large amounts of cured meats, in addition to the nitrates already in the diet, increases the risk of cancer. Nitrosamine formation in the body is decreased by vitamin C present in the stomach at the same time. Perhaps a large glass of orange juice should be the beverage with a corned beef sandwich. Nitrite can be omitted from household recipes for products that will always be stored frozen or refrigerated; however, the color and flavor will be altered.

**Q.** *Is it possible to make low-fat sausage?*
**A.** Householders may experiment safely with the amount of fat in a recipe; however, the quality will vary.

**Q.** *How long should meat be cured?*
**A.** For dry cured bacon, 7 days per inch of thickness at 36–40F. For dry cured ham, count on a month for a 15-pound ham or approximately 7 days per inch of thickness or 2 days per pound. Combination cure hams and partially skinned short-cut hams take approximately 5 days per inch or 1 1/2 days per pound. Mark your curing process and date of treatments on the calendar.

**Q.** *I've heard that there are earthworms in hot*

*dogs and bologna.*

**A.** Stories of ground earthworms in hot dogs circulate in the media periodically. Perhaps the reason behind the myth is the similarity of the word erythorbate to one of the scientific names for earthworm. Consumers can be assured however, that worms are not incorporated in these meat emulsions because worms wholesale for much more per pound than either beef or pork. Cost-effectiveness controlled the meat industry before the FDA did and will continue to be a powerful influence in the future.

# 7 Freezing

Freezing as a preservation method probably was observed by prehistoric people during cold weather; and, until frozen storage cabinets were developed in the late 1800s, naturally occurring snow and ice were used to freeze foods outside. There is a story of Sir Francis Bacon in seventeenth-century England getting fatal pneumonia while attempting to freeze chickens by stuffing their cavities with snow. The quality of such frozen foods was low compared to today's. Clarence Birdseye was one of the first to experiment with quick freezing as a way to retain fresh taste and texture. In the 1930s, his products were introduced to U.S. consumers. He initially provided free freezer cabinets to encourage retail stores to carry his frozen items. The increased quality and variety which freezing provided householders made frozen foods popular throughout the country in less than 10 years (Trager, 1970).

When compared to most other food preservation methods, freezing requires the least amount of food preparation before storage and under optimum conditions it has the best nutrient, flavor, and texture retention. This good quality along with widespread availability of household freezers and microwave ovens for defrosting has maintained the prominence of freezing as a preservation method. This chapter focuses on quality deterioration that occurs during frozen storage, but it needs to be remembered that most frozen foods are the best quality stored foods available.

Freezing preserves food through temperatures too low for microbial growth and so low that chemical reactions in the food proceed at a very reduced rate. Microorganisms are more resistant to low temperatures than they are to heat. At household freezer temperatures (0F, −18C), microorganisms have reduced activity but they are not killed. Generally, food will be as safe to eat when it is thawed as it was immediately before freezing. Trichinae and fish parasites are exceptions since they are killed during frozen storage and such infected foods become safer to eat.

Since food remains microbiologically safe during freezing, its shelf life is determined by chemical and physical changes that occur during storage. Rancidity—oxidative with and without enzyme involvement—and tissue damage from ice formation are responsible for most of the quality deterioration in frozen foods. At 0F (−18C) fruits can usually retain good quality for 12 months, vegetables for 8–12 months, beef 9–12 months, and poultry

8–12 months. Increasing storage temperature results in shorter shelf lives. For each 18F (10C) increase in temperature, the storage time is approximately cut in half. Sliced foods (increased surface areas), cured foods (low $a_w$), and fatty foods (rancidity) lose quality more rapidly (Table 7.1).

**Table 7.1. Recommended storage times at 0F (−18C), based on quality retention**

| Food | Time (mo) |
|---|---|
| Beef and lamb | |
| Roasts | 12 |
| Ground | 8 |
| Pork | |
| Roasts | 6–8 |
| Ham | 5–7[a] |
| Bacon | 3 |
| Poultry | 6–12 |
| Fish | |
| Lean (cod, flounder) | 6 |
| Fat (salmon, swordfish) | 3 |
| Shellfish | 3–4 |
| Eggs, whole, beaten | 12+ |
| Bread | |
| Yeast | 6–12 |
| Quick | 2–4 |
| Cakes and cookies | 4–6 |
| Fruits | 12 |
| Vegetables | 8–12 |

[a]Based on quality deteriorations from rancidity. Adverse texture changes occur immediately.

## Initial Freezing Process

Water is the major component of most foods (Table 7.2) and, since it changes both its state and volume during the freezing process, understanding some of its physical properties is necessary for comprehending household frozen food storage. During freezing, water is redistributed in food by the formation of ice crystals. This alters the characteristics of a food upon thawing since separated water usually does not return to its original position. Ice crystals themselves do not preserve the food but in fact damage it. When water changes state from liquid to solid, there is a 9% increase in volume that is responsible for many of the inferior textural characteristics of frozen food. Undesirable texture changes in thawed tomatoes and potatoes are extreme examples.

**Table 7.2. Composition and characteristics of perishable foods**

| Food[a] | Water[b] (%) | Freezing[c] Temp. (F) |
|---|---|---|
| Vegetables | | |
| Broccoli | 90 | 30.9 |
| Lettuce, head | 95 | 31.7 |
| Peas, green | 74 | 30.9 |
| Peppers, sweet | 92 | 30.7 |
| Potatoes, white | 78 | 30.9 |
| Squash, winter | 85 | 30.6 |
| Fruits | | |
| Apples | 84 | 30.0 |
| Peaches | 89 | 30.4 |
| Pears | 83 | 29.2 |
| Strawberries | 90 | 30.6 |
| Fish | | |
| Halibut | 75 | 28.0 |
| Salmon | 64 | 28.0 |
| Poultry | | |
| Chicken | 74 | 27.0 |
| Meats | | |
| Beef, round | 67 | 28.0 |
| Ham, light cure | 57 | 29.0 |
| Pork sausage, country | 50 | 25.0 |
| Dairy | | |
| Milk | 87 | 31 |
| Ice cream (10% fat) | 67 | 21 |

[a]Average or values for the most common form of the food are given.
[b]Data from USDA, 1990.
[c]Data from ASHRAE, 1986.

Plant and animal tissue cells react similarly to freezing conditions but, for the purpose of this discussion, a plant model will be used. A typical plant cell has a semiflexible cellulose wall, inner membrane(s), and a fluid center that is largely water. The plant's tissue is composed of groups of these cells attached to each other. As ice forms and the center expands, the walls bulge as much as their cellulose structure allows, and then with continued expansion rupture may occur. At the same time, as the water outside the cells freezes, the extracellular ionic (salt and sugar) concentration increases, which draws water out from within the cell and dehydrates it. Ice crystals formed on the outside may pierce cell walls, which further damages them. A similar scenario occurs in freezing gel matrices such as starch-thickened sauces, pie fillings, and gravies.

In water's solid state (ice), molecules are held together with hydrogen bonds. Hydrogen bond structure among water molecules varies with the temperature. Ice can exist in 11 different crystalline structures; however, only hexagonal ice is stable at 32F (0C) and normal pressure. Extremely rapid commercial freezing can result in other crystalline forms that cause less physical damage to the food's original structure but hexagonal is the form of importance in household foods. This hexagonal form is the most ordered type of the solid state and contains more spaces between the molecules than are found in the liquid state (water) (Fig. 7.1). Between 32 and 24F (0 and −4C), ice is least dense (i.e. it has the greatest volume). Since adverse physical changes are most likely to occur in the food while it is passing through this range, rapid freezing is desirable.

Ice which forms in the presence of solutes may contain crystals other than the hexagonal types. Frozen water with large amounts of gelatin contains ice structures with greater disorder. Apparently the large gelatin molecules impair hexagonal formation. The difference in

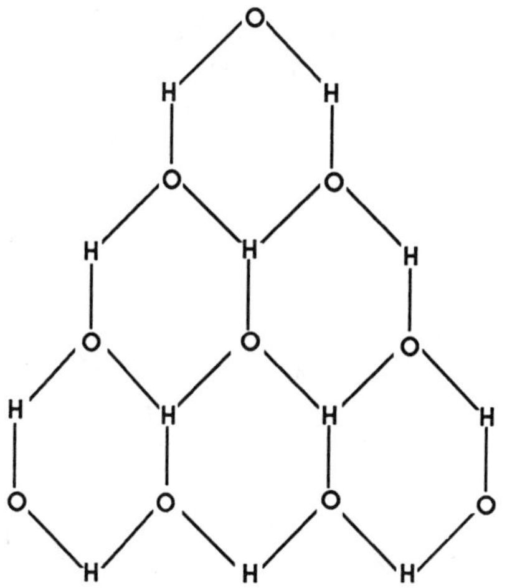

7.1. Proposed hexagonal arrangement of water molecules in ice.

ice crystal size of frozen desserts made with gelatin and those without can be detected with the tongue.

The aqueous portion of foods usually contains dissolved and dispersed substances (sugars, salts, fats, proteins, carbohydrates); however, during freezing, the ice crystals formed are essentially pure water. This leaves the lipid phase and a small amount of unfrozen water that contains a high concentration of solutes. This concentrated solution has a different freezing point, pH, $a_w$, ionic strength, and viscosity than the original aqueous portion of the cell. Gases such as oxygen and carbon dioxide may come out of solution and large molecules are forced closer together, which makes chemical interactions between them more likely. Concentrating substances (reactants) increases reaction rates which offsets some of the desirable effects of the lowered reaction rate achieved by lowering the temperature ($Q_{10}$). Oxidation of ascorbic acid and the fat in butter, cooked meats, and milk is slightly accelerated at 32F (0C) due to this concentrating of substrates. However the net rate of oxidation at 0F (−18C) is less than at room temperature (Fennema, 1985). This is the main reason behind the recommendation that home freezers be kept at 0F (−18 C) instead of higher, even though at 20F (−7C) most foods are solid (see Table 7.2). Standard home freezers are not designed to be cost effective at lower temperatures so the recommendation for most householder freezers is not the −22F (−30C) or −58F (−50C) used commercially even though oxidation would proceed even more slowly at these temperatures than at 0F (−18C). Several new, very expensive models designed for home kitchens can effectively hold −22F (−30C). These are excellent for long-term storage but not available to most householders.

Location of ice crystals in the food is a function of freezing rate, initial food temperature, and the nature of the cells in each food. Slow freezing, at the rate of 1 degree per minute or less, usually results in ice crystals being formed exclusively in extracellular areas (Fig. 7.2). These are the most damaging to structure of surrounding cells. Slow freezing also favors the formation of large ice crystals

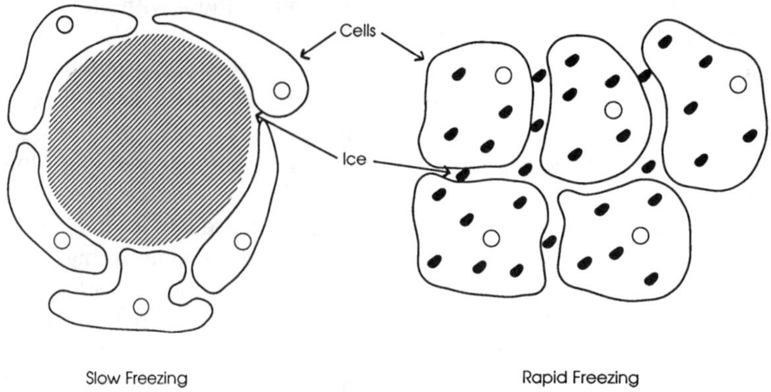


7.2. Large extracellular ice crystals in slow freezing disrupt cell structure.

and maximum dislocation of water. With extracellular ice, the cells appear shrunken while they are frozen, and the quality of the thawed product is usually inferior to quick frozen. Rapid freezing results in the formation of numerous small crystals inside and outside of the cells. The cells themselves are not squeezed by large crystals in the extracellular spaces between cells.

Rapid freezing rates are achieved commercially in a variety of ways. Individually quick frozen vegetables are frozen by blowing cold air (air blast freezing) as vegetables pass through the freezer on a belt or by "tossing" the vegetables with cold air as they pass through the freezer on a mesh screen (fluidized bed freezing). Plate freezing is also used. In this method packaged vegetables are placed on a metal surface that is cooled by refrigerants and another plate may be placed on top of the package. Vegetables frozen by blast methods are in discrete pieces and usually cooked without thawing but solid masses of foods such as pureed squash may be thawed before cooking.

Household freezers usually achieve slow freezing rates, but initially putting packages of food in single layers in the coldest part of the freezer speeds the rate slightly and may result in better quality retention. Freezing pieces of vegetables or whole berries on a metal cookie sheet is a home adaptation of blast freezing.

## Mechanical Freezing Injury

Mechanical stress can result from the dislocation of water that accompanies slow freezing, from recrystallization (many small crystals combining as one large crystal), from temperature gradients within the food (which occur as the temperature fluctuates), and from volume increases (9% expansion) which occur with the liquid-to-solid transformation (Fennema, 1985). Poor quality food due to mechanical damage is more likely in plant tissue because of its rigid structure and poorly aligned cells while muscle tissue is more pliable and tends to have cells arranged in parallel fibers. Delicate plant tissue such as that found in strawberries is particularly subject to mechanical damage.

Householders may not have the option of quickly lowering the temperature of their food during freezing, but they do have packaging options. Freezing smaller samples minimizes mechanical stress. In samples of 3 ounces or more, the outer surfaces become solid before the inner areas do. Then when the inner portions freeze and expand, the outer solid shell is ruptured. In large samples of beef, internal pressures of 200 psi can build up. This is referred to as Langham-Mason stress after the researchers who first documented it in studies of pure water.

At the household level, Langham-Mason stress occurs, but when other factors are taken into consideration, freezing large quantities such as roasts is still recommended since there is less surface area than with steaks. Airtight packaging of food is not always achieved in homes and small amounts of air typically surround meats wrapped in freezer paper making quality deterioration of the surface possible with long-term frozen storage. Mechanical stress damage in frozen rump roast compared to frozen steak is not noticeable to consumers but quality changes due to air exposure are. Conversely small quantities of frozen fish may be preferable as mechanical stress is apparent in such delicate tissue.

## Changes during Storage

Chemical and physical reactions that decrease overall quality continue even in foods stored at 0F (−18C). The losses that occur during a normal storage period usually exceed any damage that occurs to the food during the initial freezing or thawing.

### Recrystallization

Recrystallization is a physical change in which many small ice crystals combine to form a smaller number of large crystals. Temperature fluctuations during storage and longer storage times enhance recrystallization. Recrystallization also occurs in the early stages of thawing where it often damages plant cells that were left intact during the initial freezing process. The end result is decreased quality of products that had been properly frozen (Bevilacqua and Zaritzky, 1982). Recrystallization frequently spoils ice cream. Large molecular weight compounds such as gums and modified cellulose may physically inhibit the growth of ice crystals and are added commercially for this reason. Householders add gelatin and fat—cream instead of milk—to reduce recrystallization.

### Sublimation

When water goes from the solid to the gaseous state without passing through the liquid phase as opposed to the way ice would normally melt if placed in a glass of water, it is called sublimation. Freezer burn is caused by sublimation of ice from the surface of the food into the air inside the freezer. Freezer burn can affect the quality so adversely that the food is discarded. Sublimation is possible when the water vapor pressure of the ice is higher than the vapor pressure of the surrounding air. This vapor pressure difference occurs between all frozen foods and the air.

The surface of the freezer-burned food may have barely visible mat white patches when slightly burned or large areas of grayish white blemishes when the food has been stored longer. The affected tissue is spongy due to many microscopic cavities. These cavities were previously fully occupied by ice crystals (before the ice evaporated) and then the cavities enlarged when the meat fibers and bundles surrounding them shrunk as the meat proteins were dehydrated. The change in appearance is the result of light scattering by the cavities (Kraess and Weidemann, 1967 and 1969). The fibers at the surface of the freezer-burned area are disorganized; presumably the fibrils become detached from each other. Just below this region a dense mass of fibers is formed by the displacement of tissue into the spaces originally occupied by ice crystals. Such severe disruption of the original tissue structure produces noticeable texture changes.

There are many variables in freezer burn—beginning meat quality, rate of freezing, and storage temperature—though at the household level, wrapping to form moisture-proof barriers is the most effective preventive measure. Vapor-resistant materials and tape and containers without excess space reduce the amount of air-to-surface exposure. Products labeled specifically for use in home freezers are capable of providing adequate moisture barriers, and ice glazes are used effectively on fish. Foods wrapped in single layers with larger surface area:volume ratios are more susceptible to freezer burn and the accompanying quality loss, though single layers freeze more rapidly so ice crystal size is smaller and more evenly distributed, which is desirable. Perhaps package shape should be determined by expected storage time since ice crystal damage can occur immediately. Freezer burn is more likely to occur with increased lengths of time, so labeling packages with storage dates and keeping a list on the outside of the freezer to aid inventory control reduces the chance of freezer burn.

### Denaturation

Protein alterations responsible for a decrease in solubility are collectively known as denaturation. In unfrozen muscle, the proteins are grouped together as filaments. As the ionic concentration increases in freezing, through a reduction in the water content, the proteins form bonds with each other instead of water, and insoluble complexes result. This change (protein denaturation) may be responsible for increased toughness in frozen muscle. Householders also notice increased drip when this meat or fish thaws since protein-protein bonds replaced protein-water bonds. The unbound water is no longer held in the food.

Egg yolk does not perform well after being frozen and thawed due to similar protein changes. Combining it with salt or sugar helps curb this water migration. Freezing pure yolk results in a gel that does not combine with other ingredients for use in food preparation.

### Chemical Changes

Microbial growth ceases at temperatures below 28F (−2C) for most foods, but chemical reactions continue in foods even at 0F (−18C). Such reactions include lipid oxidation, Maillard and enzymic browning, flavor deterioration, protein insolubilization, and degradation of chlorophyll and vitamins. Quality changes, not safety changes, are the reason for frozen food discard and should be addressed.

When food is placed in the freezer for storage, the rates of nonenzyme catalyzed reactions decrease as the temperature is lowered to about 28F (−2C). During this time much of the water in the food is in a supercooled state—cooled below freezing, but not yet crystalline ice. As the temperature continues to decrease, the reaction rates actually increase until reaching a maximum—often 21 to 18F (−6 to −8C). The increased concentration of solutes in the unfrozen portions may be responsible for this occurrence as they are closer to each other for reactions. The rates then decline again with further cooling. Oxidation, including loss of vitamin C and E, goes through such a freeze-induced reaction rate increase. When household freezers fluctuate, they often go from 0F (−18C) to 21F (−6C) to 28F (−2C). With each fluctuation, the quality deteriorates and shelf life is shortened significantly. Air temperature in frost-free freezers may vary as much as 9F (5C) when defrost cycle heating occurs, so keeping the freezer fully loaded is important to decrease temperature fluctuations in the stored food.

The increased ionic strength (salt or sugar content) with lower $a_w$ as ice crystals form enhances protein denaturation and usually decreases fat stability. It is theorized that such a changing salt and nitrite concentration in frozen cured products is responsible for these becoming rancid much faster than frozen uncured meat. Reaction rates that increase during frozen storage cannot be calculated by the usual straight line relationship of the $Q_{10}$ doubling for every 18F (10C) increase in temperature since the composition of the food changes variably as the temperature changes.

Vitamin E and ascorbic acid—themselves antioxidants—are often destroyed by oxidation first, leaving tissue lipids vulnerable to further oxidation. Fortunately, the unfrozen portions with their lower $a_w$'s are more viscous, which decreases the mobility of rancidity catalysts; the overall effect is that lipid oxidation is retarded but not prevented during frozen storage.

Off-odors that develop in frozen foods routinely cause consumer discards. Volatile compounds can accumulate in the tissues during storage which are related to off-odors in thawed raw and cooked vegetables. Off-odors may also be the result of oxidation of lipids. Those due to chemical changes in the food have been described as alfalfalike in green beans and peas, as composted grass in spinach and asparagus, as stale cabbage in Brussels sprouts, as cardboard in ice cream (not related to the storage container), and oxidized oil in lima beans.

In unblanched or underblanched vegetables, enzymes remain active during the freezing process, frozen storage itself, and thawing to catalyze reactions that produce off-flavors and off-odors. Enzyme activity may increase or decrease depending on the enzyme as the food chills from 32 to −14F (0 to −10C); however temperatures below this result in greatly decreased enzyme activity.

Color retention in frozen foods is more easily achieved than with canning, but problems still occur. Some color-degrading reactions are catalyzed by enzymes and some are nonenzymatic. Blanching green vegetables slows the enzyme-catalyzed conversion of chlorophylls a and b (green pigments) to pheophytins (olive green) during frozen storage (Diehl et al., 1936). This is not, however, a perfect treatment since during blanching heat and acid from the plant tissues also cause a significant amount of this conversion. Blanching at a high temperature, 199–212F (93–100C), for a short time of 1–2 minutes causes less chlorophyll conversion than blanching at lower temperatures, 189–194F (87–90C), for a longer time of 3–5 minutes (Olson and Dietrich, 1968). The recommendation to add vegetables to water (or steam) only after it has reached a full rolling boil addresses this phenomenon. Chlorophyll conversion can also occur during frozen storage. The rate is slowed with lower temperatures. Broken plant tissue (cut, torn, bruised) accelerates chlorophyll degradation, thus frozen chopped spinach has poorer quality color than the whole leaves. When green vegetables are stored for periods longer than 1 year, nonenzymic oxidation of chlorophylls to pheophytin usually occurs (Lee et al., 1955). Blanching does not have an affect on this reaction. When dry white sugar is added to strawberries prior to freezing, the anthocyanin pigment (red) becomes more stable and browning reactions are also diminished (Wrolstad et al., 1990).

Enzyme-catalyzed oxidative browning also decreases the appearance of frozen food. Raw plant foods such as apples, peaches, pears, yellow cherries, mushrooms, potatoes, cauliflower, and beets are particularly affected since their color shows the defects well. When ice crystals disrupt the cell's natural barriers, the browning enzymes and substrates are free to react. Oxidative browning is most severe near the surface where oxygen is readily available; however, in foods with many intercellular air spaces such as apples, the browning may be throughout. This oxidative browning can be retarded by heat inactivation of the enzymes, adding antibrowning agents such as ascorbic acid and sulfite to packing syrup, and excluding oxygen. When mature-green bananas are peeled, blanched for 10 minutes in boiling water, then sliced and packaged, the end product shows little browning (Cano et al., 1990). Some household freezing methods such as solid side freezer boxes with headspace left at the top for expansion have greater browning potential because of the availability of oxygen than do flexible bags and vacuum packing methods.

Ascorbic acid is the most difficult nutrient to preserve in frozen storage because it can be oxidized at low temperatures. This reaction requires oxygen, is catalyzed by enzymes, and its rate is temperature related. The enzymes which catalyze ascorbic acid oxidation can be heat inactivated, but some ascorbic acid continues to be lost with storage (Table 7.3). If fruits and vegetables are packaged in oxygen permeable containers such as plastic coated cardboard boxes, there is significant loss of ascorbic acid. Foods with lower pH (fruits as opposed to vegetables) and those containing less copper lose less ascorbic acid during storage.

Chemical changes such as insolubilization or gelation of proteins, lipid oxidation, and degradation of vitamins and pigments occur fairly slowly at 0F (−18C), but at faster rates as storage temperature is increased. A good home freezer setting is 0F (−18C).

## Thawing

Plant and animal tissues and gels—this encompasses most foods—thaw more slowly than they freeze. It takes longer to thaw a sample than to freeze it because the heat

**Table 7.3. Ascorbic acid retention of raw, water blanched and frozen vegetables for a short time, and water blanched and frozen for 12 months**

| Vegetable | Blanch time | Raw[a] Asc. Acid | Freshly Frozen Asc. Acid Retention | | Frozen a Year Asc. Acid Retention | |
|---|---|---|---|---|---|---|
| | (*min*) | | | (%) | | (%) |
| Asparagus | 2 | 9.4 | 9.1 | 97 | 8.2 | 87 |
| Peas | 1 | 24.2 | 19.8 | 82 | 16.6 | 68 |
| Broccoli | 4 | 114 | 56.0 | 49 | 36[b] | 32[b] |

[a]mg ascorbic acid per 100 g plant tissue. Data from Batchelder et al., 1947.
[b]frozen 9 months

transfer is by conduction. During freezing, the outer surfaces contain ice crystals first, followed by the center portions. In thawing, the outer portions first change from solid ice to liquid water followed by melting of the central portions. Initially the temperature rise is rapid, but this only occurs before much of the outer surface has changed to the liquid state; most foods remain solid until 23F. Nonflowable water is a better insulator than ice. A plateau stage follows which is the result of this unflowing water impairing heat transfer. When thawing is done with microwaves, conduction is not the major method of heat transfer and this scenario does not apply.

During thawing, the food passes slowly through temperatures near the melting point where chemical reactions and recrystallization can occur. Thus, there is more opportunity for the quality of the food to be decreased during thawing than freezing. This is particularly the case when householders exercise little care in thawing the food they so carefully blanched, wrapped, and quickly froze. Rapid thawing minimizes recrystallization and limits the time cells are exposed to detrimental concentrations of solutes at high subfreezing temperatures.

Rapid and safe thawing at the household level requires some judgement decisions. A single method is not best for all situations. Thawing cooked foods at room temperature may be more rapid than in the refrigerator, but it may pose a health hazard when the outer portions are warmed beyond safe temperatures. Thawing with microwaves is efficient but often uneven and, because parts may become warm, it should be done just before cooking or serving. Thawing wrapped foods under cool running water is a rapid and safe method, but due to high water waste, it is not commonly used. Placing wrapped foods in still water may not be safe if forgotten. Cooking vegetables directly from the frozen state minimizes thawing changes, but will increase the cooking time. For solid blocks of food, this may result in overcooking outer layers. For frozen ground meats, the undercooked central portion may pose a risk of the survival of pathogenic bacteria.

## Microorganisms

Frozen food does not deteriorate from a safety (microbial) standpoint, but the freezing process cannot be counted on to render the food safe to eat if it contained viable microorganisms that can cause foodborne illness. As a general rule, the food should be considered as safe to consume after freezing as it was before freezing.

Laboratory results of survival of microorganisms are difficult to evaluate in food systems because microorganisms injured in freezing may be able to recover in the food but not when placed in media commonly used for microbial analyses. During freezing, the microbial cell goes through a transition similar to that of the plant or animal cell—cooling, extracellular ice, intracellular ice, solute concentration. Freezing injury may occur at any phase of the storage depending upon the organism. This injury is usually due to damage to the microorganism's cell membrane or to its enzyme and metabolic processes. In general, death rate of bacteria during storage of frozen foods is greatest at temperatures between 28 and 14F (−2 and −10C). This is also the temperature range that causes excessive quality changes in the food. As with other types of preservation, starting with top quality fresh food is important. Highly contaminated foods will have more organisms after freezing than if a fresh food with a low concentration of organisms were used.

The most common bacteria in meats are gram negative and freezing reduces their numbers significantly during storage. However, the thawed product is a suitable substrate for their growth. Freezing poultry reduces but does not eliminate salmonellae. Salmonellosis has occurred when thawed poultry was not properly handled or cooked. Fish contain a large proportion of gram-negative bacteria and often their total numbers are reduced by at least 50%. However, viable *C. botulinum* type E has been recovered from samples of raw, previously frozen flounder and polio virus can survive in oysters during frozen storage. Trichinae are killed by holding 20 days at 5F (−15C), −10F (−23C) for 10 days, and −20F (−29C) for 6 days. However, an Alaskan trichinosis outbreak from frozen bear meat suggests that there is survival in bear meat or that the Arctic strain may be different. Fish parasites are killed by holding 24 to 48 hours at −4F (−20C) or lower; a longer time will be needed for larger fish so that they are completely frozen.

The microbial flora of frozen prepared foods depends upon the microorganisms in the raw ingredients, microorganisms introduced during manufacturing from equipment surfaces and personnel, the amount of heat treatment before freezing, extent of recontamination before packaging, and the conditions of the frozen storage. The occurrence of organisms which can cause foodborne illness in frozen foods that will be consumed without further heating is of special concern. It is variable depending upon all of these factors. Commercial manufacturers that prepare frozen food are subjected to periodic government inspections for sanitation and food safety. Freezing, whether at the commercial or household level, cannot be relied upon to render unsafe food safe; proper handling before and after frozen storage is still needed.

## Controlling Enzymes

Enzymes may catalyze quality deteriorations such as undesirable changes in color, flavor, texture, and nutritive value (vitamins A and C) in frozen plant foods. Vegetables are frequently blanched to inactivate enzymes but many frozen fruits are ultimately consumed uncooked; thus, blanching is not a practical control in them. Peaches, apricots, and apples do blanch well so batches which will be used in cooked recipes—pies, cobblers, cakes—can be heat treated to inactivate enzymes before freezing. Blanching also removes oxygen from the plant's tissues.

Blanching treatments for vegetables range from a few minutes at 190F (88C) to less than 1 minute at 212F (100C). Blanching times are calculated for each food to allow heat penetration to the center of the product. The product must be cooled promptly to prevent cooking the food. This is best done by cold water. The postblanch equipment and water must be kept clean to decrease the chance of high bacterial counts.

Early research noted advantages of blanching as a pretreatment before freezing (Diehl et al. 1933). Diehl et al. (1936) found that scalding of English (green) peas for at least 30 seconds in boiling water resulted in a product with better color and odor after 4 months frozen storage at 20F (−7C). As the effects of different scalding methods were explored over the years, nutrient retention became a concern. Boiling water destroys enzymes in snap beans more rapidly (3 min) than does steam (5 min); however, more of the water soluble nutrients are lost when the beans are blanched by immersing in the boiling water (Melnick et al., 1944). Some vegetables are also more palatable when steam blanched. The better method to use will vary with the vegetable so tested procedures should be followed.

Additives which inhibit enzyme activity, alter enzyme substrates, or limit contact of oxygen with phenolic substances are also used to decrease browning during frozen storage. Dipping the fruit before packaging is the most common method of applying these additives.

Pretreatments of chloride ion (1–3% salt in water), sulfite, citric acid, and malic acid act by inhibiting the activity of the enzyme o-diphenol oxidase, which causes browning of phenol compounds in the plant tissue. Ascorbic acid, erythorbic acid, and dihydroxymaleic acid keep the phenolic compounds in a reduced (colorless instead of brown) state. Packing fruits in syrups (sugar and water) acts as a barrier to oxygen which decreases enzymic browning, adds sweetness, helps retain volatile aromas, and decreases the amount of ice crystals at any given subfreezing temperature.

### Other Additives

Sugar, sorbitol, salt, and low methoxyl pectinic acids can be added to increase firmness of fruit. Salt is especially effective in preserving apple texture, but since householders usually serve apples sweetened and because of recommendations to decrease sodium intake, it has limited use. Salt does not have this firming effect on peaches or strawberries. Low methoxyl pectinic acids are used commercially for firming berries. The mechanism is discussed in Chapter 11.

## Vegetables and Fruits

Plant tissues may be irreversibly damaged during freezing and be of unacceptable texture after thawing. The extent of softening is related to the degree of tissue disruption which is dependent upon ice crystal location and size. Blanching that is necessary to inactivate enzymes also alters the quality. Adding sugar to plant tissue before freezing lessens freeze damage, as does forming small intracellular ice crystals by rapid freezing, low storage temperature, and rapid thawing. Householders should strive for the most optimum conditions (rapid freezing, airtight containers, constant temperatures, short storage times) their equipment allows but be aware that some softening will occur, as it does even under commercial frozen storage conditions. The rate of freezing which is a major quality factor can be influenced by freezer temperature, amount and placement of unfrozen packages, size and shape of package, and air circulation.

The length of storage time may produce unexpected changes in fruit firmness. A study using sour cherries found the firmness increased steadily during 5 weeks storage in 60% sucrose syrup at 20F (−6.7C) before leveling off. This increase in firmness is most likely due to changes in pectin (Guadagni et al., 1958). Demethoxylation of pectin in the plant cell wall structure means that more calcium bridges form, and this increases cell cohesion.

## Meats and Poultry

Most meats and poultry can be frozen with minimal preparation since enzymes do not need to be inactivated. Wrapping the meat in vapor-proof packaging and limiting the storage time to 12 months are the primary precautions householders need to take. There are some chemical and physical changes that occur during frozen storage that decrease quality.

The size and distribution of intracellular (within fibers) ice crystals in meats, poultry, and fish is governed by freezing rate (Love, 1966). Rapid freezing to 0F (−18C) is recommended.

Juiciness is an important quality characteristic of cooked meats. It is related to both the cooking temperature and the water holding capacity of the raw muscle tissue (meat). Water-holding capacity decreases with slow freezing, with higher frozen storage temperatures, and with longer storage times. Loss of this water-holding capacity results in excessive drip during thawing and cooking.

Tenderness is affected slightly, if at all, by freezing rates. There are conflicting reports on the effect of time of storage on tenderness, perhaps because so many factors affect tenderness. However, the temperature of storage is a major factor in texture. Samples become tougher at higher temperatures.

Breakdown of fats through oxidation is responsible for off-odors and -flavors in frozen muscles, though at a much slower rate than occurs in refrigerated muscle. Under refrigeration, changes occur within a few days. Temperature is a critical factor in rancidity reactions. Frozen storage for periods of 2 months at 5F (−15C) or higher results in rancid flavor in fish; in beef, rancid odor develops after 3 months storage at 18F (−8C) but not at 5F (−15C). Beef stored for 6 months at 5F (−15C) and for 12 months at −7.6F (−22C) does develop rancid odors (Awad et al., 1968). The general recommendation that beef can be frozen for 12 months may not be acceptable to all people.

High-salt and low-water content of cured pork products (ham, bacon) prevent them from solidly freezing at temperatures of 0F (−18C), but ice crystals which damage texture are present. Thus, the large areas of muscle in cured ham do not freeze well because of excessive texture deterioration. Pork contains moderately high levels of fat; rancidity in frozen pork is the primary limiting factor in its storage life.

## Fish

The quality of frozen fish is dependent upon many factors including the species, individual fish, water environment, harvest conditions, and state of rigor; conditions most consumers cannot control. However, the handling of fish after purchase will markedly affect the acceptability as oxidative deterioration such as rancidity, dehydration, loss of juiciness, excessive drip and toughening can occur during storage, even with top quality raw fish.

Slow rates of freezing can cause great mechanical stress due to size and location of the crystals. A study of codfish found that times of 20 minutes or longer for the center to reach 32F (0C) resulted in mechanical tissue damage from ice crystals (Love, 1966). This freezing rate is typical of household conditions. As with other foods, fast freezing results in many small ice crystals which cause minimum dislocation of the delicate structural components of the fish tissue.

Oxidative deterioration in fish can be decreased by removing much of the oxygen (vacuum packing, freezing in water, and wrapping tightly in plastic films instead of rigid boxes), preventing entrance of more oxygen (vapor-tight seals and protective coatings), using antioxidants (ascorbic acid), and using the lowest storage temperatures and time. Dehydration can be reduced significantly by protective coatings (ice and/or wrapping well), and drip can be decreased by storing and thawing properly. Phospholipids in fish are hydrolyzed rapidly during the first 5 months at 6.8F (−14C) (Olley and Lovern, 1960), thus recommended storage times for fish are short (see Table 7.1).

In some fish (gadoid) the texture becomes tough and rubbery during freezing due to the conversion of trimethylamine oxide in the frozen fish to dimethylamine and formaldehyde. The formaldehyde serves as a cross-linking agent for the protein (Kelleher et al., 1981). Cross-linked protein is tougher than unconnected protein strands.

The same principles for freezing fruits, vegetables, and meats apply to freezing fish but the delicate structures of fish tissue are more readily damaged so following these principles is crucial.

## Dairy Products

Large quantities of milk are discarded at the household level due largely to usage miscalculations and errors in estimating the speed of microbial growth at refrigerator temperatures (VanDeRiet and Woodburn, 1987). Freezing the surplus before spoilage occurs could save money for the average household; however, freezing dairy products poses some problems. Disruption of the original distribution of fat, water, carbohydrates, and protein in the food during freezing affects the quality of dairy foods too.

Foods chill by conduction so initially ice crystals which are pure water form next to the container walls, and lactose (a sugar) and soluble salts concentrate in the central, unfrozen phase. The fat does not move. Higher percentages of fat droplets in the food inhibit the movement of sugar and salt to the center. When heavy cream (25% to 35% fat) is frozen, the composition is uniform throughout. Homogenization also inhibits this movement of sugars and salt to the center.

Most dairy products such as milk, cream, and cheese are emulsions. During freezing, alteration of the lipid components often occurs and the fat droplets may coalesce (the emulsion is broken), giving the liquid a nonuniform appearance which very likely will result in consumer rejection if used for other than baked products. Dairy products with broken emulsions usually have granular fat particles or oil floating on the surface. The exact mechanism of fat globule damage during freezing is not clear; however, the destruction of lipoprotein layers—important to stabilize emulsions—around lipid droplets may precede the droplets coalescing to form granular, plastic fat particles. If the temperature of the thawed product is increased to 100F (38C), the solid fat particles melt and form an oil layer on the surface. Thawed cream added to hot coffee results in such an oil layer on the surface. If the thawed product is not heated above 77F (25C), there is much less fat separation (Doan and Keeney, 1965). Commercially, rapid freezing (within a few minutes) and storage at −20F (−29C) has been found to give a more stable thawed dairy product. Householders do not have this rapid freezing capability so must consider whether the quality of the thawed dairy product will be acceptable.

Oxidation of unsaturated fatty acids with accompanying off-flavors and off-odors may occur during longer storage times or when dairy products are stored at higher temperatures. The oxidized flavor is often described as "tallowy." Oxidation of lipids is catalyzed by small amounts of copper and iron. These metals may be naturally present in milk or enter from the equipment during processing. Pasteurization and homogenization both inhibit oxidative deterioration of lipids (Schwartz and Parks, 1965).

The stability of the major protein in milk—caseinate micelles; caseinate is a protein, micelle is a type of particle—depends upon the physical properties of micellar (particle) hydration and upon the net surface charge. The surface charge is influenced by ionic (salt) concentration. Any changes in these characteristics may cause protein aggregation (to clump together) and subsequent flocculation (the clumps settle). As the water portion freezes, the caseinate particles become more closely spaced and the ion concentration increases. This is a more unstable arrangement than in the original fresh milk, and clumping is likely. If the clumping proceeds to flocculation, the thawed product appears curdled. Householders associate curdled milk with microbial activity, but in this case it is a quality spoilage. Evaporated milk that has been taken through the freeze-thaw cycle several times will form a three-dimensional gel network of the protein. Milk proteins are destabilized gradually during frozen storage. It may take 1–2 months at 10F (−12C) before a noticeable amount of precipitate is formed in skim milk and 3–6 weeks in evaporated milk. Although homogenization has a stabilizing effect on fat emulsion and rancidity in frozen milk, it decreases the protein stability.

Consumers trying to freeze liquid dairy products at home may find that thawed milk has enough quality changes to be rejected by their families as a beverage. The cost-effectiveness of long-term storage of milk should be weighed carefully for each household situation.

When hard cheeses are frozen, the emulsion is often broken and a crumbly thawed product results. Grating the cheese before freezing and then using it on top of heated foods (casseroles, pizza) may produce an acceptable quality since the broken emulsion is much less noticeable than with other serving methods.

## Eggs

Backyard chickens tend to lay more in the winter months and the price of commercial eggs may be low around the Easter season. Some householders like to store their surplus eggs for later use or purchase large quantities during these sales. Eggs store well at refrigerator temperatures for 5 weeks, but will keep for 6 months when frozen. Eggs, like most foods, have water as their main component. This water expands when frozen and would split the shell if whole eggs were stored, thus all freezing recommendations start with removing the shell.

Separated whites can be frozen by simply

packaging in airtight containers; however, freezing egg yolks poses special problems. During frozen storage the yellow color which is due to carotenoid compounds is retained, and oxidation of lipids and browning reactions do not adversely affect the product during short storage periods; however, the texture changes dramatically. Yolk stored below 21F (−6C) becomes irreversibly pasty due to gelation of the proteins and then does not combine well in recipes. The addition of salt or sucrose inhibits this gelation. Sugars such as fructose, arabinose, galactose, and glucose also have been found to restrict gelation (Meyer and Woodburn, 1965). Addition of salt as an anti-gelling agent increases the viscosity and translucency at the 5 to 10% levels used commercially. These changes may be attributed to the disruption of the granules within yolks.

The recommended amount of sugar or salt varies, and householders achieve good results within the range. Addition of 1/2 teaspoon salt or 1–2 tablespoons granulated sugar per cup of yolks is recommended as an antigelling additive depending upon the use to be made of the thawed yolks. Whole eggs are cracked and scrambled with 1 teaspoon sugar, honey, or salt per cup of eggs and packed in freezer containers or initially frozen in ice cube trays (1/4 c = 1 large egg) and then packed in freezer bags. Packages should be labeled with the type of additive used, as tasting raw eggs to determine if they are salty or sweet is a potential health hazard. Enough time must also be allowed, usually overnight, to thaw eggs in the refrigerator before using since they are a good medium for bacteria growth at room temperature.

## Starches

Freezing starch-thickened foods (pudding, gravy, thick soup and sauce) results in poor quality due to retrogradation which increases at low temperatures. During retrogradation starch-starch bonds are formed and the starch gel network becomes more dense, squeezing water out. The result is a tough, rubbery gel with separated liquid. Thawed starch gels have unusually large amounts of syneresis (weeping liquid). Many households find these products unacceptable. Cooked pasta is a firm starch-water product. It also looses quality when frozen due to retrogradation.

Some starches are modified in such a manner as to slow retrogradation. Modified phosphate starch is such an example. The large phosphate groups along the molecule produce stearic hinderance (get in the way) which decreases the number of new starch-starch bonds that form. Modified celluloses, alginates, and gums are also used successfully in frozen gel products. These gums and modified starches are generally not available to consumers but may be to institutions as well as to the food industry.

## Household Freezers

There are four main types of freezers used at the household level for food storage. Two involve the freezer portion of refrigerators—above refrigerated section and side-by-side, and two are free-standing freezing-temperature-only boxes—upright and chest.

The most common type of household refrigerator has a freezing compartment above the refrigerated section. The refrigerator controls adjust the amount of cooling in the freezer compartment and the excess cold air drifts downward (cold air sinks, hot air rises) to cool the refrigerated section. Side-by-side refrigerator-freezers have difficulty in achieving a good balance between desired refrigerator and freezer temperatures. Since the cold air does not simply drift down to the refrigerated section, balancing side-by-side models can be time consuming; adjustments are made during several days of measuring and observation. Usually with balancing, a good temperature in the refrigerator section is the goal. Lowering the controls to achieve 0F (−18C) in the frozen section may result in temperatures below 32F (0C) in the refrigerated portion so that the fresh foods freeze. Opening the door frequently further complicates temperature control as the small mass of cold food and air in these boxes means greater fluctuations and slower recovery. Storing food in refrigerator-freezers is acceptable for the short term but quality is not expected to last as long as the times indicated in Table 7.1.

Upright freezers consist of one box to be maintained at freezing temperatures; 0F, −18C, is ideal. Some more expensive models are designed to hold temperatures lower (−20F, −29C) and also have sections for rapid initial freezing of food. Standard upright freezers provide acceptable storage for most households when the principles of wrapping, loading, and storage times are followed. Selecting appropriate size is important. Full freezers use the least electricity and maintain temperatures best. Household needs vary widely, those storing large amounts of bulky baked goods have different needs than households storing only meats.

Chest freezers also consist of a single box for frozen storage. They have the advantage over uprights of not losing as much cool air when the door is opened—cold air sinks—so usually require less energy to operate. Some householders find loading and unloading chest freezers more difficult than uprights.

Self-defrost models of upright freezers and combination refrigerator-freezers are popular time-saving additions to the household. However, they require more energy to operate than properly defrosted (usually every 6 mo) manual models and the quality of stored food may deteriorate more rapidly if foods are not carefully packaged, since excessive moisture is removed from the freezer's air and from any improperly packaged items. During the defrost cycle which occurs every 8 to 16 hours, the coils, which usually cool the box, heat and any accumulated ice melts off. De-iced coils result in more efficient cooling but the overall operation costs are usually higher for self-defrost models. The temperature of the air surrounding the coils often rises from 0F to 27F (−18 to −3C) but temperature of the stored food does not always rise since the unit is in defrost for a short time (15 to 20 minutes). The surface of stored foods may experience some temperature fluctuations with accompanying ice crystal distribution changes but most consumers do not find that self-defrost models produce food of unacceptable quality.

Alarm systems for household freezers alert people to accidental unpluggings and flipped circuit breakers. They are relatively inexpensive and easily installed by nonprofessionals. These devices are particularly useful when the freezer is in an infrequently used portion of the house.

Widespread power outages pose a more serious problem as they may last several days and the option of using a shelf in the neighbor's freezer is usually eliminated. Simply waiting for power to be restored is the only recommended procedure for the first 24 hours. The freezer box is well insulated and opening the door for curiosity's sake worsens the situation by raising the temperature inside. If the freezer is in a room not occupied by people and the power is expected to be off for an extended period during the winter, chilling the room may be an option. Recommendations for the second day are to mentally (without opening) assess the contents of the freezer. If it was full, simply wait another day. If it was less than half full and in a heated room, the food has probably thawed or is starting to thaw. This problem does need to be addressed as many of the foods will produce drip when they thaw. This drip will spoil along with the food but it has the ability to seep into any cracks in the lining or around the seal to give a permanent odor to the freezer.

Putting Dry Ice in the freezer before the foods thaw works well, but Dry Ice is not adequate for refreezing large amounts of food. Two or three pieces of Dry Ice (10-lb pieces) should hold the food below 32F (0C) for another 24 hours. Dry Ice that comes in contact with skin immediately freezes it leaving a "burn." Some grocery stores and ice cream parlors (such as the Baskin Robbins national chain) carry Dry Ice. When the power returns, the Dry Ice should be removed from the freezer and placed in an ice chest until one is confident there will be no more immediate outages. Dry Ice evaporates quickly in a freezer that is running.

Decisions regarding the use of thawed foods are sometimes not clear-cut as the householder may not know how long perishables have been thawed and at room temperature. Potentially hazardous foods (see Chapter 2 for definition) that have completely warmed to room temperature for an unknown period of time should be discarded. Other foods may have quality deterioration from the freeze-thaw cycle but are safe to eat and even refreeze. (Note: some less-perishable foods such as raw vegetables are partially cooked during blanching so now are potentially hazardous foods.)

Breads (including egg bread), cookies, cakes (with nonperishable fillings and frostings), raw fruits, fruit pies, freezer jam, and foods that ordinarily could be held at room temperature for several days are safe to eat even though they have completely thawed and been at room temperature over 3 hours. Pathogenic bacteria do not grow in these foods but quality changes will make them undesirable after several days unfrozen.

Cooked meat, poultry, fish, combination dishes, and cooked vegetables should be discarded if they were above 40F (4C) for over 3 hours. If only the surface was warmed above 40F (4C), the entire package should still be discarded. These foods are an excellent environment for growth of bacteria which can cause foodborne illness. When householders do not have a thermometer to test the temperature of the food, squeezing unopened softsided packages to test for the presence of ice crystals is advised. When ice crystals are present, the temperature is usually below 40F (4C) even in the outer portions of wrapped packages.

Raw meat, poultry, and fish are safe to eat when below 40F (4C). If these raw products are refrigerated and they are *thoroughly* cooked (to 165F [74C] for a margin of safety), they are safe to consume. Pathogenic bacteria do not compete well with all the other organisms present in uncooked foods so their growth is delayed. The quality spoilers outgrow the pathogenic bacteria and produce off-odors which may cause these products to be of too poor quality to use. Thorough cooking before eating kills the bacteria present, both those causing spoilage and pathogenic types.

## Summary

Frozen foods (*all*)

**SAFE IF:**

Safe to eat when frozen

**TOP QUALITY IF:**

**a.** Best maturity or quality for eating.
**b.** Blanched if vegetable other than green pepper, onions, or herbs.
**c.** Antioxidant added such as ascorbic acid if a light-colored fruit.
**d.** Packaged in moisture vapor-resistant packaging.
**e.** Frozen quickly.
**f.** Stored at 0F (−18C) or below.
**g.** Used within times recommended in reliable source.

How thawed:

**a.** By cooking—safe, top quality. Note longer time, higher energy use.
**b.** In refrigerator—safe, good quality.
**c.** In refrigerator, refrozen—safe, poorer quality.

After thawing if potentially hazardous food:

At room temperature
2 hours or less above 60F (16C)—safe.
3 hours or longer above 40F (4C)—unsafe, poor quality.

## References

ASHRAE, 1986. ASHRAE Handbook, refrigeration systems and applications. Atlanta: ASHRAE Publ.
Awad, A., Powrie, W.D. and Fennema, O. 1968. Chemical deterioration of frozen bovine muscle at −4C. J. Food Sci. 33:227.
Batchelder, E.L., Kirkpatrick, M.E., Stein, K.E. and Marron, I.M. 1947. Effect of scalding method on quality of three home-frozen vegetables. J. Home Ec. 39:282.
Bellow, R.A., Luft, J.H. and Pigott, G.M. 1982. Ultrastructural study of skeletal fish muscle after freezing at different rates. J. Food Sci. 47:1389.
Bevilacqua, A.E. and Zaritzky, N.E. 1982. Ice recrystallization in frozen beef. J. Food Sci. 47:1410.
Cano, P., Marin, M.A. and Fustu, C. 1990. Freezing of banana slices. Influence of maturity level and thermal treatment prior to freezing. J. Food Sci. 55:1070.
Diehl, H.C., Campbell, H. and Berry, J.A. 1936. Some observations on the freezing preservation of

Alderman peas. Food Res. 1:61.
Diehl, H.C., Dingle, J.H. and Berry, J.A. 1933. Enzymes can cause off-flavors even when foods are frozen. Food Ind. 5:300.
Fennema, O.R. 1985. *In:* Food chemistry, ed. O.R. Fennema, 24. New York: Marcel Dekker.
Fennema, O.R., Powrie, W.D. and Marth, E.H., eds. 1973. Low temperature preservation of foods and living matter. New York: Marcel Dekker.
Guadagni, D.G., Nimo, C.C. and Jansen, E.F. 1958. Time-temperature tolerance of frozen foods XIII. Effect of regularly fluctuating temperatures in retail packages of frozen strawberries and raspberries. Food Tech. 12:306.
Hagen, P.O. 1971. *In:* Inhibition and destruction of the microbial cell, ed. W.B. Hugo, 36. London: Academic Press.
Joslyn, M.A. 1966. *In:* Cryobiology, ed. H.T. Meryman, 565. New York: Academic Press.
Kelleher, S.D., Buck, E.M., Hultin, H.O., Parkin, K.L., Lucciardello, J.J. and Damon, R.A. 1981. Chemical and physical changes in Red Hake blocks during frozen storage. J. Food Sci. 47:65.
Kraess, G. and Weidemann, J.F. 1969. Freezer burn of animal tissue. 7. Temperature influence on development of freezer burn in liver and muscle tissue. J. Food Sci. 34:394.
Kraess, G. and Weidemann, J.F. 1967. Freezer burn as a limiting factor in the storage of animal tissue. V. Experiments with beef muscle. Food Tech. 21:143.
Lee, F.A. , Wagenknecht, A.D. and Hening, J.C. 1955. A chemical study of the progressive development of off-flavor in frozen raw vegetables. Food Research 20:289–297.
Love, R.M. 1966. *In:* Cryobiology, ed. H.T. Meryman, 317. New York: Academic Press.
Melnick, D., Hochberg, M. and Oser, B.L. 1944. Comparative study of steam and hot water blanching. Food Res. 9:148.
Meyer, D.D. and Woodburn, M. 1965. Gelation of frozen-defrosted egg yolk as affected by selected additives: Viscosity and electrophoretic findings. Poultry Science, 44:437.
Olley, J. and Lovern, J.A. 1960. Phospholipid hydrolysis in cod flesh stored at various temperatures. J. Sci. Food Agr. 11:644.
Olson, R.L. and Dietrich, W.C. 1968. *In:* The freezing preservation of foods, 4th ed. eds. D.K. Tressler, W.B. VanArsdel and M.J. Copley, 2:83. Westport, Conn.: AVI Publ.
Rhee, K.S. and Watts, B.M. 1966. Lipid oxidation in frozen vegetables in relation to flavor changes. J. Food Sci. 31:675.
Schwartz, D.P. and Parks, O.W. 1965. *In:* Fundamentals of dairy chemistry, eds. B.H. Webb and A.H. Johnson, 170. Westport, Conn.: AVI Publ.
Trager, J. 1970. Foodbook. New York: Grossman Publ.
USDA, 1990. United States Department of Agriculture Handbook No. 66, Washington DC: U.S. Govt. Printing.
VanDeRiet, S.J. and Woodburn, M.J. 1987. Food discard practices of householders. J. Am. Dietetic Asso. 87:322.
Winter, J.D., Hustrulid, A. and Noble, I. 1952. The effect of fluctuating storage temperature on the quality of stored frozen foods. Food Tech. 6:311.
Wrolstad, R.E., Skiede, G., Lea, P. and Eneisen, G. 1990. Influence of sugar on anthocyanin pigment stability in frozen strawberries. J. Food Sci. 55: 1064.

## CONSUMER QUESTIONS

**Q.** *What should be done with food in the freezer past recommended storage times?*
**A.** It is safe to eat, but quality may be unacceptable for some uses to some householders.

**Q.** *What causes frozen fish to have patches of white?*
**A.** This is freezer burn. It is a quality change, usually caused by packaging that allowed air to enter or excessively long storage times. Off-flavor and -odors may accompany freezer burn and discard may be appropriate in some cases. If the burn is slight, removing the affected areas and preparing the rest of the fish may be acceptable. Use of packaging materials designed especially for freezing will provide airtight conditions which slow freezer burn.

**Q.** *What may cause off-flavors, -colors, or -odors in thawed food?*
**A.** If these occur in vegetables, it may be due to action of enzymes. Review blanching procedure and times. These changes may also be the result of excessively long storage times, lack of pretreatment in fruits, poor packaging, fluctuating freezer temperatures, and storage above 0F (−18C).

**Q.** *After a power outage is the food safe?*
**A.** Do not open door of freezer first day of outage. If power is only off for a few hours the food will remain frozen in closed freezer. For extended outages, adding Dry Ice into the freezer is recommended.

If food became somewhat thawed but still

contains ice crystals (determine by squeezing unopened package), it is safe to consume now or refreeze for later. If no ice crystals are present, but all parts of the food are still below 40F (4C), it is safe to consume now or refreeze for later. Refrozen food may be of lesser quality when finally used. If it is a potentially hazardous food and warmed above 40F (4C), then discard is recommended for safety. See Chapter 2 for list of perishable foods.

**Q.** *How long will food keep at 10F (−12C), the temperature of the freezer compartment above the refrigerator?*
**A.** Do not expect food stored in this manner to remain top quality for the recommended storage times. Those times assume temperatures at least as low as 0F (−18C). The exact length foods can be stored at higher temperatures varies with the packaging, pretreatments, and quality expectations of the householder. Each household will be able to make these determinations through experience.

Microbial growth is not possible so food remains safe, but browning reactions and reactions that produce off-flavors and off-odors will continue. The quality is poorer than at 0F (−18C), and the longer the storage time, the more quality deteriorates. However, this is the freezing condition at which much household food is stored and families routinely consume it without noticing quality defects.

# 8 Refrigeration

Ice and water are used in many parts of the world to chill foods, but in the United States gas-cooled (usually freon or ammonia) refrigeration systems are widely available. Refrigerators in households have dramatically changed shopping habits and food consumption, and widespread commercial refrigeration systems provide fresh food variety throughout the year even in northern areas.

Foods can become spoiled due to chemical reactions, physical reactions, or the action of microbes. Lowering the temperature slows the rate of these processes. This is the basic principle behind prolonging shelf life through refrigerated storage. Refrigeration temperatures do not slow the reaction rates as much as freezing temperatures so are not used for long-term storage of most foods. However refrigerated foods do not suffer from ice crystal damage and they do not have to go through a thawing time before preparation as do many of their frozen counterparts. For these reasons, refrigeration is the storage method of choice for many situations. Chilled foods are increasing their share of the convenience food market.

Reduction in temperature will give some reduction in the rate of change of any given parameter whether it is plant respiration, change in texture, loss of vitamins C and E, or decreased microbial growth (Fig. 8.1.). This is called the rule of $Q_{10}$.

$Q_{10}$ measures the sensitivity of chemical changes to temperature change. $Q_{10}$ can be calculated from shelf life as:

$$Q_{10} = \frac{\text{shelf life at Temp C}}{\text{shelf life at (Temp C + 10C)}}$$

The range of $Q_{10}$ for most biological systems such as fruits and vegetables is 1.5 to 2.5. When $Q_{10} = 2$, the rate doubles for each 10C increase in temperature. However, the effect of reducing the temperature is not uniform for all physiological factors. The action of enzymes, which is responsible for many color, texture, and odor changes, is not uniformly slowed when temperature decreases and activity ceases when the enzyme becomes heat denatured.

A temperature of 37–39F (3–4C) is ideal for a home refrigerator. Higher temperatures result in decreased shelf life for the majority of foods, and temperatures between 32F (0C) and 37F (3C) do not preserve most foods enough better to justify the additional energy costs. The maximum temperature in the refrigeration range is 41F (5C). This is based on food safety. The

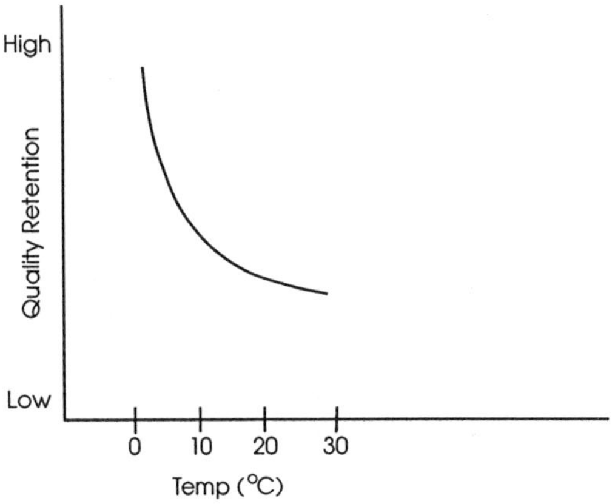


**8.1.** Reaction rates are decreased by lowering the temperature.

lowest reported temperature for growth of *Staphylococcus aureus* and *Salmonella*, which are responsible for much of the bacterial intoxications and infections in the United States, is 44F (6.7C)—above the recommended home refrigerator temperature.

## Microbial Growth

Foods with pH and $a_w$ levels within the ranges for pathogenic bacterial growth are considered potentially hazardous. The temperature at which these foods are held will determine the rate of bacterial growth and, for toxin producers, the level of toxin production. It has been recognized since the early part of this century that cold temperatures are bacteriostatic (stop multiplication) rather than bactericidal (kill), and many studies have been conducted to determine food storage temperatures at which pathogen growth and toxin production cease. There are interrelating factors that affect the temperature at which pathogenic bacterial growth is possible; however, the growth temperature ranges used in consumer education materials are 50 to 120F (10–50C) (Bryan, 1988). This range is often referred to as the danger zone. To add a margin of safety, 40 to 140F (4 to 60C) is more generally used. It is recommended that potentially hazardous food not remain in this temperature range for longer than 3 hours.

The exceptions to the range for pathogenic bacterial growth are *C. botulinum* nonproteolytic types B, E, and F which have been found to grow and produce toxin slowly at temperatures as low as 37F (3C) (Speck, 1976), *Yersinia enterocolitica* which has been found to grow slowly in nutrient broth at temperatures as low as 34F (1C) (Swaminathan et al., 1982), and *Listeria monocytogenes* which can multiply at any temperature above freezing, although more slowly as the temperature is lowered. These organisms and *C. botulinum* toxin are heat sensitive. They would not present a health hazard in foods thoroughly heated after storage and before consumption. Their more common substrates include raw and cooked fish, poultry, meats, and unpasteurized milk. Botulism poisoning is rare, but it is an important concern because the disease is so serious. *C. botulinum* multiplication is slow at refrigerator temperatures; food quality generally becomes unacceptable in a shorter time. The exceptions are partially preserved foods such as lightly smoked fish in which the spores are not killed during the preservation process—but many bacteria are, so quality deterioration occurs slowly. It is recommended these foods not be refrigerated for longer than 3 weeks.

The rate of cooling of hot foods is very important since refrigeration is effective only when the food has cooled below 40F (4C). Foods must be precooled if quantities are too large to cool quickly in the refrigerator. It is possible for *C. botulinum* and *C. perfringens* to multiply in the center of large food items such as casseroles when the center is in the growth zone for over 3 hours. Precooling is best done by placing the container of food in a sink of cold water because the heat transfer is faster than if cold air were the surrounding medium. Some dishes may be stirred during precooling to speed the process. Hot foods less than 3 inches in depth will cool rapidly enough for safety during refrigeration in modern systems.

For an extra margin of safety, many authorities recommend reheating leftovers thoroughly before serving. However, this is often not achieved in all parts of the food when portions are heated in the microwave range. To further compound this problem, many parents of young children reheat leftovers in the microwave range until they are just at a slightly warm serving temperature. Food that has been kept above 140F (60C) for serving, then cooled quickly is acceptable to eat without being reheated, except for those people at high risk should they contract foodborne illness.

Time limits for food safety when potentially hazardous foods are held below 40 or 45F (4 or 7C) have not been set, except for those packaged in a reduced oxygen atmosphere. Expected shelf life of refrigerated foods takes into account quality changes which can be brought about by nonpathogenic bacteria which are able to grow under refrigeration (Rutgers, 1971) or are the result of physical changes in the food, such as drying. The approximate storage life times in Table 8.1 are determined by quality changes due to both nonpathogenic microbiological and chemical changes such as rancidity.

Ground meats have more surface area than roasts so spoil more quickly due to surface microorganisms and reactions with oxygen. Gravies and sauces develop off-flavors rapidly and become less smooth. Fish develop off-odors rapidly from the activity of low-temperature-tolerant enzymes and bacteria, so have a very short storage life. Manufacturers of perishables designed for home refrigerated storage take all these factors into account when determining the package pull date. Each may use different criteria for the expected time of storage after this pull date. Consumers should view pull dates as only a guideline, not a determining factor for discard.

Bacteria able to grow at refrigerator temperatures are classified as psychrotrophic. Cold temperatures are not their optimum range for rapid growth, but their numbers continue to increase at 40F (4C). Most psychrotrophic bacteria belong to the genus *Pseudomonas* and to a lesser extent *Achromobacter*, *Flavobacterium*, *Alcaligenes* and *Arthrobacter*. Common low-temperature yeasts include species and strains of *Debaryomyces*, *Torulopsis*, *Candida*, and *Rhodotorula*. These are not pathogenic but

**Table 8.1. Expected shelf life of refrigerated foods for best quality**

| Food | Storage Time[a] |
|---|---|
| Fruits | |
| Apples | 3–8 mo |
| Blueberries | 2 wk |
| Cherries | 2–3 d |
| Lemons | 3 wk |
| Melons | 3–5 d |
| Oranges | 3 wk |
| Peaches | 3–5 d |
| Plums | 3–5 d |
| Grapes, table | 3–5 d |
| Strawberries | 1 d |
| Vegetables | |
| Asparagus | 2–3 d |
| Beans, snap | 7 d |
| Broccoli | 3–5 d |
| Brussels sprouts | 3–5 d |
| Celery | 1 wk |
| Cucumbers | 1 wk |
| Lettuce | 1 wk |
| Mushrooms | 1–2 d |
| Peppers | 1 wk |
| Squash, summer | 3–5 d |
| Meats and poultry | |
| Roasts[b] | 3–5 d |
| Steaks, chops | 3–5 d |
| Ground, stew | 1–2 d |
| Variety | 1–2 d |
| Cooked meats | 3–4 d |
| Gravy, broth | 3–4 d |
| Bacon | 7 d |
| Frankfurters | 7 d |
| Ham, whole | 2 wk |
| Poultry, raw | 1–2 d |
| Poultry, cooked | 3–7 d |
| Fish | |
| Fish, marine | 14 d |
| Shrimp | 7 d |
| Oysters | 14–21 d |
| Clams, live | 3 d |

Data from ASHRA, 1986 and Labuza, 1982.

[a]Approximate household storage times. Total refrigeration time in distribution system is longer. Some consumers may find quality changes in refrigerated meats, poultry, and fish objectionable after only 1 day household storage. The lower the refrigeration temperature, the longer the quality remains fresh.

[b]Average range for raw beef, lamb, pork, and veal. Sensitive consumers will notice rancidity after only 1 day storage.

illness has been reported when refrigerated foods are consumed that contain large enough populations of nonpathogenic microorganisms to result in visible slime. Consumption of such spoiled food does not occur frequently in the United States. Molds able to grow at low temperatures are many but the most commonly found spoilers of household refrigerated foods are penicillia and aspergilli (VanDeRiet and Woodburn, 1987). It is possible for certain types of these molds to produce toxins, but toxin production is not favored at refrigerator temperatures. It has not been found on refrigerated moldy cheese. Since most bacteria are not able to grow well at refrigerator temperatures, the food often eventually spoils due to molds.

Reduced oxygen packaging (ROP) has offered a wider variety of ready-to-eat foods but caused a concern for storage limits since typical spoilage will not occur and growth of pathogenic bacteria may (FDA, 1993). Such foods will usually have safety barriers incorporated as a precaution if the food is stored at higher temperatures. The package will have a "use-by date" not longer than 14 days after packaging.

## Fruits and Vegetables

The cells of fruits and vegetables must be kept alive for the food to remain acceptable in quality during storage. Lowering the temperature generally slows the metabolic rate of cells and they use substrates such as nutrients, oxygen, and water more slowly. After harvest, these substrates cannot be replenished by the roots and leaves; conserving them by slowing the rate of their consumption increases shelf life. When the fixed supply of nutrients is depleted, the cells die and their membrane systems break down, releasing enzymes which brown and soften tissue. Apples removed after 6 months of storage often contain such brown and soft portions. These storage disorders increase in frequency with increased storage time.

Lowering the temperature is effective only to a certain point, as freezing kills the cells, and the cells of some tropical fruits are unable to survive cold below about 45F (7C). Their death results in quality spoilage of the food referred to as chilling injury. Bananas, mangos, tomatoes, cucumbers, and most tropical plant foods have a longer shelf life when stored at low room temperature than in the refrigerator.

Refrigeration of some plant cells alters their metabolism in addition to slowing it down. (See root cellaring in Chapter 10.) Storage of some vegetables, such as white potato, sweet potato, peas, and sweet corn, at low temperatures alters the starch-sugar balance. At room temperature, white potatoes and sweet potatoes are heavily biased towards starch accumulation. At low temperatures of 50–59F (10–15C), the conversion of sugar to starch decreases and sugar accumulates in the tissues. A high sugar content is desirable in sweet corn and peas. These vegetables are commercially harvested at the immature stage when the sugar content is highest and then stored at low temperatures to retard the conversion of sugar to starch, but in white potatoes it is not desirable. Refer to root cellaring in Chapter 10 for handling potatoes that have been stored cold.

Dehydration is a problem in home refrigerators. It is particularly pronounced in those with self-defrosting cycles because moisture is removed from the refrigeration coils and also from the air and any uncovered foods. Moisture loss can cause both quality and nutritive losses in commodities. Leafy vegetables with a large surface area lose water at a rapid rate. The thickness and nature of the protective waxy coating on produce determines the amount of water loss through the skin but this varies with the commodity, cultivar, and growing conditions. For example, carrots have less of this protective waxy covering than apples or pears, so lose water faster. Unfortunately, the amount of wax on a given apple is variable and cannot be determined by the average consumer (USDA, 1990). Some vegetables such as cucumbers are waxed by food wholesalers to decrease moisture loss. Moisture loss is usually more rapid in fruits and vegetables when they are stored at higher temperatures. The vapor-pressure deficit between the vapor pressure of the food and the vapor pressure of the surrounding air is greater at the higher temperature, given constant relative humidity for the storage temperatures.

Shrivel is noticeable when grapes lose 5–6% of their original weight and when apples lose 5–8%. Water loss not only decreases the aesthetic quality of the food, it may be associated with nutritive losses. In kale, slight wilting caused 90% of the original vitamin C to be lost in 3 weeks at 32F (0C), 4 days at 50F (10C), and 2 days at 70F (21C) (Ezell and Wilcox, 1962).

Packaging that allows gas exchange while maintaining a moist environment is best at retaining the texture and nutritive value. The open-ended plastic bag the food was purchased in is frequently used. The 1993 introduction of a perforated self-closing bag gives more control. Some household refrigerators have produce storage compartments designed to provide this environment, but the most efficient systems tend to be in the highest-priced models. Moist environment compartments frequently have mold problems when used in household settings so diligent observation by householders is required. Cleaning them thoroughly with a 10% bleach solution is often necessary once mold appears.

A few organisms such as *Colletotrichum* are able to penetrate the skin of healthy produce but if fruit is bruised or cells die because of unfavorable conditions, molds and bacteria are better able to invade the tissue and grow. Such damaged produce should be stored separately and consumed first.

## Cooked Foods

Heating food to adequate temperatures—165F (74C) or between 145 and 164F (63–73C) and holding for a period of time—kills vegetative cells; however, the spores are not destroyed at usual cooking temperatures. These spores may germinate if conditions are favorable. Organisms may also contaminate the food after it has been cooked. This growth after cooking is particularly dangerous, because the pathogen's competitors were destroyed by cooking. Therefore, the food should always be refrigerated promptly. Reheating thoroughly gives a margin of safety when preparing for those with increased health risks.

In cooked foods, reactions are also slowed at refrigerator temperatures. The growth of pathogenic microorganisms is eliminated or slowed. Others slowly multiply and cause spoilage. Enzyme action is not of much importance in the cooked food, as enzymes have usually been inactivated by the heat used in preparation. Rancidity is significantly slowed for increased shelf life of the food. As householders use less oils, many find they must store the bottles in the refrigerator instead of the cupboard to prevent rancidity.

Changes such as dehydration, broken emulsions, and weeping occur with time in refrigerated foods. Dehydration may be a problem with improperly packaged cooked foods. There are many excellent containers and wrapping materials on the market that provide moisture barriers and since the plant cells of cooked foods are no longer alive, these packages need not be oxygen permeable. When householders take the time to properly store cooked foods, quality deterioration due to water loss is usually not a problem. Salad dressings, sour cream, yogurt, and cottage cheese may acquire a layer of thin liquid on top during storage. This is due to physical changes and not microbial activity so the food is safe to eat.

## Shelf Life

While products do not usually become unsafe when stored at recommended refrigerator temperatures, the quality may be affected and the total shelf life time limited. As household refrigerator temperature increases, the amount of discards also increase significantly (Table 8.2). In one study 21% of the household refrigerators surveyed were ≥50F (10C). At these temperatures, pathogen growth becomes a possibility (VanDeRiet and Woodburn, 1987).

Householders often find time:temperature relationships confusing and determining the safety of refrigerated foods is also unclear for many. When asked all of the ways they evaluate food safety; 87% said time unrefrigerated,

**Table 8.2. Discards increase with increasing storage temperatures**

| Storage Temp | Amount Discarded Weekly | |
|---|---|---|
| | (g) | ($) |
| ≤ 40F (≤4.5C) | 553 | .74 |
| 41–45F (5–7C) | 578 | .87 |
| ≥50F (≥10C) | 1,128 | 1.93 |

Adapted from VanDeRiet, S.J. and Woodburn, M. 1987.

87% said smelling, 71% said looking, and 43% said tasting. More experienced householders are less likely to taste a food to determine safety and the younger are more concerned with the length of refrigerated storage, which should not be a factor in the safety if the refrigerator is at the recommended temperature.

Bacterial spoilage of meats in the home is mostly due to psychrotrophs and, in unground products, the effect is surface growth with production of slime, off-odors, and off-colors. When the shelf life of refrigerated raw beef is evaluated by odor—odor is related to bacterial growth—it is generally 3 to 4 days and has a $Q_{10}$ of 3.2 (Labuza, 1982). Beef shelf life evaluated by color is usually less than 24 hours and has a $Q_{10}$ of 750 (MacDougal and Taylor, 1975).

Dairy products may spoil due to reactions of lipids (autoxidation or enzymic) or from the presence of bacteria. Control of bacteria depends upon preventing contamination before and after pasteurization and creating conditions unfavorable for microbial growth. Storing milk in the purchased container and discarding leftover milk from serving pitchers instead of returning it to the carton lessen contamination.

## Open Shelf Life Dating

Uncoded dates that consumers can easily read are now commonly used on many products. The dating may indicate when the product was packaged, when it should be sold, or the date by which it should be discarded if uneaten. The arguments for increased open dating include informing consumers if the product is overage even though it is not unsafe. However, a direct relationship between open shelf-life dating and the actual freshness of the food products when sold has not been established so Congress is reluctant to make the dating mandatory. Since the freshness of perishable foods is so temperature dependent (see Fig. 8.1), some consumer advocates are asking for a temperature-sensitive tag on containers that will show a color change if held above refrigerator temperatures for a significant period of time.

It is common for consumers to base a discard decision on an uncoded pull date without opening the container to examine the food, and for householders to attempt to unscramble a manufacturing code for the container and then discard the food because they believe the date is past (VanDeRiet and Woodburn, 1987). Certainly there is a need for more consumer education.

## Summary

Refrigerated foods (except lightly smoked fish or other cooked foods*)

SAFE IF:

**a.** Handled safely before refrigeration.
- Potentially hazardous foods, whether hot or cold, should not be left at room temperature longer than 2–3 hours.
- Keep cold foods cold by refrigerating promptly.
- Cool hot foods quickly.
- Refrigerate small amounts of food while still hot (example, as soon as a meal is over for leftovers).
- Refrigerate in small quantities or in shallow (3-inch) layers.
- Cool pots of hot food by putting pan into sink of ice water and stirring occasionally until cooled (about 30 minutes).

**b.** Kept at refrigerator temperatures of 40F (4C) or below.

**c.** Protected from contact with drip of raw meat, poultry, or fish.

**TOP QUALITY IF:**

**a.** Refrigerator temperature is 33–40F (1–4C). The lower the temperature within this range, the longer dairy products and other highly perishable foods remain at top quality.

**b.** Packaged so food does not dry out.

**c.** Acidic foods are removed from tin cans and put into glass or plastic containers.

**d.** Used as soon as possible, i.e., not forgotten.

**e.** Not moldy. If a solid food molds, trim the mold and 1/2 inch of food beneath it. If a liquid molds, discard it.

*Cold-tolerant strains of *Clostridium botulinum* may grow in lightly smoked fish and other cooked foods at refrigerator temperatures. These products should not be stored in the refrigerator longer than 3 weeks.

## References

Angelotti, R., Wilson, E., Foter, M.J. and Lewis, K.H. 1959. Time-temperature effects on Salmonellae and Staphylococci in foods. I. Behavior in broth cultures and refrigerated foods. Cincinnati: U.S. Dept. of Health, Education, and Welfare, Robert A. Taft Sanitary Engineering Center.

ASHRAE, 1986. American Society of Heating, Refrigerating and Air-Conditioning Engineers Handbook, refrigeration systems and applications. New York: ASHRAE.

Bryan, F.L. 1988. Risks of practices, procedures and processes that lead to outbreaks of foodborne disease. J. Food Prot. 51:663.

Ezell, B.D. and Wilcox, M.S. 1962. Loss of carotene in fresh vegetables as related to wilting and temperature. J. Agric. Food Chem. 10:124.

Fennema, O.R., Powrie, W.D. and Marth E.H., eds. 1973. Low-temperature preservation of foods and living matter. New York: Marcel Dekker.

Food and Drug Administration. 1993. Food Code 1993. Washington DC: U.S. Dept. Health and Human Services, U.S. Public Health Service.

Labuza, T.P. 1982. Shelf life dating of foods. Westport: Food and Nutrition Press.

Lyons, J.M. 1973. Chilling injury in plants. Annual Rev. Plant Phys. 24:445.

MacDougal, P.B. and Taylor, A.A. 1975. Color retention in fresh meat stored in oxygen—A commercial scale trial. J. Food Tech. 10:339.

Rutgers University Food Science Department. 1971. Food stability survey I and II. Washington DC: U.S. Government Printing Office.

Shear, C.B. 1975. Calcium related disorders of fruits and vegetables. HortSci. 10:361.

Swaminathan, B., Harmon, M.C. and Mehlman, I.J. 1982. A review: *Yersinia enterocolitica*. J. Applied Bact. 52:151.

United States Department of Agriculture. 1990. The commercial storage of fruits, vegetables, and florist and nursery stocks. Washington DC: Author.

VanDeRiet, S.J. and Woodburn, M. 1987. Food discard practices of householders. J. Am. Dietetic Asso. 87:322.

Vanderzant, C. and Splittstoesser, D.F. 1992. Compendium of methods for the microbiological examination of foods. 3rd. ed. Washington DC: American Public Health Asso.

## CONSUMER QUESTIONS

**Q.** *How can I determine if refrigerated food is safe to eat?*

**A.** Keep a thermometer in the refrigerator so the temperature is known to be 40F (4C) or below; 37–39F (3–4C) is ideal home refrigerator temperature.)

Odor and appearance of the food will determine the acceptability; most microorganisms that cause illness cannot grow at such low temperatures, so the quality of the food rather than the safety will deteriorate first. If a refrigerated food has off-odors, discard is advised. Some microorganisms that ordinarily do not cause illness can bring about digestive upsets when their numbers are large enough to produce surface slime, so slimy food should be discarded.

Rancidity is the most common spoilage of high-fat foods such as salad dressings, ham, butter, margarine, and sausages. These foods are often in the refrigerator for long periods of time. Typical off-odors result.

**Q.** *How long can leftovers be safely stored in the refrigerator?*

**A.** If leftovers were safe to eat when they reached refrigeration temperature then shelf life will depend on the acceptability of their quality. Cold temperatures do not make spoiled food safe to eat. A maximum of 3 hours above 50F is a good rule-of-thumb for keeping food safe. Since they have been heated, leftovers have an additional problem of the natural competitors for pathogenic bacteria not being present. They were destroyed in cooking. If pathogens are then introduced in the food by a human (hands, sneezing, etc.) or utensils, they are able to grow without competition for space or nutrients. This allows more rapid growth and possible rapid toxin production when storage temperatures allow. Leftovers that have been held at 40F (4C) or below should be evaluated by the same criteria used for other foods.

**Q.** *Is moldy cheese safe to eat?*
**A.** Consumption of moldy food is controversial. Some molds are capable of producing toxins on certain foods at certain temperatures. More research is needed in this area, but the current studies indicate that toxin production is unlikely at temperatures below 40F on cheese. To further reduce the risk, it is suggested that 1/4 to 1 1/2 inches of the cheese beyond the moldy area be removed. For softer cheeses, remove the larger area. Lowering both the temperature (refrigeration) and the pH (wrapping in vinegar-soaked cheesecloth) of the cheese for storage usually gives a longer shelf life than lowering only the temperature.

# 9 Canning: Preservation by Heat and Vacuum Seal

Early humans discovered the preservative effects of drying, salting, pickling, cold storage, and freezing through trial and error and by observing food during naturally occurring weather cycles. However, sealing hot food in an airtight container and, later, heating the food in the container to destroy microorganisms for preservation was truly an invention.

In the late 1790s, France, under Napoleon's leadership, was at war and having great difficulty feeding its troops. Salted and dried foods were the army's staples. An award of 12,000 francs and fame was offered by the government to an inventor of a useful method of food preservation. Nicolas Appert was awarded the prize in 1809 for what he called "the art of Appertizing." Both glass and metal containers were used. The early tin-plated metal containers were called canisters and the term "can" is assumed to be derived from them. The canisters had a hole in the top which was covered with a loose lid as the filled container was heated in a waterbath—cans placed in boiling water up to the lid, but not covering. After heating, the lid was hand-soldered into place (Desrosier, 1970), and as the contents cooled, a vacuum was formed inside the sealed container. The glass jars were stoppered with a cork after heating and a vacuum also formed in them as the food cooled and shrank.

Appert noted that cleanliness was an important factor in his process but prominent scientists of the time concluded that Appertizing was successful due to a magical combining of air and food in a sealed container which prevented putrefaction. Canning was practiced this way with some success for the next 50 years.

Through observations of visibly spoiled cans, Appert knew that some foods required more heat treatment than others. By 1824, Appert's canning schedules for 50 different foods were widely used. Sir Edward Perry carried some of these processed goods on his Arctic expeditions. Leftover cans from this voyage ended up in museums, and in 1938 several displayed at the National Maritime Museum in London were opened and found to be nontoxic to animals. They did, however, contain many viable (live) bacteria cells.

As with modern processing plants, decreasing production time was of critical importance to the early canning industry. Heating for shorter times at higher temperatures was the main approach. In one factory, adding calcium chloride to the waterbath raised the water temperature and reduced the holding time in boiling water from 6 hours to 1/2 hour which increased plant production from 2,000 cans per day to 20,000 cans. In 1851, Chevalier and Appert invented an autoclave—industrial size pressure cooker—to further increase temperatures and decrease time the cans needed to be held in the heat. It was a dangerous contraption to operate and many of the hand-soldered cans burst when their contents reached 240F (116C), but this was the first canning process that was potentially capable of destroying botulism spores. However, the effects of microorganisms were not to be discovered until the 1860s by Louis Pasteur, and scientific studies of the relationships between heat treatment of canned foods and microorganisms were not carried out until the 1890s. In 1874 Shiver's invention of a steam retort in which the pressures could be controlled paved the way for mechanized canning plants. The first canning plant opened in the United States in the 1920s and by 1940 they had appeared throughout this country.

The early 1900s was a time of many great processing discoveries. Can manufacturing and can sealing were mechanized. In the 1920s researchers accumulated accurate processing-time data based on the heat resistance of bacterial spores and heat penetration for a variety of foods, and C. Olin Ball developed a mathematical formula for calculating time-temperature processing schedules (Desrosier, 1970). By World War II, a tin-coated steel can called the "sanitary can" was in general use. During World War II, the U.S. troops consumed a variety of safely canned foods and the population as a whole had the best nutritional status of any before it, partially due to the availability of canned fruits and vegetables in the winter months. Canning has changed the diet of industrialized societies and still has a significant affect today.

## Containers

### Glass Jars

Glass containers are inert, nonmetallic material. They do not react with the food they are holding and are transparent so householders can easily examine the contents. Glass jars manufactured specifically for home canning are the containers used most frequently by householders; however, some also use jars in which they have purchased commercially canned foods. In a survey of household canning (Fields et al., 1977) 133 of 292 tomato jars sampled were reused commercial types and of 290 jars of green beans sampled, 100 jars were reused commercial types. Some commercial jars are made to withstand large pressures induced by mechanical pumping from a sealing head over the container when the cap is set (Jackson and Shinn, 1979), and others are not as strong as the jars designed specifically for home use. Householders experience excessive breakage when they use the weak jars for canning. Several finishes—top portion which contains threads for the lid to screw onto—are used commercially. The continuous thread finish for screw caps may fit the rings used in home canning and be acceptable for such use. The commercial push-on, turn-off finish appears similar to continuous threads, but household rings will not screw on properly and hold the flat lid in place for a good seal.

Closures for glass containers have come a long way since Appert's corks. The wire-and-dome-type glass lids which seal with a disposable rubber ring were used almost exclusively a generation ago. These seals needed to be completed by tightening down the top lid with the wire bail as soon as the jars were removed from the processing pan. The rings are no longer being manufactured and these canning jars have become collector's items. The glass-lined zinc cap which was also used with a rubber ring followed these, and then the currently used flat lid has been marketed with relatively few changes since the 1950s. The flat lids (also called two-piece lids) put out by several manufacturing companies have a rim that is usually first softened in hot water and then establishes an absolute barrier when the jar seals so microorganisms cannot enter, and there is no gaseous exchange between the jar interior and the environment. This rim is made of material specifically designed to withstand processing temperatures and hold a seal for years. Householders suggest substitutions such as epoxy-type glues and denture adhesives so they can reuse flat lids, but these would not be effective. The flat lids complete their seal during the cooling after processing. If the lid is tightened after the jars are removed from the canning kettle, as required with the old rubber rings, this seal is broken.

A disadvantage to using glass jars in canning is their vulnerability to thermal shock breakage (Fig.9.1). Thermal shock occurs when there is a large temperature differential between various parts of the jar as in the case of adding hot food to a cold jar or placing a cold-packed jar into the boiling-water canner. Some jars are manufactured to withstand temperature differentials of 150F (66C); however, it is advisable that the differential not exceed 80F (27C) between the outer and inner surfaces of the jar. Round jars, of uniform thickness, made of thin glass are the most resistant to thermal shock. Jars may also be weakened by a scratch from a table knife used to remove air bubbles when the jar is filled or a spoon used in removing food from the jar. Thus, reused jars are more prone to thermal shock breakage.

**9.1.** Typical thermal shock breakage originates at the base of the jar and few pieces result.

## Rigid Metal Containers

In 1810 the "tin canister" was invented for preserved foods by Peter Durand. These cans were made by hand; an expert could make five to six per hour. In the early 1900s the process was mechanized and canned foods became much more affordable. Metal cans were used almost exclusively by industry until the late 1980s when it became popular to market sauces and gourmet canned foods in glass jars.

Home canning in metal cans has been used; however, in their 1988 revision of processing times, the United States Department of Agriculture (USDA) did not test the processing time recommendations for tin cans and, therefore, they currently do not have recommended processing times. At the household level, cans were closed by double seaming the lid onto the filled can with a special can sealer. Such sealers currently retail for over $100 so householders who have made this purchase do not easily cease using them. In many parts of the United States, cans are not available for home canning, but some householders with can sealers are able to purchase cans through mail order suppliers or directly from commercial canneries, and can sealers are still promoted in mail order catalogues.

Those who canned outside of their household kitchens, where container breakage during transport was a problem, found metal cans desirable. It is still common for pleasure-fishing boats off the western United States' coastline to be equipped with a hot plate, pressure cooker, and a metal-can sealer for preserving tuna as it is caught. The cans had the advantage of stagger-stacking in the pressure cooker instead of only a single layer of jars on the bottom of the pressure cooker. This made canning fish with 90-minute processing times much more rapid through fewer batches. Cans also protected the food from light which can accelerate some quality deteriorations.

Tin and aluminum cans are still used by commercial canneries but most modern cans are made of steel with a thin (0.25%) coating of tin or completely without tin. Tin is fairly inert to corrosion but damage is possible by interaction with the food over time. When the steel is exposed, hydrogen gas is evolved from reactions between the can and the food and a "hydrogen swell" results. This food is not microbiologically unsafe to eat, but because it is difficult to distinguish between this gas and gas produced by pathogenic bacteria, all canned foods swollen with gas should be discarded. Aluminum cans have the disadvantage of bleaching some products, such as red fruits. Enamels were originally developed to coat the inside of cans for cherries and berries to prevent this color fading. Now most aluminum cans are enameled inside, as are tin cans.

Many vegetables such as peas and corn contain sulfur compounds—sulfur in the amino acids of protein—which react with tin and iron to produce dark-colored metal sulfides. These sulfides are harmless to consume but have an objectionable appearance. In C-enameled cans, a zinc-oxide pigment in the coating traps the sulfur compounds before they cause discoloration. Epoxy-based coatings are often used for fish, meats, and dairy products to protect the contents from color change. Release agents are sometimes added to enamels for canned meats so contents do not stick to the container. Uncoated cans are used for applesauce and grapefruit sections because the reducing action of tin salts slows darkening and flavor changes that may occur with storage.

## Canning Procedures

Clean, sound canning jars should be selected. A smooth, un-nicked sealing surface at the top is imperative for an airtight seal to form with the lid. In the past, jars were "sterilized" by immersing in boiling water immediately before filling. This step was not a true sterilization process because microorganisms could enter as soon as the jar was removed from the water. Washing the jars thoroughly in a dishwasher provides adequate sanitation. The "sterilization" step is not necessary.

Foods are usually thoroughly heated, then added to jars (a hot pack). However, products may also be packed raw (formerly called cold pack) and covered with hot liquid. Air bubbles trapped between pieces can be removed with a rubber spatula or knife.

Home canning recipes will often specify "fill to within 1/2 inch of the top of the jar" or give other instructions indicating the amount of air to be left in the container. This air at the top is called the headspace. The headspace is important for pulling a vacuum and completing the seal, and also decreases the chance of food particles getting between the sealing surface and the lid during processing. Air in the food's tissues expands greatly when it is heated. The headspace provides room for this expansion. Current USDA recommendations are 1/2-inch headspace for tomatoes and fruits processed in boiling-water baths, 1- to 1 1/4-inch for low-acid foods processed in pressure canners, and 1/4-inch for jams and jellies.

Hot packs are popular because, for some foods which heat chiefly by conduction, the already hot food requires less time in the processor to reach sterilizing temperature. During preheating, the food becomes softer and air is expelled from plant tissue. If these changes do not occur during preheating, they occur when the jar is being heat processed. The result is jars that are partially empty. A hot pack reduction in bulk for spinach has been measured at 50 to 60%, and in peas 6 to 15% (Andress and Kuhn, 1983).

Most raw packs (filling jars tightly with unheated food then adding hot liquid) have the advantage of time-saving convenience on canning day. Some foods have been found to have superior quality when raw packed; but for most fruits and vegetables, the palatability is similar or inferior to hot packing. Typically, with raw packs, fruits tend to float and are then particularly susceptible to discoloration from the entrapped air; unacceptable browning may occur after only 2 months of storage. Using hot liquid in raw packing is necessary to remove some air from the plant tissues.

When the jar is capped and heated, the contents expand and their vapor pressure rises. The amount of pressure depends upon the amount of headspace and the temperature. Air escapes from the jar during heat processing. This is called venting or exhausting the jar. Closures designed for home canning vent at relatively low pressures so that dangerously high pressures do not build up within the jar.

The flat self closing lid vents at lower pressures than did the older 3-piece glass lids.

The venting or exhausting of jars causes some concern for home canners because liquid may be lost, and food particles may move between the lid's sealing compound and the jar's sealing surface. Such food particles may prevent a seal and vacuum from forming or cause a weak seal which fails during storage. Exhausting is important because it expels air from the plant tissues which helps reduce oxidation reactions. Metal cans which are tightly sealed before processing do not vent so need to be exhausted by heating until the contents reach 170F (77C) before sealing. In some foods, exhausting metal cans can be eliminated by using hot packs. Fluctuating process pressures, as is common with dial-gauge pressure canners, and rapid cooling of the canner—common with the newer thin-walled pressure cookers—increase liquid losses. Allowing sufficient headspace for expansion of the food during processing (35 mL minimum for pints, 70 mL minimum for quarts according to 1949 research) also reduces liquid loss.

Canning fruits and vegetables without adding sugar or salt is a matter of personal preference and does not affect processing times or the microbiological safety of the food. The amount and type of sugar used are also a matter of personal preference. Using light syrups or water for fruits instead of the traditional heavy syrups will result in a less firm product since the sugar and plant pectin bonds will be less (see Chapter 11 for exact mechanism). Soft fruits such as peaches may even lose particles into the surrounding liquid during processing in light syrups. Most recipes are formulated for table sugar (sucrose). Some fruits contain larger proportions of fructose, as does honey. Using these juices or honey will change the flavor of the products and may result in more browning as fructose participates more readily in the Maillard reaction than does sucrose.

## Alternative Canning Methods

Steam canning is not a new method, but it has gained recent popularity with the marketing of steam canners. Steam-air mixtures are less efficient at heat transfer than water or saturated steam environments, as in a pressure canner. Most of the modern steam canners do not have a vent at the top so air pockets exist. It is impossible to keep the percentage of steam surrounding the jars constant; the air pockets have very slow heat transfer. One study did find steam canning of tomatoes acceptable (Collins et al., 1982); however, further research which includes heat penetration data and jars inoculated with organisms that grow readily in the test food is needed before steam canning can be recommended. An additional problem with some of these canners is a design without handles which is an attractive addition to the kitchen cupboards but likely to give the user steam burns.

Oven canning recipes are still found in nonreliable publications and practiced by some householders. Oven canning was extensively researched in the 1930s and concluded to be unsafe due to low interior jar temperature, large fluctuations in heat penetration among the jars, and fairly common jar breakage (Andress and Kuhn, 1983). No new developments refute this research.

Microwave ovens and dishwashers have the ability to heat the contents of jars so that a vacuum is produced as they cool which completes the seal. However, heating of the jars is variable and heat penetration data and researched processing times do not exist for either method. Cold spots in microwave ranges and uneven heating of the food is a major problem, and the hottest cycle on dishwashers is not high enough to kill bacteria inside a sealed container.

In the open kettle method, foods are heated in a kettle and poured into "sterilized" jars. The lid is then placed on top. As the food cools and shrinks, a vacuum forms. This method gained popularity in the early days of home canning because it was so convenient. It is now discouraged for all products because of the risk of contamination during filling and sealing of the jars. Problems with this method were observed early. In 1944, open kettle canning was a recommended method only for relishes, pickles, preserves, jams, and jellies. It was considered wasteful for fruits and tomatoes, and dangerous for low-acid foods (Andress and Kuhn, 1983). A 1979 USDA survey of home canning practices found 70% of the households using open kettle canning for at least some products (Davis and Page, 1979). Eighty-five percent of those using this method used it for jams and jellies and experienced only 1 to 2% failure rates. However 44% used it for fruits, 35% for tomatoes, 43% for tomato sauce, 14% for vegetables, and 57% for pickles, and observed greater failure rates.

A process similar to open kettle canning, but under very strictly controlled conditions and with sealing in a sterile environment, is the commercial aseptic packaging. The product to be canned is heated in a high-temperature unit for the same effective heat treatment as required for conventional sterilization, then used to fill a sterilized container and sealed under sterile conditions. This is currently used for many dairy and juice beverages.

Commercial pouch packaging is canning without cans or jars. The food is sealed in a flexible-wall container such as a pouch and then heat sterilized. These foods have similar shelf lives to foods canned in metal or glass. Pouch packages and equipment are not available for household use. Devices developed for vacuum sealing dried and frozen foods in pouches cannot safely be used for homestyle packaging for room-temperature storage.

## Heat Sterilization

Developing the processing time for a food product to be canned is a complex process. The lethal effect of heat on bacteria depends upon:

**A.** The number and thermal resistance of the organisms present, including the effect of the composition of the food.

**B.** The rate of heat transfer through a given food in a given size container.

**C.** The temperature and the time of the heating.

The initial number of bacteria present in a food is an important factor because, when heat is applied, the bacteria are killed at a rate proportional to the number present. This phenomenon is called a logarithmic order of death, which means the death rate is constant. With a given time of heating, the same percentage of the bacterial population will be killed regardless of the initial numbers present. Therefore, if the number of bacteria in a food is large, the heating time required to sterilize the food increases. The 1917 Farmer's Bulletin recognized this fact and stated that "stale" vegetables could be home canned if the processing time were lengthened (Andress and Kuhn, 1983). However it is impossible to give accurate, general advice for processing foods with unusually high microorganism loads. This is why starting with sound produce of optimum ripeness is stressed in canning instructions.

To arrive at a processing time for each food, sample jars are inoculated with bacteria which are more heat-resistant then *C. botulinum*. Usually flat-sour bacteria are the most heat-resistant that are likely to be in low-acid foods so they are used. The jars are heated for a variety of processing times, then they are held and observed for spoilage. A Thermal Death Time (TDT) curve can be made from data such as that in Table 9.1 by plotting the heating temperature as arithmetic values and the time as logarithms (Fig. 9.2). A single line is drawn and all points above the line represent processing conditions which destroy the organisms, and points below are processing conditions which allow organism survival. The straight line indicates that the order of death with heating is logarithmic; the death rate is constant.

Home canners and food professionals seeking information on a specific new food or process turn to research articles. In research articles on canning, the z and F values are

**Table 9.1. Food inside jars becomes safer with increasing heat treatments**

| Temp (F) | Heating Time (min) | Samples Heated | Positive[a] Samples |
|---|---|---|---|
| 230 | 100 | 6 | 6 |
| | 200 | 6 | 4 |
| | 300 | 6 | 0 |
| | 400 | 6 | 0 |
| 240 | 30 | 6 | 6 |
| | 50 | 6 | 2 |
| | 80 | 6 | 2 |
| | 100 | 6 | 0 |
| 250 | 3 | 6 | 6 |
| | 5 | 6 | 5 |
| | 10 | 5 | 5 |
| | 15 | 6 | 2 |
| | 20 | 6 | 0 |
| | 40 | 6 | 0 |

[a]Samples with viable spores or cells after the heat treatment are positive. Jars in which all microorganisms are killed are negative. Data in table are hypothetical and presented only as an example of actual research findings.

that have a known quantity of bacteria added (inoculated packs) and a margin of safety is added.

Heat penetration throughout a jar is important for all parts of the food to receive a thorough heating. The time that the slowest heating portions are exposed to the lethal heat treatment is most critical. This is the time used in calculating the processing time recommendations.

Heat transfer from the outside of the jar to the interior of a solid canned food, such as pureed winter squash, is by conduction. This molecule-to-molecule transfer is slow. Note: It is now recommended to process squash as cubes surrounded by liquid instead of solid pack for this reason. The last portion heated, the cold spot for these foods, is usually the geometric center of the container (Fig. 9.4). Unusually shaped jars with large diameters will have a cold spot that requires longer to heat so they are not recommended for canning. Good

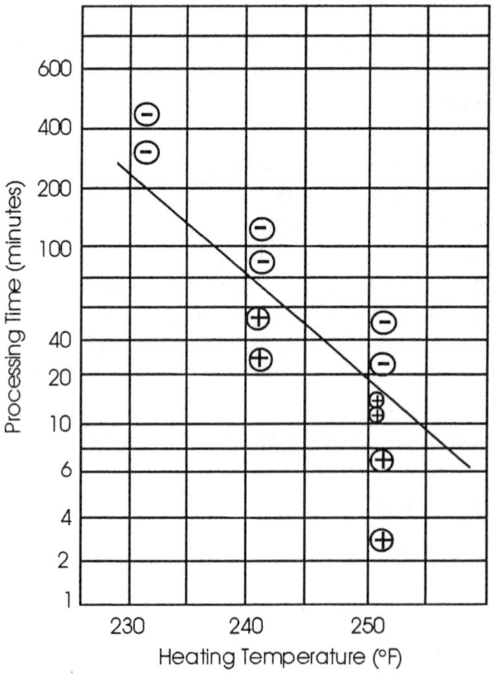


**9.2.** Thermal death time (TDT) curve calculated from data in Table 9.1.

referred to frequently. The slope of the TDT line is called the z value. It is graphically represented as the degrees Fahrenheit required for the line to pass through 1 logarithmic cycle (100 to 10) on semilogarithmic paper (Fig. 9.3). The z value is the number of Fahrenheit degrees required to reduce the TDT 10-fold. In Figure 9.3, the z value is 8. The F value is the minutes required to destroy the organism in a given medium at 250F (121C) processing temperature. The F value is found by reading the time which corresponds to the intersection of the TDT curve and the 250 vertical line. In Figure 9.3, it is 7.5. Both the z and F values will vary with the organism's heat resistance, the number of organisms present, and the medium the organisms are heated in. These variables will be stated in published research so comparisons can be made. Processing times are calculated from z and F values, taking into account heating (come-up) and cooling (come-down) times and temperatures for household pressure canners. The calculated processing time is always verified by actual tests with jars

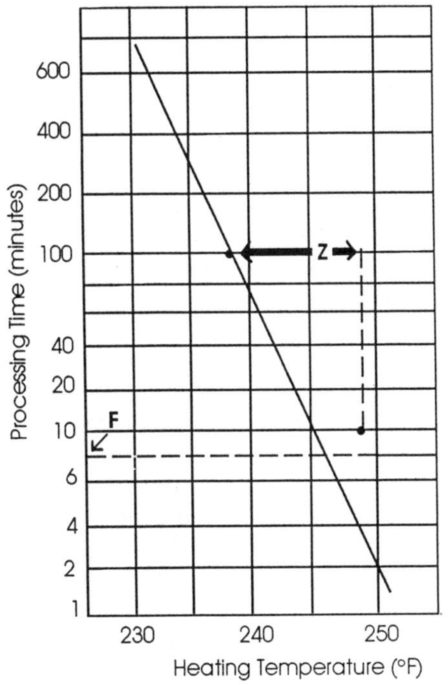


**9.3.** The slope of a TDT curve is called the z value.

canning recipes specify the size of jar the processing time is calculated for. Meats, cream-style corn, white potatoes, sweet potatoes, beets, and fish are heated primarily by conduction. Most often home canned foods are heated by a combination of conduction and convection currents. For example, whole kernel corn and beef stew are heated by both methods (Nordsiden et al., 1978).

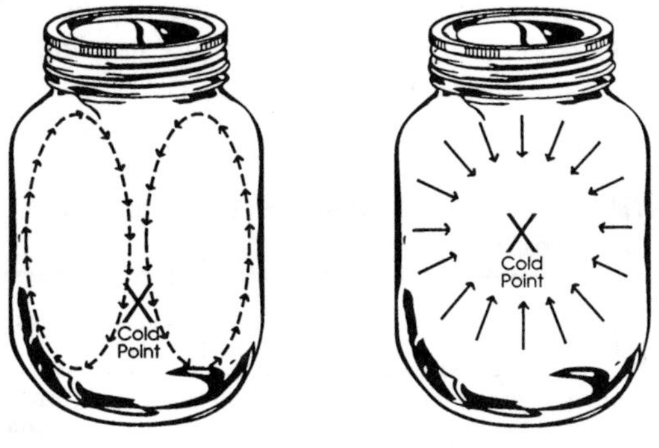


**9.4.** Transfer of heat energy in food in a jar during processing—heat evenly surrounding the jar.

Heat transfer in liquids is by convection currents; the heat is transferred by movement of the fluid itself. Convection type products are in continuous motion during heating and cooling due to the currents caused by temperature differences in the food. Since convection is the faster method, it is important that the food be in pieces and surrounded by adequate liquid to allow these currents. Changing the style of pack, i.e., whole, vertically packed green beans instead of 1-inch pieces, will alter these currents and greatly affect heat penetration. Foods that layer horizontally such as spinach interfere with convection currents so they heat by conduction and require unusually long processing times, as do solidly packed foods. Thick sauces or foods that increase the particle concentration of their surrounding liquid by becoming mushy also inhibit convection currents and heat primarily by the slower conduction method. Foods which change their physical state with heating such as some starchy products or starch-thickened foods that change from a liquid state to a more solid phase present special heat penetration problems. Therefore, canning a vegetable in a sauce instead of in water will result in underprocessing.

An example of the need to determine processing times for specific foods is research which has shown that pint jars of smoked salmon must be processed at 10-lb pressure (11 with a dial gauge type) for 110 minutes as compared to 100 minutes for fresh salmon. Heat penetration was affected by the percentage of moisture remaining in the fish and the tightness of pack (Raab and Hilderbrand, 1993).

## Food Composition and Processing Temperature

### pH

Acid foods are those with a pH at or below 4.6 (see Table 2.4) (Food and Drug Administration, 1979). Fruits, such as apples, with an average pH of 3.4, apricots 3.9, blackberries 3.5, cranberry sauce 2.6, peaches 3.8, pears 4.1, and tomatoes 4.3 fall into this category. Acid foods can safely be processed at the temperature of boiling water because the goal is to kill all vegetative cells that could grow under the storage conditions. Since spore formers such as *C. botulinum* cannot grow in acid foods, spore-killing temperatures of 240F (116C) are not needed for safety. The goal for most foods is to reach 185F (85C) in the coldest spot. Processing times in the boiling-water canner are calculated to hold this temperature until vegetative cells are killed.

Though acid foods are not as subject to microbial growth as low-acid foods, *Bacillus thermoacidurans*, a flat-sour spoilage organism which causes quality loss of tomato products, can be a problem. Flat refers to the nongas producing characteristics of this organism and sour refers to its acid production. Since the container does not bulge from excess gas pressure (i.e., flat lids), spoilage from this organism is apparent only after opening. The off-odor is often more apparent during heating. *Bacillus thermoacidurans* is a spore-forming bacteria that tolerates heat (a thermophile) so it is not killed with boiling-water processing and its growth is favored by holding the jars at warm temperatures after processing—slow cooling or storage in a warm place. Canning directions often state to space the jars apart until completely cool. Householders who pack jars in boxes while still warm are likely to have flat-sour spoilage. Canned foods should be stored in a cool, dry place. Basements are common storage areas in all parts of the country, but in the areas with mild climates, jars are sometimes stored in garages and attics where an unusually hot summer generates flat-sour spoilage.

Most spore-forming bacteria have maximum heat resistance at neutral pH values, and are not as resistant at low pHs. Adding acid to tomato products decreases the heat resistance of *B. thermoacidurans*. Lactic acid is most effective, followed by citric and acetic; however, householders have access to citric in lemon juice and acetic in vinegar so those are used in recipes.

Low-acid foods have a pH above 4.6 (see Table 2.4). Meats and vegetables (except tomatoes) usually fall into this category. Asparagus has an average pH of 5.5, baked beans 5.9, green beans 5.4, beets 5.4, carrots 5.2, corn 6.3, mushrooms 5.8, peas 6.2, white potatoes 5.5, yellow squash 5.1, and spinach 5.4. If *C. botulinum* spores are present and germinate, the organism could grow and produce toxin in these low-acid foods. Therefore, a "botulism cook" of temperatures in the 240–260F (116 to 127C) range for calculated periods of time to kill any spores in the food is required. Since 212F (100C) is the maximum temperature reached by boiling water, steam under pressure is needed; a pressure canner achieves this (Table 9.2).

Development of safe processing recommendations took several decades. The recommended fractional sterilization canning process in the 1909 USDA Farmer's Bulletin recognized that spores are more heat resistant than vegetative bacterial cells and may germinate upon cooling. Fractional sterilization involved heating the jar to boiling and holding it for 1 hour on each of 3 successive days (Andress and Kuhn, 1983). This process would be successful in killing the bacteria that grew from the spores which had germinated after heat shock (the heat they received during processing) and was a slight improvement over only water bathing 5 hours the first day which was previously recommended. Now it is felt that 7 to 11 hours in a boiling-water canner would be required instead of pressure (USDA, 1988) but research has not been done to determine the specific times since quality of the product would be poor and energy use high. Pressure canning is the only recommended way to can low-acid

**Table 9.2. Temperatures achieved with various pressures**

| PSI | Temperature (F) | Temperature (C) |
|---|---|---|
| Boiling water vapor[a] | 212 | 100 |
| 5 | 227 | 108 |
| 10 | 240 | 115 |
| 15 | 250 | 121 |

[a]Sea level.

foods. It was not until 1917 that the Farmer's Bulletins included steam-pressure processing and explained the difference between making the food free from spores, not just free from vegetative cells. The most recent change was from 10-lb pressure with a dial gauge canner to 11 lb to insure that a safe processing temperature was reached (USDA, 1988).

### Effect of Compounds in the Food

As the concentration of sugar in solution increases, so does the heating time required to kill yeasts, molds, and bacterial spores. It is theorized that the higher sugar concentration partially dehydrates the cell protoplasm which protects the microorganism's proteins from coagulation. An example of this phenomena is also seen in baked goods such as angel food cakes when sugar is added to egg albumen—protein in the egg white—and the coagulation temperature is increased.

Sodium chloride in solutions up to 4% increases the resistance of spores to heat; above 8%, resistance is decreased. Table salt added to tomato juice reduces flat-sour spoilage (1 tsp salt per qt).

Starch does not increase the heat resistance of spores, however it permits growth of more organisms if the spores germinate because of adsorption of inhibitory compounds such as $C_{18}$ unsaturated fatty acids. Fats and oils greatly retard moist-heat sterilization of bacterial spores, vegetative cells and yeasts. This is a problem with beef meat balls, chicken pieces, and some canned fish. Protein foods also provide some heat resistance. These foods have longer processing times.

## Nutrients

Research on the nutritive value of canned foods started in the 1920s. It was greatly accelerated by the National Nutrition Conference for Defense in 1941 and continues today primarily to evaluate new processes and foods.

The cultivar and growing conditions can affect total nutrient content of plant foods but householders should strive to retain the maximum amount in each product. Harvesting at optimum maturity, using promptly after harvest, and keeping processing losses at a minimum are important.

From a practical standpoint, the changes in meat and fish proteins in canned products are insignificant in the average U.S. diet, even though bio-availability of the amino acids lysine and methionine is adversely affected (Poling et al., 1944).

Losses of ascorbic acid occur during precooking for hot packs and processing, and significant amounts also leach from plant tissues into the liquid in the jar. Heat-labile thiamin may be substantially lost in pressure canned foods but this is variable. Peas lose approximately 50%, lima beans 80%, corn 80%, carrots only traces, and canned meats approximately 67% (Jackson and Shinn, 1979). Thiamin is more heat stable under acidic conditions and acidic foods receive less heat treatment in canning so their losses are not as great. Retention of riboflavin is quite good (ranging from 70 to 90%) due to its heat stability (Desrosier, 1970). However, it is light sensitive so storage conditions of glass jars are important. Niacin is stable to heat and light. The acid form found primarily in plant tissues is less soluble than the niacinamide found primarily in animal tissues. Overall, the retention of niacin in canned foods is excellent. Vitamin $B_6$ is water soluble and sensitive to light and oxidation. Additionally, the pyridoxal form is heat labile. The retention of vitamin $B_6$ in canned foods has not been thoroughly researched using modern assays. Carotenoids in plant tissues and vitamin A in animal tissues remain relatively unchanged with thermal processing. Their retention averages 80 to 98% (Cameron et al., 1949). Losses of vitamin D are negligible. A study of vitamin losses in canned foods during storage (Elkins et al., 1976) found only trace losses of thiamin, riboflavin, and niacin; however, ascorbic acid content decreased from 5.8 mg/100g in freshly canned green beans to 5.6 after 6 months storage, 3.6 after 12 months, and 3.1 after 18 months.

Minerals can be lost if they leach into the blanching or cooking water which is then not consumed. For most vegetables, blanching losses of minerals rarely exceed 10% (Kramer,

1974). The exception is water blanching of spinach in which phosphorus can be reduced as much as 40% (Jackson and Shinn, 1979). Analyses of the liquid surrounding vegetables in canned foods after processing has found significant amounts of calcium, phosphorus, and iron leached from the plant tissues (Cameron et al., 1949). Therefore, it is desirable to incorporate this liquid into dishes served.

## Quality Changes

### Inactivation of Plant Enzymes

Enzymes are proteins that allow specific reactions to occur much more easily. They lower the energy of activation. It is important that these enzyme-catalyzed reactions occur in living plant tissues for their growth and for their desirability as human food stuffs. However, after harvest these reactions spoil the food quality so it is necessary to stop them. This is most easily done by inactivating the enzymes with heat.

In acid foods, the goal of heat processing is to inactivate enzymes, destroy vegetative cells, and heat the food thoroughly to produce a good seal. The 185F (85C) food temperature achieved with current boiling-water canner processing times can adequately do this in most foods. Most enzymes are irreversibly destroyed within a few minutes at 175F (79C) in neutral pH foods. In canned acidic foods such as pickles, peroxidase (the most heat-resistant enzyme, so its presence is used as an indication of the efficiency of the heat-processing step in industry and research) can withstand heating to 185F (85C).

### Color

The bright colors of fruits and vegetables make them appealing. These colors change during canning due to the effect of heat and also the presence of plant acids that were kept separated from the pigments in the live plant. Chlorophyll is affected by these acids and converts to the olive-green compound pheophytin. The color of canned peas and spinach is remarkably different from the color of the frozen.

Carotenes may fade slightly; but, overall, the color of yellow and orange vegetables such as canned winter squash and carrots is good. The lycopene pigment in tomatoes and watermelon is stable during canning processes, though it is water soluble and leaches into the liquid surrounding the food. Anthocyanin pigments—raspberries, blackberries, blueberries—are also water soluble so they tint the canning liquid, but the overall berry color remains good because so much anthocyanin was initially present.

Color changes during processing of canned beet products are a concern for commercial processors who need to grade their finished products, but at the household level, it is an interesting phenomenon. The betacyanin pigment is sensitive to heat and some converts to nonred compounds. Immediately after canning, 41% of the original betacyanin amount is present—approximately half is lost during blanching and half during pressure canning. After the heat treatments cease, the betacyanin pigment regenerates and the red color is restored. Within 10 hours the regeneration process is usually complete (VonElbe et al., 1981).

### Texture

The texture of canned foods is usually softer than their frozen or fresh counterparts due to the heat treatments required for safety. Consumers in their first canning season may complain about poor quality. Upon examination by those familiar with canned foods, the products are usually found to be top quality canned vegetables, but the family was accustomed to the characteristics of frozen.

### Flavor

Changes in flavor compounds also occur during canning. The flavor of canned spinach, vegetables from the cabbage family, and meats are examples. Generally, if household members are accustomed to the flavor of commercially canned vegetables and meats, they find these home canned foods acceptable too.

## Householder Practices and Errors

Improperly home canned foods may be the result of using inadequate processing methods (boiling-water canner instead of pressure cooker), inadequate processing times for a given product, or contamination of the food after processing. If a toy or appliance were marketed with the same hazard as improperly processed low-acid foods, consumers would be outraged; however, some householders knowingly have this danger in their homes. One such person processed half of her asparagus in a boiling-water canner and half in her only pressure cooker to save time on canning day. The jars were stored on separate shelves and she felt there was no hazard as she planned to thoroughly heat the underprocessed jars before serving. Unfortunately, one meal another household member brought the asparagus out of the pantry and mistakenly brought the wrong batch. It was only warmed and *Clostridium botulinum* poisoning was the result.

Processing at inaccurate pressures and incorrect timing of processing are the major sources of household variation. Insufficient heat treatments carry with them a risk of foodborne illness and spoilage. Underprocessing by 30 seconds at 15 psi results in a 100-fold increased probability of *C. botulinum* spore survival. Underprocessing for 1 minute at 10 psi results in a 10-fold increased probability. Overprocessing is also ill advised. Excessive heat treatments increase the loss of thiamin, chlorophyll, and overall quality. Strictly following USDA processing recommendations is important, but the equipment must be accurate to do this.

For 240 pressure canner dial gauges that were brought in for testing at a Minnesota USDA-staffed center (Thompson et al., 1979), workers found that when the gauges read 10 psi, the average pressure was actually 10.8 psi. When the gauges read 15 psi, the average pressure was actually 15.7 psi. Since the average of the readings was high, the use of the majority of these gauges resulted in safely processed food. Some however were under the expected pressure since the standard deviation was 1.62 psi. Annual testing of gauges is recommended. This testing may be available without charge through the Cooperative Extension Service. In some counties testing booths manned by master food preservers—volunteers trained by the Extension Service—are set up in local businesses that sell new gauges for several weekends each summer. The testing device is a plug-in appliance that builds pressure. A valve releases some of this pressure into the lid gauge of the unit to be tested. Both the gauge of the testing device and the gauge of the lid being tested are then read to determine accuracy.

Household timers are another source of error. The Minnesota survey found that 20% of the timers on kitchen ranges were inaccurate. Comparing the accuracy of the timer to a known accurate clock is recommended.

Human error in placing the jars in the processor, when to start timing, and using a boiling-water canner when a pressure canner is called for also occur frequently. Home economists regularly answer processing questions such as the batch of tuna that was processed 15 minutes short of the recommended time because the trailer ran out of propane, the last batch of salmon that was left in the closed canner all night after the burner was turned off at 1 A.M. and the householders fell asleep during the come-down time, and the corn that was processed 5 minutes less than recommended and then processed again for 5 minutes later in the day when the error was discovered. In addressing these problems, a firm rule is that underprocessing is never safe. If the error is discovered within 24 hours (USDA, 1988), the food may be processed again for the total length of time with new lids, otherwise discard is prudent. In the case of the salmon that was not cooled properly, the salmon is safe but spoilage may occur due to spore-forming bacteria that multiply at warm temperatures. If gas production occurs during storage and the seal is lost, discard is recommended.

Common recipe book instructions recommend placing a 10-pound weight on the canner. In this instance, the weight is a small metal piece with a hole to place over a steam nozzle on the canner. The weights of most pressure canners can be chosen to result in 5, 10, or 15 inside the canner. Since pressure is measured in pounds exerted per square inch (psi), the term "weight" for the pressure regulator seems an appropriate term to some people; to others, it is a source of confusion. Some novice canners want to know if the same effect can be achieved without the expense of purchasing a pressure canner by placing 10 pounds of rocks or books on the lid of a boiling-water canner.

Contamination of the food after processing is a rare occurrence with cans and usually involves a faulty seam or problems with the sealer. If cans are allowed to freeze, expansion of the contents may cause the seam to split. Microorganisms can then enter through these areas.

With glass jars, problems with flat top lids or the sealing surface are usually apparent when the jar fails to develop a vacuum seal. However, it is possible to have a weak seal immediately after processing that later fails. The appearance of molds or other visual defects may occur, or the poor seal may be noticed when the jar is opened without resistance. It is important to check jar seals 3–4 hours after removing from the kettle. Those poorly sealed should immediately be reprocessed, or stored in the refrigerator or freezer. Jars which were sealed when checked but are later found to not have a vacuum are suspect. Seal failure during storage may be the result of gas production from microbiological spoilage or a hydrogen swell. These should be discarded without tasting.

## Altitude Adjustment

A kettle of water boils when the individual water molecules have enough energy—heat energy from the electric unit or gas burner—to escape into the atmosphere. Atmospheric pressure is less at high altitudes. Less atmospheric pressure pushing down on the top surface of water in a kettle means less energy will be required on the part of the water molecules to escape into the air as a vapor. Thus, water heated at high altitudes boils at lower temperatures than the 212F (100C) at sea level. Lower temperatures are less effective at killing microorganisms so high-altitude adjustments must be made for safety. Processing times for boiling-water canners are calculated at sea level with boiling water being 212F (100C) so canning at altitudes of above 1,000 feet requires increases in the processing times (Table 9.3).

Some high-altitude home canners prefer to process fruits under pressure to bring the temperature up to 212F (100C). The actual effective pressure is atmosphere pressure plus that added in the pressure canner. When the temperature surrounding the jars is 212F (100C), then the times are the same as in boiling-water processing.

Little research has been reported on home canning at high altitudes. Inoculated pack studies of tomatoes and green beans in Laramie, Wyo. (elevation 7200) just a few miles from the continental divide (Williams and Maki, 1980) found spore survival with the 1976 USDA recommended processing. The researchers concluded that high-altitude processing times need to be researched under specific altitude conditions; simply adding calculated extra times or pressures is not always effective. The most recent comprehensive USDA canning publication (USDA, 1988) gives processing information for canning at altitudes above 1,000 feet for the foods included. Heat treatments for some of the more commonly canned foods appear in Table 9.3.

## Canning Tomatoes

Tomatoes produce abundant crops in backyard gardens and tomato texture decreases significantly when frozen, so tomatoes are one of the most commonly canned foods. By horticultural definition, tomatoes are a fruit, but they are served as vegetables because of their low-sugar level. The acid level is usually below 4.6—to classify it as an acid food. However, because *C. botulinum* does not grow in tomatoes below pH 4.7, this is the maximum pH

**Table 9.3. Processing treatments vary among foods and increased time or temperature is needed at high altitudes**

| Food | | Altitude (ft) 0–1,000 | 1,001–3,000 | 3,001–6,000 | >6,000 |
|---|---|---|---|---|---|
| | | | (*min*) | | |
| ***Boiling-water canner*** | | | | | |
| Applesauce, qt | | 20 | 25 | 30 | 35 |
| Peaches, hot, pt | | 20 | 25 | 30 | 35 |
| Peaches, hot, qt | | 25 | 30 | 35 | 40 |
| Peaches, raw, qt | | 30 | 35 | 40 | 45 |
| Pears, hot, qt | | 25 | 30 | 35 | 40 |
| Tomatoes, whole, water, qt | | 45 | 50 | 55 | 60 |
| Tomatoes, whole, juice, pt or qt | | 85 | 90 | 95 | 100 |
| | (*min*) | | (*lb*[a]) | | |
| ***Pressure canner*** | | | | | |
| Snap beans, pt | 20 | 11 | 12 | 13 | 14 |
| Snap beans, qt | 25 | 11 | 12 | 13 | 14 |
| Beets, pt | 30 | 11 | 12 | 13 | 14 |
| Carrots, pt | 25 | 11 | 12 | 13 | 14 |
| Corn, knl, pt | 55 | 11 | 12 | 13 | 14 |
| Corn, knl, qt | 85 | 11 | 12 | 13 | 14 |
| Spinach, hot, pt | 70 | 11 | 12 | 13 | 14 |
| Chicken, bnd, pt | 75 | 11 | 12 | 13 | 14 |
| Chicken, bnd, qt | 90 | 11 | 12 | 13 | 14 |
| Salmon, pt | 100 | 11 | 12 | 13 | 14 |
| Tuna, pt | 100 | 11 | 12 | 13 | 14 |

All processing recommendations from USDA, 1988.
[a]Pressures are given for dial gauge canners only. Recommendations for weighted gauge types are the same processing times and pressures of 10 lb for 0–1,000 ft and 15 lb for above 1,000 ft, for all foods.

approved by FDA (1979) for processing at atmospheric pressure. Since the pH of tomatoes may vary with the cultivar, growing conditions, and maturity, the general practice is to slightly acidify by adding an acid.

In testing of 107 cultivars significant differences were found as ripening progressed (Wolf et al., 1979). Underripe red color tomatoes averaged pH 4.19; ripe, 4.35; and overripe, 4.38. Underripe yellow color tomatoes averaged pH 4.02; ripe, 4.20; and overripe, 4.33. Underripe orange color tomatoes averaged pH 4.17; ripe, 4.25; and overripe, 4.36. The tomato cultivars advertised as "low acid" are genetically designed to contain more sugars so they taste less tart. Their pH is still below 4.6 in the firm-ripe stage. Orange and yellow tomatoes have the reputation of being less acid in flavor, but the pH is not high enough to cause safety concerns. The type of cultivar is not as important for pH as the stage of ripeness. Tomatoes infected with plant pathogens and those from dead—including frost killed—vines have abnormally high pH values (Sapers et al., 1978). These are not recommended for canning.

Since some householders use overripe tomatoes and cultivars are occasionally developed which have a pH above 4.7—the accepted minimum for safety for boiling-water processing of tomatoes, current USDA recommendations acidify canned tomato products. For whole, crushed, or juiced tomatoes, the USDA recommendation is 2 tablespoons of bottled lemon juice, 1/4 cup commercial vinegar, or 1/2 teaspoon of citric acid per quart. These amounts should be halved for pints (USDA, 1988).

Tomatoes which have been processed in the boiling-water canner may have surviving spores, and if growth conditions exist, these may germinate and multiply. In doing so, the bacilli bacteria may use the acids of the tomato as an energy source. This decreases the acidity to above pH 4.7 which permits subsequent growth of *C. botulinum* and toxin production. A similar metabiosis (one organism changing the environment which then enables others to grow) effect has been observed with molds (Anderson, 1984 and Montville, 1982). Acidifying decreases the chances of this sequence of spoilage. Most important is obtaining a vacuum seal since bacilli growth is favored by air and mold growth requires air. The seal should be checked on each jar before it is stored.

Keeping jars in a cool place also decreases the chance of spore formers (*Bacillus coagulans* and *B. licheniformis*) growing. Examining the jars before consumption for quality spoilage is a good way to detect spore outgrowth as gas production, off-odors, and/or off-colors are usually present. Contents of jars with these problems should be discarded with care that an animal does not have access to them. Garbage disposals make this task much simpler than the old instructions of burying the spoiled food.

In actual practice, using household equipment, even when processing at boiling is carefully timed, the total heat treatment is variable. In a study of the survival of *B. licheniformis* spores in tomatoes (Montville et al., 1983), use of large boiling-water canners resulted in significantly greater reductions in spore numbers than smaller canners. The mass of water being heated is thought to be the major factor in these differences because a small quantity of water comes to a boil more quickly and heat input during come-up time is less. Current boiling-water canner steps advise preheating the water to 140F (60C) for raw-packed foods and to 180F (82C) for hot-packed foods, adding the jars, covering with hot water and then heating to boiling. Large kettles which allow the user to cover quart jars with at least 2 inches of water can provide adequate heat treatment. These are often difficult to find. Smaller kettles in which quart jars can be barely covered are displayed with most canning supplies and thus frequently used by householders. In recent years the label on these small kettles has been changed to picture pint jars of canned foods instead of quarts.

## Frequency of Spoilage

In a survey of home canned tomatoes in Georgia, Powers and Godwin (1978) found that only 7 of 387 jars of tomato products canned at home showed microbial growth indicative of underprocessing or seal failure. Of the 380 sound jars, only 1 had a pH greater than 4.6; it was 4.65. (Note: 4.7 is the maximum for tomatoes.) In a study of canned tomatoes from a variety of Missouri households, 116 out of 292 tomato jars contained viable *Bacillus* organisms (Fields et al., 1977). Pressure processed jars of tomatoes accounted for 30% of those with viable microorganisms. However, survival of bacilli spores would be expected and would not cause spoilage under usual storage conditions. The pressure processes used varied from just bringing up to 5 pounds to holding at 15 pounds pressure for 30 minutes.

In 1951, 1965, 1969, and 1974, five outbreaks of botulism attributed to tomatoes were recorded. The 1974 outbreak involved open kettle canning. Some of the early reports of botulism outbreaks from home canned tomatoes are now questioned. It was commonly thought that botulism poisoning came only from home canned foods, and other possible sources—prepared low-acid foods improperly stored in the home—were not suspected by the investigator. Since tomatoes are sometimes the only home canned food in a household, they were listed on the report without having been tested for the presence of toxin.

The outbreaks that have occurred probably represented underprocessing—vegetative cells initially present were not killed—not growth because of high pH. Adequate processing destroys viable cells and heat damages the spores so that they cannot grow out. The problem is not product pH which receives a lot of press, but one of education—getting householders to follow important directions. A 1980 survey of householders in eight locations in Canada found 88% did home preserving. Sixty-six percent of these used boiling-water processing for low-acid foods and word-of-mouth advice was commonly followed (Andress and Kuhn, 1983).

A rule of thumb for canning mixtures of acid and low-acid foods is to process for the longest time at the temperature required for the most severely processed ingredient. Surveys of home practices (Andress and Kuhn, 1983) indicated that 20% process these mixtures inadequately. Unfortunately there is currently no accepted method for determining the equilibrium pH of these combination foods other than controlled research for each recipe. Sapers et al. (1982) have studied the problem of processing times for tomato-based combination foods. Their survey developed the following data (Table 9.4) which may help to determine appropriate canning methods. This method is too recent to have stood the test of time so is usually used only to screen a recipe that the table predicts would have a pH above 4.6. The adequacy of process for these still needs to be verified in the laboratory.

**Table 9.4. Range of pH for tomato-based combination foods**

| Product | No. Recipes Tested[a] | pH Range |
|---|---|---|
| Juice blends | | |
| Tomato veg. | 23 | 4.1–5.1 |
| Soups | | |
| Tomato broth | 13 | 4.5–4.8 |
| Tomato veg. | 18 | 4.1–4.6 |
| Minestrone | 9 | 5.2 |
| Manhattan clam | 8 | 5.0–5.3 |
| Chili con carne | 34 | 4.9–5.1 |
| Chili w/o beans | 6 | 5.2 |
| Tomato-meat sauces | | |
| Spaghetti | 43 | 4.6–5.0 |
| Red clam | 5 | 4.8–5.1 |
| Tomato sauces w/o meat | | |
| Tomato puree | 6 | 4.4–4.5 |
| Tomato paste | 9 | 4.4 |
| Spaghetti | 16 | 4.3–4.5 |
| Marinara | 7 | 4.4–4.5 |
| Mexican salsa | 40 | 3.9–4.6 |
| Barbecue | 22 | 4.0–4.4 |
| Tomato mushroom | 16 | 4.3–5.0 |
| Creole | 20 | 4.2–5.2 |
| Tomato-veg. dishes | | |
| Stewed tomatoes | 16 | 4.4–4.5 |
| Tomato celery | 6 | 4.5–4.6 |
| Tomato okra | 12 | 4.4–4.9 |
| Tomato bean | 12 | 4.9–5.3 |
| Tomato zucchini | 3 | 4.7–4.8 |

[a]Data from Sapers et al., 1982.

## Identifying and Preventing Spoilage

Consumption of spoiled canned foods is not recommended even though the usual spoilage organisms are nonpathogenic. No organisms should be growing in properly canned foods, unless stored at higher temperatures, so the presence of any type of organism indicates improper processing. When the processing has not been adequate, there is always the potential for pathogenic bacteria to be present, even in acid foods. Jars showing any type of spoilage should be discarded, and the rest of that batch checked carefully. Understanding the type of organism causing the spoilage and its prevention is important for future canning (Table 9.5). Using only the most recent USDA recommended processing treatments and starting with low initial numbers of organisms (top quality products and clean equipment) is important. The following description of common spoilages and causes can be used to define canning errors (Dunn, 1974).

### Fermentation

Fermentation by yeasts may occur in canned fruits and fruit juices (not tomatoes). Bubbles; a cheesy, alcoholic odor; and a sour taste are characteristic. During fermentation, both alcohol and carbon dioxide are produced; as carbon dioxide builds up, it may break the seal on glass jars or split the seam on metal cans. When the consumer breaks the seal on a fermented jar, there is an outburst of gas and spurts of liquid. Yeasts are very easily destroyed by heat. Boiling-water canning for recommended times prevents fermentation. Fermentation usually develops quickly in underprocessed jars. The fermentation yeasts are not pathogenic, but such randomly fermented jars will be of unacceptable quality and discard is recommended.

### Swells, Fruit

In canned acid foods, swells are characterized by gaseous and frothy fruits, fruit juices, and tomato products. The gas often will bulge (swell) the lid or metal can. Acid-tolerant, spore-forming bacteria produce swells. The more heat-resistant kinds may not be destroyed by boiling-water canner processing for marginally adequate times. Starting with low initial numbers of these bacteria is the main prevention method. This is achieved by using top quality fruits and thoroughly washing jars and equipment before using. Swells develop within a few days after processing. These bacteria are nonpathogenic, but the food in not palatable and should be discarded.

### Swells, Vegetables

Greens, mature peas, shelled beans, and corn have more swell spoilage than other

**Table 9.5. Identification and prevention of canned food spoilage**

| Food | Spoilage | Organism | Prevention |
|---|---|---|---|
| Fruits, juices | Fermentation | Yeast | B-w proc[a] |
| Fruits, juices, tomatoes | Swells | Bacteria, usually nonspore forming | B-w proc |
| Vegetables | Swells | Bacteria, spore forming, thermophilic, anaerobe | Press proc[b] |
| Tomatoes | Flat sour | Bacteria, spore forming, 80–100F[c] | Keep jars cool |
| Vegetables | Flat sour | Bacteria, spore forming, 100–130F[c] | Press proc, keep jars cool |
| Mostly fruits | Mold | Mold | Vacuum seal |
| Vegetables | Sulfide | Bacteria, 100–130F[c] | Press proc |
| Low acid | Putrification | Bacteria, spore forming, anaerobic | Press proc |
| Low acid | Botulism | Bacteria, spore forming, anaerobic | Press proc |

[a]Boiling-water canner processing as recommended by USDA, 1988.

[b]Pressure canner processing as recommended by USDA, 1988.

[c]Growth temperature range. Mesophilic bacteria favor temperatures under 100F (38C); thermophilic bacteria have optimum growth temperatures of 100–130F (38–54C).

canned vegetables. As with fruit swells, large amounts of gas are produced which break the vacuum seal or may even burst the glass jar or metal can. This spoilage also has a faint odor of rancid butter. These bacteria are thermophilic, so storage temperatures under 100F (38C) discourage their growth, as does rapid cooling after processing. Just a few weeks of storage at warm temperatures can result in swell spoilage. These bacteria are not pathogenic, but USDA recommended pressure processing times are calculated to destroy these organisms so swells in jars of vegetables indicate underprocessing and the entire batch is suspect. Underprocessed vegetables should be discarded because pathogenic bacteria may also have not been destroyed.

### Flat Sour, Tomatoes

This spoilage occurs in tomatoes and tomato juice (not other fruits) fairly often. No gas is present, the lids are flat, and there is no noticeable change in appearance. The tomatoes will have a medicinal, sour, or bitter flavor. Sometimes there is also a sour odor. The bacteria that cause flat-sour spoilage have very heat-resistant spores that are not inactivated by boiling-water processing. Warm storage temperatures make germination of these spores more likely. The primary preventative measure is storing the canned food at cool temperatures, but other steps also help. Hot packing, adequate processing in a boiling-water canner, cooling quickly after canning, and storing in a cool location decrease the chance of flat-sour spoilage. Flat-sour bacteria are nonpathogenic, but such spoiled jars are unappetizing and should be discarded.

### Flat Sour, Vegetables

Legumes (shelled beans, peas), corn, pumpkin, greens, and mature snap beans are most susceptible. No gas is produced, so the jar lids are flat and seams of metal cans do not bulge. In flat-sour spoilage of vegetables, cloudy liquid and softer textures are often visible. When the jar is opened, there is an offensive odor. Since flat-sour bacterial spores are encouraged to germinate with warm temperatures, prevention of this spoilage in vegetables also emphasizes lower storage temperature (below 100F, 38C) of the jars. Pressure processing times are calculated to destroy these very heat-resistant spores, so following USDA recommended procedures with accurate equipment should eliminate flat-sour spoilage in vegetables. Flat-sour bacteria are nonpathogenic, but their presence in low-acid foods indicates underprocessing so the food should not be consumed.

### Molds

Molds require air to grow, so their presence indicates a poor seal. Canned food without a strong vacuum seal should be discarded because pathogenic organisms may enter the jar and grow. Molds may use acids and increase the pH to a level where pathogens may grow in previously acid foods. Mold presence is visible before opening jars as a fuzzy, gray or white growth on the surface. Fruits may be slimy if mold enzymes have softened the flesh. There is usually a musty odor and off-flavors will penetrate the entire jar. Mold prevention is by boiling-water processing to heat-inactivate mold spores and obtain a strong vacuum seal. Moldy fruit jams and jelly do not have to be discarded for safety because they are preserved by both acid and sugar, but simply removing the top moldy layer will not remove the off-flavors throughout the jar.

#### Sulfide Spoilage

This occurs most often in corn, mature peas, and beans. It is characterized by grey or black discoloration throughout the jar and a rotten-egg odor (hydrogen sulfide is produced as proteins are broken down). Gas is not produced; the lids are flat. The bacteria responsible are thermophilic so storage at warm temperatures (above 100F, 38C) or slow cooling after processing is necessary for sulfide spoilage to occur. Pressure processing times are calculated to inactivate these bacteria, so following USDA recommended procedures with accurate equipment is the best prevention. Sulfide spoilage indicates underprocessing, and all jars of the batch should be examined carefully. Sugar provides some protection to this bacteria during processing, so sugar should not be added to vegetable recipes; the processing treatments are calculated without added sugar. Sulfide-spoiled jars are too foul to consume and are discarded without question by householders.

#### Putrefaction

This is most likely to occur in meat, greens, corn, and mature shelled beans and peas. The putrefactive bacteria grow anaerobically (without air) so the seal initially is good, but gas is produced as the bacteria break down proteins and the gas may break the seal. This spoilage develops within several weeks. The food has a foul, sewagelike odor which becomes very strong if the food is heated. The food may also be soft or slimy and darker than normal. Putrefactive food may cause illness. It should be discarded in such a manner that animals do not consume it. A garbage disposal is ideal, or boiling for 20 minutes then burying is also safe, but the odors during this heating will be extreme. Adequate pressure processing will destroy putrefactive bacterial spores. Putrefaction indicates inadequate processing and all jars of that batch are suspect.

#### Botulism

Botulism spoilage is possible in all low-acid foods that are not adequately processed. Some types of the botulism bacteria produce gas and the food is readily recognized as spoiled, but some types do not produce gas or other quality changes in the food. Sometimes the appearance of food in which *C. botulinum* has grown is similar to putrefied food. Off-odors such as rotten, cheesy, or rancid butter may be present and become more pronounced with heating. The liquid may be cloudy. The food may be soft or slimy.

*Clostridium botulinum* bacteria form heat-resistant spores and grow anaerobically (without air), so their growth is possible in jars with a good vacuum seal. However, pressure processing treatments are calculated to destroy botulism spores, so botulism is not a concern if USDA recommended procedures and a correctly functioning pressure canner were used. Two to 3 weeks storage is usually required for botulism spores to germinate and produce toxin. All suspect food should be discarded in such a manner that animals do not consume it. If underprocessing occurred in one jar, the entire batch is suspect and should be discarded.

## Storage

If a jar fails to seal and it is noticed within 24 hours, it may be safely reprocessed (USDA, 1988). Check the sealing surface for defects before reusing the same jar. Reprocessing must be done using the same time and temperature as was originally required. Alternatively, enough food may be removed from the jar to give 1 1/2-inch headspace, an airtight seal placed on top, and the jar frozen or refrigerated for short-term storage.

Adequate processing produces jars in which biological foodborne illness spoilage will not occur. Thermophilic organisms—heat loving, can withstand many heat treatments—may be present, but unless the jars are stored under warm (95F, 35C) conditions, spoilage by them is unlikely. However, chemical reactions that affect flavor, texture, color, and nutritive value may occur. The speed of these reactions is directly related to storage temperature. Summer storage at 50F (10C) or below is recommended. Increasing this temperature to 68F (20C) will about double the speed of reactions in the food and 86F (30C) will quadruple the reaction rate from the 50F (10C) level. The jar stored at 86F (30C) will have the same quality loss as the one stored at 50F (10C) in one-quarter of the time. The reaction rate of oxidative rancidity in canned meats often determines the shelf life. In winter, storage conditions should be such that the product does not freeze. Freezing may cause quality deterioration, and lost seals and broken jars from expansion of water in the food. Additionally, glass jars should be stored in the dark as light catalyzes reactions that bleach food pigments, destroy certain vitamins, and alter flavor compounds. The storage area should be dry and have adequate air circulation to remove excess moisture. This decreases the chance of mold growth on the outside of the containers, rusting of metal cans, and broken seals.

Proper labeling with the date processed and a number code of the batch is helpful for inspecting the entire batch when spoilage occurs in one jar. Spoilage is more likely with increased storage times. Canning only the amount needed for one season is recommended for top quality, but canned foods are safe for an unlimited time if the seal remains strong.

Before consumption, the jar should be inspected for loss of seal and for signs of spoilage. If either is suspected, the contents should be discarded. Low-acid food should be detoxified before discarding by boiling for 30 minutes and then disposed of in such a way that pets and wild animals will not have access. If the consumer has any question about the accuracy of the processing or the jar is a gift and the processing time is unknown, the USDA (USDA, 1988) recommends that low-acid and tomato foods be boiled for 10 minutes and an additional minute of boiling be added for each additional 1,000 feet elevation above sea level. An alternative oven-heating method for foods such as fish that lose palatability after a 10-minute boil has been developed (Woodburn et al., 1979). The opened jar is heated in a 350F (85C) oven until a thermometer inserted into the center of the contents reads 185F (85C). This oven method has the additional advantage of using a thermometer to determine the heat treatment instead of visually determining boiling since spoiled foods often foam at low temperatures as gas escapes, which may be interpreted as boiling. Since this could be a serious judgment error, foods which are boiled are held for 10 minutes after the start of boiling. For thick mixtures, 20 minutes of boiling is recommended since the bumping of food from pockets of steam may occur before the temperature has reached the boiling point. At the end of this time, temperatures at which toxin is inactivated will have been reached throughout the food.

## References


Anderson, R.E. 1984. Growth and corresponding elevation of tomato juice pH by *Bacillus coagulans*. J. Food Sci. 49:647.

Andress, E.L. and Kuhn, G.D. 1983. Critical review of home preservation literature and current research: Home canning. University Park: Pennsylvania State Univ.

Banwart, G.J. 1989. Basic food microbiology. 2d ed. Westport: AVI Publ.

Cameron, E.J., Pilcher, R.W. and Clifcorn, L.E. 1949. Nutrient retention during canned food production. Am. J. Public Health 39:756.

CDC. 1974. Botulism in the United States 1899–1973. Atlanta: Centers for Disease Control.

Collins, J.L., Man, Y., Draughon, F.A. and McCarty, I.E. 1982. Modification of the processing method for home preservation of tomato juice. J. Food Prot. 45:580.

Davis, C.A. and Page, L. 1979. Practices used for home canning of fruits and vegetables. USDA Home Econ. Research Report. No. 43.

Desrosier, N.W. and Desrosier, J. 1977. The Technology of food preservation. 4th ed. Westport: AVI Publ.

Dunn, C.M. 1974. The food spoilage chart—how to identify and prevent spoilage in home canning. Madison: Univ. of Wisconsin.

Elkins, J.R., Kemper, K. and Lamb, F.C. 1976. Investigations to determine the nutrient content of canned fruits and vegetables. Washington DC: Natl. Canners Asso. Research Foundation.

Esleen, W.B. Jr. and Fellers, C.R. 1948. Effect of different processing procedures on venting and loss of liquid from home canning jars. Food Tech. 2:222.

Fields, M.L., Zamora, A.F. and Bradsher, M. 1977. Microbial analysis of home-canned tomatoes and green beans. J. Food Sci. 42:931.

Food and Drug Administration. Title 21. Code of federal regulations low-acid canned foods. From 1979 Federal Register 42(1433A): Section 113.3.

Frazier, W.C. and Westhoff, D.C. 1988. Food microbiol. 4th ed. New York: McGraw-Hill.

Jackson, J.M. and Shinn, B.M. 1979. Fundamentals of food canning technology. Westport: AVI Publ.

Karmis, N. and Harris, R.S. 1987. Nutritional evaluation of food processing. 3rd ed. New York: Van Nostrand Reinhold.

Kramer, A. 1974. Storage retention of nutrients. Food Tech. 28:50.

Montville, T.J. 1982. Metabiotic effect of *Bacillus licheniformis* on *Clostridium botulinum*: Implications for home-canned tomatoes. Appl. Env. Micro. 44:334.

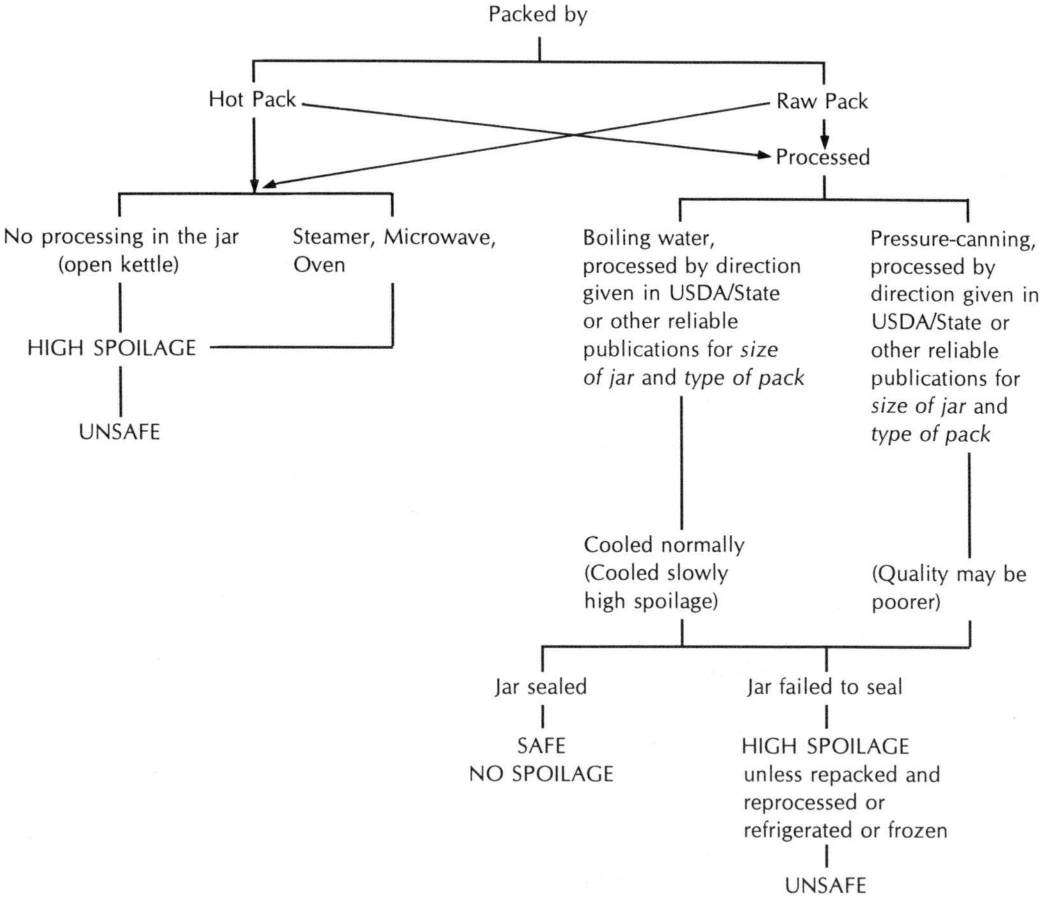


**Summary Chart I.** Canned fruits (except figs), tomatoes (except overripe), quick pickles made with vinegar, at least 1:1 proportions.

Montville, T.J. and Conway, L.K. 1982. Oxidation-reduction potentials of canned foods and their ability to support *Clostridium botulinum* toxigenesis. J. Food Sci. 47:1879.
Montville, T.J., Conway, L.K. and Sapers, G.M. 1983. Inherent variability in the efficacy of the USDA

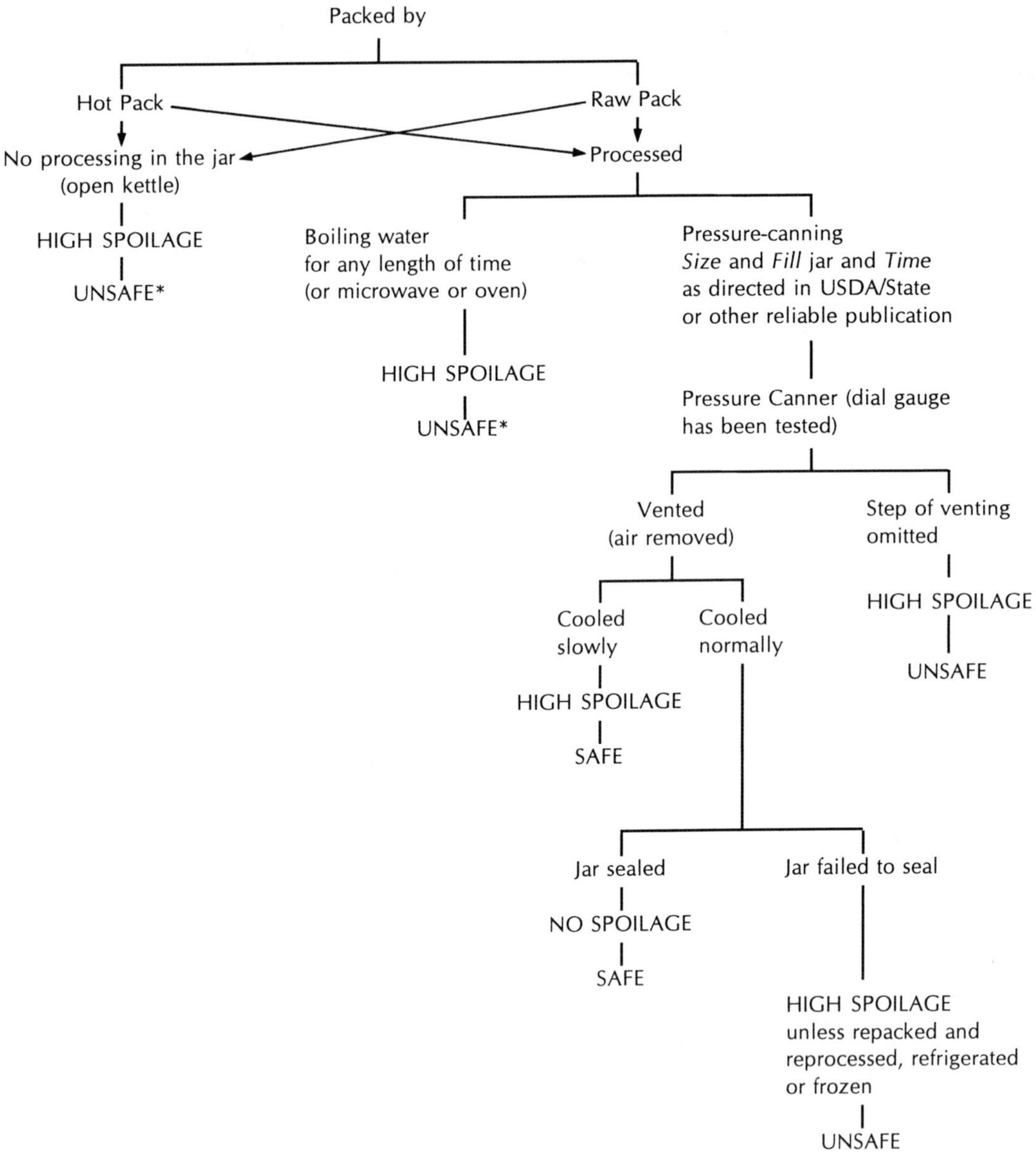


*Boiling for 10–20 minutes or heating fish to 185F (85C) in oven will destroy botulinum toxin. Not safe method because others opening the jar may not remember or know food is not safe until boiled and may taste before boiling. Do not use unsafe processing methods.

**Summary Chart II.** Canned vegetables (except tomatoes), meats, fish, poultry, and mixtures containing any one or more of these foods unless also acidified to pH 4.6 or below.

raw-pack process for home-canned tomatoes. J. Food Sci. 48:1591.

Nordsiden, K.L., Thompson, D.R., Wolf, I.D. and Zottola, E.A. 1978. Home canning of food: Effect of a higher process temperature on the safety of low-acid food. J. Food Sci. 43:1734.

Raab, C.A. and Hilderbrand, K.S., Jr. 1993. Home-canned smoked fish: A new processing recommendation. J. Food Prot. 56:619.

Poling, C.E., Schultz, H.W. and Robinson, H.E. 1944. The retention of nutritive quality of beef and pork muscle proteins during dehydration, canning, roasting and frying. J. Nutr. 27:23.

Powers, J.J. and Godwin, D.R. 1978. pH of tomatoes canned at home in Georgia. J. Food Sci. 43:1053.

Sapers, G.M., Phillips, J.G. and DiVito, A.M. 1982. Equilibrium pH of home canned foods comprising combinations of low acid and high acid ingredients. J. Food Sci. 47:277.

Sapers, G.M., Phillips, J.G., Panasiuk, O., Carre, J., Stoner, A.K. and Barksdale, T. 1978. Factors affecting the acidity of tomatoes. Hort. Sci. 13:187.

Sapers, G.M., Stoner, A.K. and Phillips, J.C. 1977. Tomato acidity and the safety of home canned tomatoes. Hort. Sci. 12:204.

Stimpson, J.K.H. 1981. Home canning of meat and poultry, a review of procedures. M.S. Thesis. Univ. of Idaho.

Thompson, D.R., Wolf, I.D., Nordsiden, K.L. and Zottola, E.A. 1979. Home canning of foods: Risks resulting from errors in processing. J. Food Sci. 44:226.

USDA. 1988. Complete guide to home canning. Ag. Infor. Bull. No. 539. Washington DC: United States Dept. Ag.

VonElbe, J.H., Schwartz, S.J. and Hildenbrand, B.E. 1981. Loss and regeneration of betacyanin pigments during processing of red beets. J. Food Sci. 46:1713.

Wiese, K.L. and Jackson, E.R. 1993. Changes in thermal process times ($B_b$) for baked beans based on water hardness and fill temperature. J. Food Prot. 56: 608.

Williams, J.C. and Maki, L.R. 1980. Survival of *Clostridium sporogenes* (PA3679) and *Bacillus coagulans* in green beans and tomatoes home canned at high altitude (7200 ft). J. Food Sci. 45:1452.

Wolf, I.D., Schwartau, C.M., Thompson, D.R., Zottola, E.A. and Davis, D.W. 1979. The pH of 107 varieties of Minnesota-grown tomatoes. J. Food Sci. 44:1008.

Woodburn, M. 1982. pH of Oregon-grown figs and their acidification for home-canning. J. Food Prot. 45:1245.

Woodburn, M., Schantz, E.J. and Rodriguez, J. 1979 Thermal inactivation of botulinum toxins in canned salmon. Home Econ. Res. J. 7:171.

## CONSUMER QUESTIONS

**Q.** *Are blackened jar lids a sign of spoilage?*
**A.** The deposit is due to a chemical reaction between the food and the lid. It is not harmful to consume but off-flavors may be present. This is most common with acid foods.

**Q.** *What is the problem with canned beets that have no unusual odor or gas present, but have blackened?*
**A.** This is due to mesophilic—grow best below 100F—bacteria which must have iron present to produce the blackening. The iron source may be from water or a container used in processing such as an iron kettle or a chipped enamel saucepan. If metal cans containing iron are stored for a long time, the iron in the can may become the source of iron. Black beets should not be consumed since this color change is the result of microbial action. Adequate processing eliminates these bacteria.

**Q.** *Received home canned low-acid food and do not know the processing history, must it be discarded?*
**A.** *If the food is a gift to the household*: In general the decision will depend on your assessment of the likelihood that reliable canning directions were followed. If you know the food was underprocessed, such as boiling-water processed green beans, then it should not be consumed. As a safety precaution, low-acid canned foods of unknown history should be heated thoroughly before consumption. Use in oven-baked dishes or boil for 10 minutes.

It is prudent not to distribute canned low-acid foods to others through an organization because of possible legal ramifications. Whether illness resulted from underprocessing or mishandling after opening, processing methods in the home are not documented as they are commercially.

**Q.** *Should I can products without established processing times? (I want to can vegetables in a thickened sauce, but cannot find a recipe. My mother's spaghetti sauce is better than the USDA recipe.)*
**A.** The best advice is to preserve by another

method such as freezing. A similar recipe may be found for which a process has been developed such as the USDA mincemeat recipe instead of the family's traditional recipe. In this case, switching recipes is necessary to can. Even though ingredients of other recipes are similar, establishing processing times is an involved microbiological process and cannot be done in households.

For mixtures in which the pieces stay discrete so the heat penetration characteristics do not change (green peas and pearl onions) the product may be pressure processed for the time for the product with the longest requirement.

Sometimes, this question will not be asked until several batches are already made and sitting on the pantry shelves. There may be alternatives to discarding these jars. Vegetables may be boiled and then frozen for later use. However, in the case of a variable product such as mincemeat, pureed winter squash, or macaroni and cheese, it is likely that all parts of the jar were not processed adequately and large amounts of bacteria or toxin may be present. Heating before eating may not be effective in this case and quality would be poor. Consumption is not recommended.

It is not unusual for the inquirer to ask, "How sick will I get if I do decide to eat it?" The worst-case scenario is that of botulism poisoning. It is also not unusual for the inquirer to contact you after consumption to report that no illness occurred at all. The reason for this phenomenon is that food poisoning organisms are not present in every jar; in fact they are in the minority in microorganism populations. In the early days of canning, underprocessed foods were consumed routinely and certainly if 100% or even 25% had become ill, their use would have been discontinued. However given the processing information we have today and the severity of the consequences of consuming botulism toxin, it is not wise to take a chance.

**Q.** *Jars did not seal, why?*
**A.** Check sealing surface of jar for defects. If there are many failures, perhaps the sealing compound on the lid was defective or not pretreated correctly. Check the box instructions. Purchasing only the lids you will need for each canning season reduces sealing compound failures. The contents may be saved if discovered within 24 hours, by reprocessing according to USDA recommendations.

**Q.** *What difference does the syrup make in canning fruit?*
**A.** Syrup with a higher proportion of sugar (heavier) results in more attractive, firmer products. Any type of sugar can be used, but molasses, raw sugar, brown sugar, sorghum, and most honeys are strongly flavored and may mask the fruit's natural flavor. It is safe to use plain water; however the fruit may lose sugar into the water and the result could be a less flavorful as well as softer product. Fruit juices and juice:water combinations are preferred by some householders; they have the same processing times as syrups. Each quart of fruit will require 1–1½ cup syrup or other liquid.

**Light syrup**
2 c sugar and 4 c water yields 5 c syrup
1 c honey, 1 c sugar, and 4 c water yields 5½ c syrup
1 c honey and 3 c water yields 4 c syrup

**Medium syrup**
3 c sugar and 4 c water yields 5 c syrup
1 c honey, 2 c sugar, and 4 c water yields 6 c syrup

**Heavy syrup**
4½ c sugar and 4 c water yields 6½ c syrup

**Q.** *My fruit turns dark, why?*
**A.** Antibrowning agents such as salt, vinegar, or ascorbic acid solutions can be used in dips between the time the fruit is cut and it is placed in the jar. Hot packing removes more air from the fruit's tissues so there is less for oxidative browning. Floating above the liquid is also decreased with hot packing, which decreases oxidative browning of the top layer.

**Q.** *Why can seeds in canned tomatoes sprout?*
**A.** If the heat treatment was not enough to inactivate the enzyme systems in the tomato seeds and the jars are stored at warm temperatures, the seeds may sprout. The sprouts are often mistaken for worms.

**Q.** *Is mold inside jars serious?*
**A.** With underprocessing, molds may survive and grow until they use up the oxygen in the headspace. A broken seal will permit more extensive growth. The jar should be discarded since molds may change the environment to one in which organisms that cause illness can grow.

**Q.** *How can I keep rings from rusting?*
**A.** If you remove rings on canning day after the jars have sealed, then wash, dry, and store them in a dry place they should not rust. It is much easier to examine the seal before consuming if the ring is not still in place.

**Q.** *I know now that I did not process long enough; can I redo it?*
**A.** *Discovered error within 24 hours*: Put new lids on and reprocess for entire correct amount of time. The USDA does not recommend reprocessing if the error was discovered after the jars had been stored for longer than 24 hours.

**Q.** *What about using outdated processing times?*
**A.** In December 1988 the USDA published its new *Complete Guide to Home Canning*. The new processing times for some foods in this manual are based upon new research findings. Although the risk is not high in using the prior USDA methods, there is more potential for spoilage. Therefore, currently recommended processing methods are the only safe ones to use.

**Q.** *Why are there gritty crystals in canned tuna?*
**A.** Magnesium ammonium phosphate crystals can form in canned tuna. There is no prevention. They decrease quality, but are harmless to consume.

# 10 Root Cellar and Dry Storage

Several generations ago, families were routinely fed carrots, turnips, apples, squash, and potatoes from the cellar throughout the winter months, but with modern intercontinental food transportation and storage systems, root cellars are seen by many to be unnecessary. Even though cellars are not required now to provide fresh produce outside of the growing season, household root cellars are a timeless way to store food since they are a low-technology, low-cost, low-energy preservation method. The style of the cellar has changed over the years as dirt-floor basements were zoned out of new neighborhoods and many suburban homes were constructed with no basement at all, but root cellaring-type storage is still very applicable today.

Root cellars are unheated (and often underground) storage spaces for vegetables and some fruits. Uninsulated basements, unheated garages, garden trenches, and holes dug into hillsides then lined with brick or concrete blocks are all examples of root cellars. Without a dirt floor, some vegetables are kept alive by covering with damp sawdust or burlap instead of mounding dirt around the roots.

Certain foods do well in this type of storage. Those experienced in the use of root cellars can provide their households with onions, cabbage, beets, carrots, potatoes, garlic, onions, leeks, parsnips, salsify, squash, pumpkins, rutabagas, radishes, turnips, and celeriac as late as February or March. In addition to selecting appropriate commodities for underground storage, the cultivar and soundness of the produce itself will influence success with root cellaring. Unblemished produce is desirable for all preservation methods but since the produce remains alive in the cellar, soundness is critical for this method. Wounded skins are common entry sites for plant pathogens—spoilage that not only destroys the wounded item but spreads throughout the cellar. Bruised flesh also releases ethylene.

Ethylene ($C_2H_4$) is one of the most important factors influencing storage life of root cellared produce. It is a naturally present plant compound that increases respiration, and hastens aging and senescence. Even after harvest, fruits, vegetables, stems, and roots are very susceptible to the effects of ethylene. When tissue is wounded or bruised, the large amounts of ethylene released stimulate not only healing but plant respiration. High rates of respiration shorten the life span of stored commodities as energy sources are used and can no longer be replenished in plants separated from their roots, stems, and leaves. Care must be taken not to bruise the fruit when packing in boxes for storage. Windfall fruit should be placed in a separate area and used first. Commercial produce warehouses go to great expense to remove ethylene from the storage rooms since its presence greatly decreases shelf life.

## Respiration

Many people consider photosynthesis as the main activity in plants, but aerobic respiration systems provide most of the energy for fruits and vegetables and are critical for survival; just as they are in animals. Photosynthesis provides the substrates for respiration by using energy from the sun with water from the soil and carbon dioxide from the atmosphere to produce oxygen and carbohydrates containing glucose. Most plants store carbohydrates as starch or sucrose, and convert these to glucose via their enzyme systems as needed. Respiration is essentially the reverse of photosynthesis; glucose and oxygen are converted to carbon dioxide, water, and energy. Plants, like animals, require the energy for growing; without oxygen, plants die.

$$\underset{\text{glucose}}{C_6H_{12}O_6} + 6\,O_2 \rightarrow 6\,\underset{\text{carbon dioxide}}{CO_2} + 6\,\underset{\text{water}}{H_2O} + \text{energy}$$

Plant respiration cycles are the same as those of animals. In the EMP pathway glucose is converted to pyruvate which is then broken down in the citric acid cycle to release energy and $CO_2$.

$$\text{glucose} + 2\,\text{ADP} + 2\,\text{P}i + 2\,\text{NAD} \rightarrow 2\,\text{pyruvate} + 2\,\text{ATP} + 2\,\text{NADH}_2 + 2\,H_2O$$

$$\text{pyruvate} + 3\,O_2 + 15\,\text{ADP} + 15\,\text{P}i \rightarrow 3\,CO_2 + 2\,H_2O + 15\,\text{ATP}$$

Respiration rates are critical to the shelf life of cellared foods because high rates use substrates which will shorten the life expectancy and increase the concentration of respiration products. Successful root cellaring involves controlling humidity around the produce (no drying or condensation) and decreasing respiration rates. Five factors determine respiration rate: temperature, concentration of reactants or products, ethylene, the produce itself, and enzyme activity.

### Temperature of the Storage Room

Higher temperatures increase respiration rates, so low-temperature storage is desirable within limits. Temperatures below 32F (0C) may result in ice crystal formation in the tissues and death, and temperatures below 45F (7C) may result in chilling injury for some tropical-type commodities such as bananas, green peppers, avocados, pineapples, and tomatoes. Chilling injury speeds senescence. A common chilling injury symptom is pitting of the skin which occurs when cells beneath the surface collapse. The pits are often discolored. Browning of flesh tissues is also common, as is failure to ripen (Table 10.1). The symptoms often do not appear until the produce is warmed. Then deterioration is rapid and may occur within a few hours.

Close to the freezing temperature of water, respiration is slow, but the change in rate with an increase in temperature is dramatic (Fig. 10.1). With temperature changes between 68 and 95F (20–35C) the slope of the curve is almost zero indicating little change in the respiration rate. Between 112F and 140F (45–60C), plant proteins denature. Protein systems are needed for life and with denaturation, respiration decreases sharply as cells die throughout the tissues.

For each 18F (10C) increase in temperature (Table 10.2), the reaction rate for most chemical reactions is approximately doubled. This rule is called $Q_{10}$ (see Chapter 8); when the rate doubles, $Q_{10} = 2$. Reactions such as respiration and the accompanying senescence basically follow the rule of $Q_{10}$; however in plants, enzyme systems and other factors influence respiration too. The $Q_{10}$ is not always a constant 2 in live plants because plant respiration involves many enzymic reactions and the activity of these enzymes does not vary linearly

**Table 10.1. Produce susceptible to chilling injury**

| Commodity | Lowest Storage Temp. for Best Quality[a] (F) | Injury[b] |
|---|---|---|
| Apples | 36–38 | Internal browning,soft tissue |
| Avocados | 40–45 | Flesh discoloration |
| Bananas, green | 53–56 | Poor skin color when ripe |
| Bananas, ripe | 53–56 | Blackened skin |
| Cucumbers | 45 | Pitting, water-soaked spots, decay |
| Eggplants | 45 | Rot, seeds blacken |
| Grapefruit | 50 | Pitting, watery breakdown |
| Jicama | 55–65 | Surface decay, discoloration |
| Lemons | 52–55 | Pitting, discoloration, red blotch |
| Limes | 45–48 | Pitting, tan discoloration |
| Melons | 45–50 | Pitting, decay, failure to ripen, may discolor |
| Oranges | 38 | Pitting, browning |
| Peppers, sweet | 45 | Rot, dark seeds |
| Pineapples, green | 45–50 | Dull green when ripe |
| Potatoes | 38 | Browning, sweetening |
| Squash, winter | 50 | Decay |
| Sweet potatoes | 55 | Decay, discolor, hard core when cooked |
| Tomatoes, green | 55 | Poor color when ripe, rot |
| Tomatoes, ripe | 45–50 | Softening, decay |

[a]Minimum temperature varies with cultivar and growing conditions.

[b]Description of typical injury when commodity is stored between minimum temperature and freezing. Defects are more pronounced with longer storage periods. Data adapted from USDA, 1990.

with the temperature; therefore, the $Q_{10}$ does not remain constant.

At very low or very high temperatures, respiration almost ceases (Fig. 10.2). When selecting a storage place, temperature should be given high priority. In the middle temperature ranges, the rule of $Q_{10}$ applies. At 50F (10C) the storage life is approximately twice as

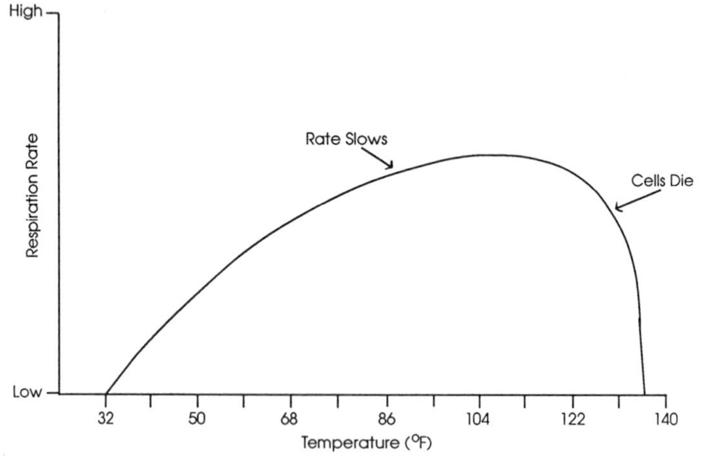


**10.1.** Respiration rate of stored vegetables in relation to storage temperature.

**Table 10.2. Storage temperatures in Celsius and Fahrenheit**

| Temperature °C | Temperature °F |
|---|---|
| 0 | 32 |
| 10 | 50 |
| 20 | 68 |
| 30 | 86 |
| 40 | 104 |
| 50 | 122 |
| 60 | 140 |
| 70 | 158 |

long as at 68F (20C).

To calculate $Q_{10}$, the respiration of a fruit or vegetable is measured at two different temperatures 18F (10C) apart. This measurement is made by placing the commodity in a closed system and then recording either the oxygen consumed or the carbon dioxide given off. $Q_{10}$ values for fruits and vegetables are usually highest between 34 and 59F (1 and 15C); they can reach 7 at this temperature range. The difference in storage life between an in-ground root cellar at 34F (1C) and a basement at 50F (10C) is significant. At temperatures above 59F (15C), the $Q_{10}$ is generally between 2 and 3. Lowering the temperature can delay ripening in climacteric fruits—such as peaches, pears, and bananas which go through a sudden ripening stage—and slow the rate of senescence in nonclimacteric produce (these gradually ripen such as carrots, lettuce, pineapple) in addition to slowing metabolic processes. At lower temperatures, plants manufacture less ethylene, and the ethylene also has less effect (activity). The cool autumn temperatures decrease respiration through both $Q_{10}$ and decreased ethylene production. Root cellaring is difficult in the summer. Potatoes, onions, and garlic are usually the only foods held in households during the hot months.

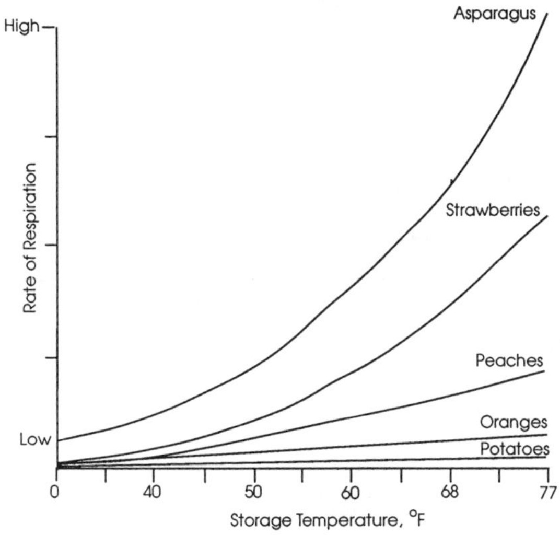


**10.2.** Respiration rates increase with storage temperature but extent of change varies.

Temperature fluctuations are also undesirable because condensation results. Excessive moisture on stored plant surfaces is an ideal environment for microbial growth. In-ground storage places tend to provide more constant temperatures than unheated garages.

## Concentration of Reactants and Products

Abundance of reactants or buildup of products influences the rate of reactions, including respiration. Increasing concentrations of reactants (carbohydrates and oxygen) increases the respiration rate. Increasing the concentrations of products (carbon dioxide) slows the rate of respiration. Fruits and vegetables are selected to have specific carbohydrate content for palatability, so this factor is not altered by the food industry to slow respiration; however, oxygen and carbon dioxide are. To have an effect, $CO_2$ concentrations must be increased 2–3%. Decreasing the concentration of oxygen is also used in commercial, controlled atmosphere storage facilities. Oxygen concentrations in air are normally 20%; a 5% reduction is needed for a decrease in ethylene activity. Special walls must be constructed for these controlled atmosphere storage rooms and special suits with breathing apparatus worn by those who enter the rooms. Controlled atmosphere storage is not practical at the household level. Householders cannot increase storage life by enclosing fruit in a bag because although carbon dioxide increases, so does ethylene, and additional oxygen needed for respiration is unavailable—the plant dies. The objective is to slow respiration, not stop it.

## Ethylene

Ethylene is a substance that can induce strong respiration responses; one ppm can stimulate respiration. Cellared foods give off ethylene which can speed the respiration and senescence of all products in the cellar. Commercially, ethylene is removed from storage warehouses by air ventilation (not just circulation), by absorbers such as potassium permanganate (Purafil pads) or bromine activated charcoal (BAC), and by hypobaric conditions. Special warehouse construction is needed for hypobaric (vacuum) storage. One-tenth to 1/12 of an atmosphere can pull internal ethylene out of the cells, not just scrub it from the warehouse air as it is released from the commodities. Since internal ethylene levels may be 100 times higher than in the atmosphere, hypobaric storage is very effective; however, because of the cost of the special warehouses, it is used only for stored fruits that bring very high prices. Hypobaric storage containers similar in size to boxcars have been used effectively to ship berries abroad. At the household level, one or two ventilation vents in a root cellar and perhaps a fan provide ethylene removal.

Harvested fruits are capable of producing the largest amounts of ethylene. For this reason, fruits are stored separately from vegetables. Many householders keep potatoes and carrots in the crawl space under their house and store apples and winter pears on the floor of their garage.

Some fruits such as apples release large amounts of ethylene during their final ripening step (climacteric), making storage with vegetables a difficult balance (Table 10.3). Commercial warehouses have sustained large financial losses from placing carrots or potatoes in the same room as apples. One such case in the state of Washington involved carrots and apples in separate rooms that shared a common warehouse wall—a wall that did not extend all the way up to the ceiling. Ethylene released from the apples diffused through the air and crossed over to the carrot's side where it increased the respiration of the stored carrots. When the carrot warehouse doors were opened several months later an unmarketable product was discovered.

Molds such as *Aspergillus* and *Penicillium* produce ethylene; another reason spoiled produce should be promptly discarded from the cellar. Householders must periodically check stored produce. Though a particular type of vegetable has an expected root cellar life of 4–6 months, all individual units of that vegetable will not usually survive for the 6-month period.

## The Produce

Stage of development and condition of the fruits and vegetables will influence respiration. Bruises and cuts release enzymes and ethylene that stimulate respiration. Fruits nearing their climacteric will have greater respiration rates so they should be harvested in the mature-green state (Fig. 10.3). At this point all of the enzymes necessary for ripening are present but the high respiration levels (the climacteric) that accompany ripening have not begun. After the climacteric, fruit respiration again returns to lower levels, but it is then overripe and the cellular structure is breaking down. Overripe produce does not have a long storage life due to enzyme-catalyzed browning and increased susceptibility to microorganisms.

Both climacteric and nonclimacteric foods continue to ripen after harvest if picked in the mature-green stage; however, climacteric fruits

**Table 10.3. Respiratory climacterics are present in some fruits**

| Climacteric Fruits | Nonclimacteric Foods |
|---|---|
| Apple | Blueberry |
| Apricot | Cherry, sweet, sour |
| Avocado | Grape |
| Fig | Citrus fruits |
| Muskmelon | Pineapple |
| Papaya | Strawberry |
| Peach | All vegetables |
| Pear | |
| Plum | |
| Tomato | |
| Watermelon | |

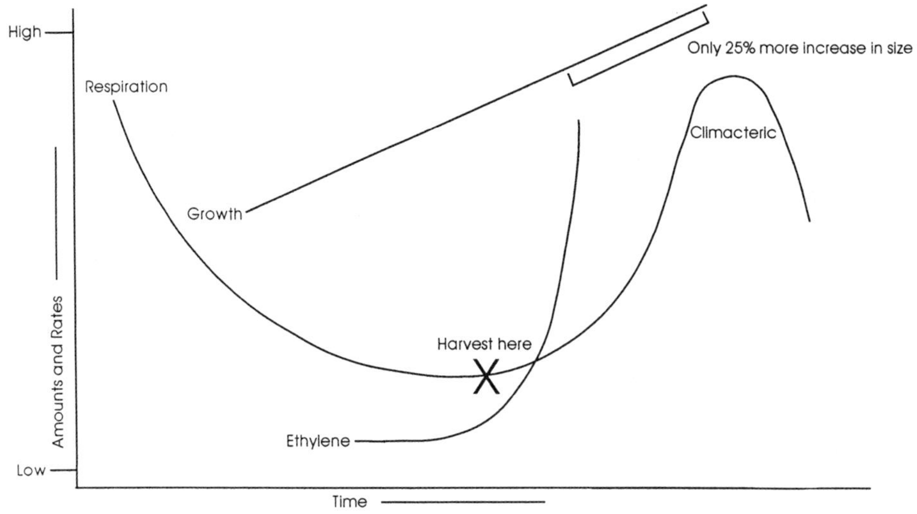


**10.3.** Fruits should be harvested before reaching the climacteric stage of maturity.

will ripen suddenly and rapidly giving off large amounts of ethylene, while the nonclimacteric ripening is a gradual steady process. Storage of climacteric fruits needs to be monitored carefully to avoid holding produce past its prime and also to remove it from storage if the climacteric begins to increase the amount of ethylene released in the storage room.

Some plant parts have naturally high rates of respiration and therefore shorter shelf lives (Table 10.4). Rapidly growing parts such as buds and flowers (broccoli, lettuce, brussels sprout, globe artichoke) and shoots (sprouts) have high rates. Slow-growing roots (carrots, ginger, beet) and stems (potato, leek, celery) have lower rates.

## Enzymes

Enzyme concentrations vs concentrations of enzyme inhibitors will influence any reaction including respiration. Enzyme concentrations are controlled at the household level by selecting fruits that are mature-green. Immature fruits do not yet contain the enzymes needed to ripen properly and mature-ripe fruit contains enzymes that greatly accelerate respiration rates and the softening of plant tissues which makes microbial spoilage more likely. Remnants of the blossom on cucumbers contain high concentrations of enzymes so should be completely removed before storing.

**Table 10.4. Respiration rates vary for plant foods**

| Food | Respiration Rate[a] |
|---|---|
| Asparagus | 400 |
| Broccoli | 320 |
| Mushrooms | 300 |
| Pea, unshelled | 260 |
| Bean | 130 |
| Banana, ripe | 142 |
| Lettuce, head | 80 |
| Strawberry | 75 |
| Pear | 70 |
| Banana, green | 35 |
| Cabbage | 32 |
| Cherries, sweet | 32 |
| Apple | 25 |
| Watermelon | 25 |
| Citrus | 20 |
| Onions, dry | 19 |
| Potato, white | 8 |

[a]Respiration measured as mL of carbon dioxide per kg hr at 68–70F (20–21C). Data from ASHRAE, 1986 and USDA, 1990.

## Root Cellar Practices

Design of the cellar is an important factor for the shelf life of produce, though the cellar construction need not be expensive. Potatoes store well in simple earth pits, but cellars with a variety of vegetables need vents and good circulation patterns to remove the heat of respiration and ethylene released from the stored, living foods. Mike and Nancy Bubel (1979) have published several dozen designs for root cellars, and the local Extension Service often has simple plans for storing local area produce very well. A building permit is required for many types of root cellars in incorporated areas.

Good root cellar management starts with selecting top quality vegetables and fruits with naturally low respiration rates, then determining and maintaining appropriate temperature and humidity.

Fruits and vegetables have different responses to low temperatures so there is no single temperature that is ideal for all of them. Selecting only the produce that will store well within the range of root cellar conditions enhances the probability of successful long-term storage. The best root cellar keepers are root vegetables, potatoes, and squash. Householders experimenting with cellaring for the first time are likely to have little spoilage if they restrict the vegetables to these items. Additionally, some cultivars in these categories are better keepers than others but experimentation is necessary for each growing area as climate and soil will influence cultivars differently. Beets, carrots, celeriac, kohlrabi, onions, parsnips, winter radishes, sweet potatoes, white potatoes, rutabagas, squash, and turnips are excellent choices for the novice. Many householders also have success each year storing apples, tomatoes, salsify, leeks, kale, escarole, endive, eggplant, collards, Chinese cabbage, celery, cauliflower, cabbage, brussels sprouts, melons, and broccoli.

Time of harvest is important for stored vegetables. Soft fruits such as tomatoes, potatoes, onions, sweet potatoes, squash, and greens should be harvested before the first frost. Freezing their tissues causes ice crystal and enzyme system damage, and results in dead portions of tissue which are very susceptible to decay organisms.

Produce is often optimally ripe for harvest in the warm days before frosts. As a result, it is warm when picked; this is called field heat. Commercially, field heat is removed with cool water dips and refrigerated trucks. Removal of field heat is difficult at noncommercial levels; however, householders have an advantage of harvesting small amounts, and, if the produce is grown nearby, extra heat from transport in a nonair-conditioned truck is avoided. Less chilling is needed before storage if these commodities are harvested in the cool of the morning. Large produce such as squash hold a lot of heat after an afternoon in the sun.

Root vegetables such as carrots and parsnips can be stored in the ground until frosts are regular and the root cellar temperature does not fluctuate with occasional warm autumn days. In locales with moderate winters such as the Pacific Northwest, commercial carrot and cabbage growers harvest directly from the field as late as January.

Special handling immediately after harvest also influences storage life. Squash should be harvested before frost and placed outside during several warm days for the skin to harden before storage. This is called curing. The warm outside temperatures accelerate healing of any skin blemishes and hardening of the shell—a necessary impermeable barrier for plant pathogens. Onions and garlic are also cured outside during warm weather before storing at cool temperatures. The green, above ground portion of these vegetables must dry thoroughly all the way down to the bulb to prevent spoilage organisms from entering the bulb in this area. Certain cultivars of onion cure well for successful long-term storage. Harvesting squash with several inches of stem that thoroughly dries when the shell is hardened similarly prevents pathogen entry via the stem scar area.

Occasionally, produce will freeze in extreme weather or inadequately insulated root cellars. The freezing point of most fruits and vegetables is 29 to 31F (−1.5 to −0.5C).

Dissolved sugars and salts give the freezing point depression. Freezing causes ice crystal damage to the cells, and usually death followed by pathogen attack of the injured tissue results. The damage may not be apparent until thawing, but the produce will remain acceptable only a few days. Due to the increased likelihood of spoilage upon thawing, frozen food should be removed from the cellar. If pathogens have not yet spoiled the frozen and thawed food, it is still consumable though the quality will be poorer.

Optimum humidity and temperature varies with the vegetable so an acceptable average is usually selected for all produce in the room. Often in household cellaring the temperature is largely determined by weather conditions, and storage disorders from improper temperatures are common.

White potatoes, sweet potatoes, peas, and sweet corn exhibit starch-sugar imbalances at low temperatures. For instance in white potatoes stored at room temperature, the reaction converting sugar to starch is favored and starch accumulates. At temperatures below 50F (10C), the rate of conversion of sugar to starch decreases and sugar accumulates in the potato tissues. If sugar has accumulated in stored white potatoes, the levels usually decrease after a week of storage at 59–68F (15–20C). Sweet corn and English (green) peas are stored at low temperatures after harvesting to delay the conversion of sugars to starch.

Relative humidity of the air surrounding the produce influences moisture loss from the vegetables and fruit. For most cellared produce, except onions, garlic and squash, optimum humidity is 85–95%. Achieving this high humidity without condensation can be difficult in autumn when atmospheric temperatures fluctuate. Dirt-floor cellars maintain this humidity level well, but other types of cellars may need water added during the storage period. Use caution when adding water if the weather is expected to become colder soon, as condensation may occur which enhances storage disorders. Hygrometers are helpful for measuring cellar humidity. Inexpensive models may be purchased from many feed stores, refrigeration supply stores, and hardware stores.

Many householders store produce in their refrigerators in plastic bags to maintain high humidity and this works well for short-term storage if the bag is not tightly closed, but when this practice is attempted on a larger scale with a garbage bag full of apples placed in the crawl space of a house, spoilage is certain. For root cellaring fruit, the environment needs to be larger than the few cubic feet that bags provide. Oxygen intake, ethylene removal, and water exchange are critical in long-term storage. Perforated plastic bags do not allow enough free gaseous exchange for cellaring. Packing root vegetables in moist sawdust is a good way of maintaining high-humidity conditions around them. Their firm structure and skin make them good candidates for such packing.

Moisture loss is greater in produce with larger surface areas such as carrots and parsnips as opposed to potatoes and squash. Leafy vegetables have the greatest surface area: interior tissue ratios and are most susceptible to moisture loss and subsequent wilting. Outer coverings also influence moisture loss; waxy apples loose less than those with less wax on the skin. Elaborate household root cellars contain several rooms with different humidities.

## Storage Disorders

Appropriate humidity decreases wilting, shriveling, and rotting. Appropriate temperature controls respiration and accompanying senescence. However, even at optimum conditions, the produce cannot be kept alive indefinitely since it is separated from the rest of the plant. Storage disorders become more frequent as storage time increases and the plant tissue breaks down.

Apples are prone to storage breakdown and most consumers are familiar with it from commercially stored apples at the retail level in later winter and spring. Pears, peaches, apricots, and citrus fruits also experience these physiological disorders that affect the skin and/or flesh, or core areas. The changes leading up to the disorders are not well understood, but those holding produce have established some cause-effect relationships.

The disorders occur only in low-temperature storage (41F, 5C). Mineral deficiencies during growth are related to internal browning. Calcium-related disorders are the most common and calcium is added postharvest commercially. Cracking, internal tissue breakdown, internal browning, and pitted flesh are some calcium storage disorders. High humidity, low ventilation, and accumulation of ethylene, a description of what frequently occurs in cellars, accelerate internal breakdown. Currently, recognition that these disorders occur with time and limiting storage life expectations accordingly is the control method for householders.

### Microbial Spoilage

Loss due to microorganisms can be high. The major storage losses are due to species of fungi which cause a variety of rots: *Alternaria* (citrus), *Botrytis* (apple, pear), *Penicillium* (apple, pear, citrus), *Rhizopus* (peach, cherry), and *Sclerotinia* (carrot). Bacteria such as *Erwinia* (potato rot) can also spoil stored produce. Most of these fungi can enter the product only in areas of damaged skin such as cuts, bruises, or physiological breakdown. Selection of top quality items, careful handling when preparing for storage, and limiting the storage time all reduce microbial spoilage.

Mold toxins are not likely to be produced on low-temperature stored fruits and vegetables, and moldy produce is routinely discarded in the United States due to poor quality and the low cost of produce, so aflatoxin poisoning has not been a concern with root cellar storage.

## Nutrients

Nutrient content of root cellared produce varies greatly because growing conditions significantly influence the vitamin content before storage, and most cellared produce originates from home gardens which have a wide range of soils. Stored greens often experience some wilting and an accompanying vitamin C loss. As much as 90% vitamin C loss is not uncommon in even slightly wilted produce. Vitamin A is relatively stable during storage of carrots and squash, and more B vitamins are usually lost during cooking than during root cellaring. Total fiber usually does not change during storage, but the texture of some vegetables may become stringy and tough.

## Specific Storage Requirements for Root Cellared Produce

### Vegetables

Listening to advice from those who root cellar in your neighborhood is an excellent way to start cellaring with locally grown produce, however the following general instructions (Bubel & Bubel, 1979 and USDA, 1990) are a good starting point. The relative humidity values refer to the humidity directly surrounding the produce, not throughout the cellar. These two measurements will be different for stored vegetables with damp burlap or sawdust covering them.

Grouping vegetables with similar storage requirements for root cellaring is not difficult. Those with the best keeping qualities tend to need the same temperature and humidity (Table 10.5). Most householders find a cool dry place for the onion family (basement, garage, closet in unheated room); a cooler, moist place for other vegetables (crawl space, in-ground cellar); and a cool moist place to store fruits by themselves adequate.

**BEETS** should be harvested before frost. Remove green tops, leaving approximately 1 inch of stem. The leaves are edible but do not store well in cellars. Removing the bottom tap root is optional but at least 2 inches of the narrow part should remain. If the tap root were to be cut close to the round beet, the scar would be larger and healing slower and pathogen entry more likely. Pack the beets, without touching, in damp sawdust, moss, leaves, or sand. Ninety to 95% humidity and 33–40F (1–4C) are ideal storage conditions. Beets have an expected root cellar shelf life of 2–4 months.

**Table 10.5. Recommended storage practices for fruits and vegetables**

| Commodity | Storage Temperature (F) | Relative Humidity (%) | Storage Life[a] (mo) |
|---|---|---|---|
| Apple, Jonathan | 35 | 90 | 2–3 |
| Apple, McIntosh | 38 | 90 | 2–4 |
| Apple, Golden Del. | 32 | 95 | 3–4 |
| Apple, most other | 32 | 90 | 4–6 |
| Blueberries | 31 | 90 | 0.5 |
| Figs, dried | 32 | 50–60 | 12 |
| Grapes, American | 31 | 85–90 | 1 |
| Grapes, Vinifera | 31 | 90–95 | 3–6 |
| Lemons | 58 | 85–90 | 1–6 |
| Melon, honeydew | 45 | 90–94 | 1 |
| Melon, watermelon | 40 | 80–90 | 0.5 |
| Peaches | 31 | 90 | 1 |
| Pears | 30 | 90–95 | 2–7 |
| Artichoke, Jerusalem | 32 | 90–95 | 5 |
| Beans, dried | 50 | 70 | 6–8 |
| Beets | 32 | 95 | 3–5 |
| Brussels sprouts | 32 | 90–95 | 0.5 |
| Cabbage, late | 32 | 90–95 | 3–4 |
| Carrots, topped | 32 | 90–100 | 4–9 |
| Cauliflower | 32 | 95 | 1 |
| Celeriac | 32 | 95–100 | 3–4 |
| Garlic, dry | 32 | 65–70 | 6–7 |
| Leeks, green | 32 | 90–95 | 1–3 |
| Onions, dry | 32 | 65–70 | 1–8 |
| Parsnips | 32 | 90–100 | 2–6 |
| Peas, dried | 50 | 70 | 6–8 |
| Peppers, dried | 32–50 | 60–70 | 6 |
| Potato, white | 38–40 | 90 | 5–8 |
| Potato, white | 40–45 | 90 | 3 |
| Radishes, winter | 32 | 90–95 | 2–4 |
| Rutabagas | 32 | 90–95 | 2–4 |
| Salsify | 32 | 90–95 | 2–4 |
| Spinach | 32 | 90–95 | 0.5 |
| Squash, Hubbard | 32 | 90–95 | 6 |
| Sweet potato | 55 | 85–90 | 4–6 |
| Tomato, mature gr. | 57 | 85–90 | 0.5 |
| Turnip, roots | 32 | 90–95 | 4–5 |
| Yams | 61 | 85–90 | 3–6 |

Data from USDA, 1990, and ASHRA, 1986.

[a]Exact storage life will vary with cultivar and growing conditions. All times are based on top quality produce and proper handling.

**CABBAGE** produces sulfur-containing volatiles during storage, so the household basement root cellars and unheated rooms in the family home are not appropriate. Firm heads have the longest shelf life. Cabbage can be stored uncovered or wrapped in newspaper on a shelf in the cellar, or layered in hay, or wrapped in newspaper and placed root side up and surrounded by at least 2 inches of dirt. Wilting of the outer leaves is likely to occur during storage so more trimming is required with cellared cabbage than with freshly picked. Chinese cabbage (and other Asian cabbages) may be harvested after several mild frosts if a sweeter flavor is desired. It may be stored surrounded with clean hay and then covered with dirt, or harvested with roots and planted in shallow containers. In the latter case, the roots need to be watered each month, but if the leafy heads become too wet, molds and rots will soon spoil them. Thirty-two to 40F (0–4C) and 90–95% humidity are recommended. Several months is a common shelf life for root cellared cabbages though some householders still have cabbages for soup from their cellars in February and March. Chinese cabbages tend to deteriorate after 1- to 2-months storage.

**CARROTS** can be stored in the garden in locales with mild winters. Other areas of the country need to harvest before the ground is solid. Some householders feel carrots that were only mulched and left in the frozen garden all winter are acceptable for spring consumption when the ground thaws enough to dig. Clinging dirt should be brushed off and the green tops trimmed to leave a 1-inch stem. The tap root should remain where it broke off when removed from the ground and trimmed just before preparing for consumption. One inch of moist sawdust, moss, sand, or leaves should surround each carrot but many carrots can be placed in the same container as long as they are not touching. Carrots have a long root cellar shelf life.

**CELERIAC** roots should be trimmed carefully and not too close to the body of the plant. The tops should be trimmed leaving a 1-inch stem and clinging dirt may be brushed off. Storage is the same as for other roots, surround each celeriac with at least 1 inch of damp sawdust, moss, or sand. Cellar temperatures of 32–40F (0 to 4C) and 90 to 95% humidity around the root is ideal. Celeriac has a long root cellar shelf life.

**CELERY** does best if stored planted in a shallow container in the cellar and the roots watered periodically (see Chinese cabbages). Its storage life is often through January in experienced hands, but celery is not an easy vegetable for the novice to root cellar. Commercially, the roots are removed at harvest. Wet burlap floors and special crates maintain moistness. One to 3 months is the commercial storage life.

**GARLIC** should be harvested when the majority of the tops start to loose their green color. Curing is necessary for the tops to dry completely down to the bulb portion so there are no damp leaves to encourage plant pathogen growth. To cure garlic, shake off clinging dirt and place bulbs in the sun for several days. The skin will harden during this time. The roots may be removed closely up to the bulb since they are such a small diameter and will not pose healing problems. The dried tops should be removed for storing the bulbs in hanging mesh sacks, or may be left on for braiding the garlic into chains. Temperatures of 35–40F (2–4C) and relative humidity of 60–70% is ideal for storage. (Note: The humidity of household refrigerators is too high for garlic storage.) Garlic is an excellent keeper in dry household basements or some attics.

**KALE** is stored unharvested in the garden. Heavy mulching with dry leaves around the sides of the plant and then a straw covering over the top usually preserves kale even during hard frosts, though the edible heads are not as large as during the summer.

**KOHLRABI** keeps fairly well when the leaves and root are trimmed and the vegetable is stored surrounded by moist sawdust or sand. Ideal temperature is 32–40F (0–4C) and humidity 90–95%. Expect kohlrabi to keep until Christmas.

**LEEKS** do best when replanted in shallow dirt (see Chinese cabbages). Cellar temperatures of 32–40F (0–4C) are ideal. In mild and moderate climates, some householders have had success storing leeks unharvested in the garden with heavy mulching.

**ONIONS** that are the hard, nonsweet types store well. When the tops bend to the ground naturally, they are mature enough for harvest. Onions, like garlic bulbs, need to cure several days in the sun to harden the skin and dry the tops. Some cultivars with large fleshy tops do not cure well and have a short cellar shelf life. After initially curing in the sun, the tops are removed, leaving a 1-inch stem, and the bulbs are spread out so they do not touch in a warm, dry, shady place for 2 weeks. Hanging onions in mesh bags at 45–55F (7–13C) and 50–60% humidity is ideal.

**PARSNIP** harvest can be delayed for a few light frosts, then these roots are trimmed and stored like carrots. Ideal temperatures and humidity are 32–35F (0–2C) and 90–95%. Parsnips are good keepers and should last until spring.

**SWEET POTATOES** are susceptible to chilling injury so harvest after the first light frost has killed the vines or in climates where fall frosts are not likely, the end of October. Let clinging dirt dry and then gently brush off without damaging the uncured skin. Curing is best done at 80F (27C) for a 2-week period. An easy storage procedure is wrapping the potatoes in newspaper and placing in nonsolid-sided baskets. Storage temperatures of 50–60F (10–16C) and 80–85% humidity is ideal. Relative humidity in household refrigerators is too high for sweet potato storage. Inexpensive sweet potato bins dug into the earth are common in areas where sweet potatoes are grown.

**WHITE POTATOES** are mature when the green tops die down. They may be harvested at this time but if the weather is hot, storage in the ground for several weeks may be advisable depending upon the likelihood of spoilage there, which will vary with locale. White potatoes must be cured in the shade—sunlight promotes solanines which are bitter and toxic—for 1 or 2 weeks, 60–70F (16–21C) is the ideal curing temperature. Potatoes require high humidity (90%) and storage temperature between 38 and 40F (3.5–4.5C). That range is low enough to slow respiration and delay sprouting but not so low that starch is converted to sugar. At this low temperature, solanine formation is also very slow. (Note: if potatoes have accumulated sugar, storage for 1 to 2 weeks at room temperature enhances conversion back to starch.) If potatoes are stored in a pile or large bin, they should be packed loosely so air can circulate. Covering the pile or bin with burlap, straw, or wood chips can help prevent water—from respiration—from condensing on the surface of the tubers. Such condensation enhances pathogen growth. Four to 6 months is average storage life for white potatoes.

**PUMPKINS AND WINTER SQUASH** belong to the same genus and species—they are different varieties. Horticulturists consider pumpkins to be those used for animal food or carving in October, and squash to be the finer grained produce used for human consumption; they have many different colors of shell. For cellaring, harvest with several inches of stem remaining, then cure outside for several warm days to heal wounds and harden the shell. The stem must dry well to prevent pathogen entry at the stem scar. Cool temperatures (above freezing) are best for slowing respiration and 60–75% humidity is ideal. Basements, unheated spare rooms and garages that do not freeze are ideal. The hybrid cultivars Sweetmeat and Delicata can be easily stored for 8–12 months and many older cultivars average a 6-month storage life, which is long enough for most households.

**RUTABAGAS** are stored similarly to carrots. Two to 4 months is an average expected storage time.

**SALSIFY** is handled similarly to parsnips. It is a good keeper.

**TURNIPS** are harvested before a hard frost. The tops should be removed without scarring the enlarged root, but the tap root is best left untrimmed. Turnips store well when handled like carrots, but also can be lightly packed in a barrel and buried.

## Fruits

Fruits pose special storage problems because they may release large amounts of ethylene, a natural plant compound, that speeds ripening and senescence of all plants. Removing ethylene from the storage area is a major concern. Commercial storage facilities have elaborate systems for doing this, but at the household level, air circulation and separation of fruits from the vegetables are the major methods.

**APPLES** that are mature and contain all the enzymes necessary for ripening, but are not yet ripe, store best. Immature apples lack a good waxy covering that is necessary to prevent excessive moisture loss and often do not develop good flavor after ripening. Wrapping apples individually in clean newspaper, placing them in a crate with ventilated sides, and adding straw on top works well. Do not stack apples more than 3 layers deep as the weight of the fruit will bruise the bottom ones, making them extremely susceptible to pathogens. Unheated basements, spare rooms, or garages that do not freeze are common storage locations. Ideal apple storage temperatures are 32–40F (0–4C) and 80–90% humidity. Quality of stored apples decreases with storage time but many householders feel their "cellared" apples are acceptable for fresh eating after the new year. Storage disorders which result in softening or browning are common in apples.

**ORANGES AND TANGERINES** are stored unwrapped at 23–40F (0–4C) and 80–90% humidity. Their life expectancy is 1–2 months.

**PEARS** are of two main types; summer and winter pears. All pears should be picked unripe, as the pear tree removes some of the sugars and acids from the fruit as ripening occurs and the result is a mushy or mealy, low-flavor fruit. Summer pears such as Bartlett can be stored similarly to apples, but the shelf life is much shorter—perhaps only a month. Winter pears such as D'Anjou and Comice are stored similarly to apples for several months. Comice will store through October and D'Anjou is usually a good Christmas pear. After removing from cold storage, winter pears ripen slowly (2 weeks) at room temperature. Winter pears require cold storage to ripen while summer pears will ripen immediately after picking.

### Nuts and Seeds

Nuts should be picked, hulled, and the drying started within 24 hours of falling from the tree for best quality; however, many householders who do not have nut trees on their own property and glean from roadside trees during an autumn outing find their products acceptable. Drying is necessary to prevent mold growth—some molds growing on nuts at room temperature may produce toxins. Nuts should be dried in a single layer or in a narrow mesh bag (nylon hose stretch too much and become a wide bag) at 80–95F (27–35C). Higher drying temperatures are associated with rancidity and excessive shriveling of the nutmeat.

**SUNFLOWER, SESAME, AND PUMPKIN SEEDS** should be dried and stored in a cold, dry place. They have long shelf lives if the humidity is low.

**PEANUTS** are dried in the sun several days after harvest and then dried several weeks in a shady, warm place. After drying, any clinging plant parts are removed and the nuts placed in burlap bags and hung in a cool, dry place with good air circulation.

## Dry Storage

Some foods have low enough moisture contents that microbial growth is not possible, and they do not contain enough sugars for water uptake from the air to be a safety concern. These foods are ideally suited for dry storage. Macaroni, flour, rice, legumes, potato chips, breakfast cereals, bread, and spices are examples. Dry stored foods usually are packaged or enclosed in canisters to discourage insects, but insect contamination of flour, rice, macaroni, or other foods that are thoroughly heated before consumption is not a safety concern. Insect bodies should be removed if possible for aesthetic reasons. Insects in ready-to-eat foods such as breads and breakfast food may have contaminated it if they have previously been in contact with human sewage (pit toilets) or animal waste (feed lots, pet feces). In most U.S. households such contamination is uncommon.

Quality changes during dry storage are responsible for significant economic losses in household budgets. Spices are very expensive and their volatile flavors and aromas decrease with storage time. Tightly closed containers that are opened only briefly are helpful. Many spices in the United States are irradiated to kill molds and insects that may be present.

Maillard browning renders much dry stored food unacceptable for consumption. This reaction between sugar and protein can proceed with small amounts of either substrate. Nonfat dry milk, white cake mixes, dry flaked coconut, and instant mashed potatoes are particularly susceptible. Heat accelerates the Maillard reaction ($Q_{10}$) so dry storage at low temperatures is recommended. Very dry products (macaroni) have slow Maillard browning, but intermediate moisture foods (coconut, dry milk) have a greater tendency to brown because reactants are close together. Maillard browning is a fairly slow reaction, taking 6–12 months to cause quality loss. Rotation of household stored goods and purchasing appropriate amounts is the best prevention measure. Store in the coolest area possible.

Oxidative rancidity is the result of changes in fats. Food does not have to be high fat for oxidative rancidity reactions to take place, actually very little lipid is needed. Shortening, oil, wheat germ, potato chips, snack crackers, and cookies are frequently discarded due to oxidative rancidity spoilage. This is a nonmicrobial spoilage, but there are health concerns about consistently consuming large amounts of rancid foods, as one of the products is potentially carcinogenic. The oxidative rancidity reaction is slowed by lower temperatures ($Q_{10}$) and oxygen is necessary. Commercial potato chips are often packed in nitrogen to decrease oxygen. However, once the package is opened, oxygen is available for rancidity reactions.

Spices, sugar, cake mixes, and powdered soup do not take up enough moisture from the air for pathogenic bacteria to grow, but they may become hydrated enough to cake. Anticaking additives such as silicoaluminate and dextrose in salt and powdered foods are routinely included by manufacturers and expected by consumers. Even so, foods such as garlic powder may be a solid, browned lump at the bottom of the jar after a year.

## Summary

For vegetables and fruits, common or root-cellar storage provides optimum conditions for keeping fruits and vegetables alive but with a slower rate of maturation.

Spoilage commonly results in loss of quality. If molds have grown, the product may not be safe to eat and should be discarded. Quality can be maintained but the length of storage time will vary with temperature, humidity, ventilation, and with the type of produce itself.

## References

ASHRAE, 1986. American Society of Heating, Refrigerating and Air-Conditioning Engineers Handbook, refrigeration systems and applications. New York: ASHRAE.

Barth, M.M., Perry, A.K., Schmidt, S.J. and Klein, B.P. 1992. Misting affects market quality and enzyme activity of broccoli during retail storage. J. Food Science 57:954.

Buble, M. and Buble, N. 1979. Root cellaring. Emmaus: Rodale Press.

Kays, S.J. 1991. Postharvest physiology of perishable plant products. New York: Van Nostrand Reinhold.

Postharvest handling of horticultural crops. 1992. Publ. 3311, 2d ed. Oakland: University of California.

USDA, 1990. The commercial storage of fruits, vegetables, and florist and nursery stock. Ag. Handbook No. 8. Washington DC: U.S. Govt. Printing Office.

Wills, R.B.H., McGlasson, W.B., Graham, O., Lee, T.H. and Hall, E.G. 1989. Postharvest. 3rd ed. New York: Van Nostrand Reinhold.

## CONSUMER QUESTIONS

**Q.** *Should all moldy produce be discarded?*
**A.** Molds that produce toxins are not likely to grow on raw fruits and vegetables; there are few known problems of toxin production. However moldy produce should immediately be separated from others in storage so the contamination does not spread. Mold spores travel through the air, so sorting should not be done in the storage room. Molds release enzymes that soften tissue, so quality spoilage has probably occurred when mold colonies are visible.

Mold on nuts is very likely to produce toxins. Moldy nuts should be discarded, even if the mold appears to be only on the shell. Patulin, a toxin produced by molds commonly involved in the spoilage of apples, has been shown to be a potential problem. Moldy apples should not be used in making cider.

Mold growth is encouraged when humidity in the cellar increases. Warm temperatures also accelerate mold growth and multiplication. Condensation on the surface of stored foods encourages molds and the spread of molds through droplets of water. Condensation is more likely to occur when temperatures fluctuate. Consistent, cool temperatures, with only enough humidity to prevent wilting is an ideal storage environment.

**Q.** *What causes texture changes during storage?*
**A.** Softening and shriveling due to wilting are the result of the humidity being too low. There is usually accompanying vitamin C loss, but the produce is safe to consume.

Mushy softness of plant tissue is usually the result of improper curing, microorganisms, or enzyme activity. Produce this badly spoiled should be discarded. Immediate discard prevents spread throughout the remaining stored produce.

**Q.** *What are the problems of insects in dry foods?*
**A.** This occurs in most households and commercial storage facilities at some point. Ants, moths, cockroaches, and earwigs are the most common in the United States. Sealing their points of entry is the best preventive. Insecticides are used in some instances, but care must be taken when applying these substances around food to be consumed. Contact a professional with a pesticide applicator license for large infestations, and the local Extension Service for advice on home-treatments. Insects usually represent only quality spoilage and economic losses. Occasionally they can bring contamination to the food if they have previously been around animal or human wastes. Some householders routinely sort through dry foods removing insects before preparing.

**Q.** *What causes yellow-brown off-flavors?*
**A.** This is especially common in stored dry milk and flaked coconut. It is the result of a reaction between sugar and proteins in the food (Maillard browning). This browning is accelerated by warm temperatures and occurs slowly with time. Shorter storage times eliminate the problem. This is quality spoilage, the food is still safe to consume.

**Q.** *Why are there off-odors and -flavors in oil and shortening?*
**A.** The product has become rancid. Using it in food preparation will result in off-flavored foods. Discard is recommended. Rancidity is accelerated by warm temperatures. If the bottle of oil is used slowly, as in the case of some seniors living alone, storage in the refrigerator and purchasing smaller amounts should eliminate the problem.

# 11 Preservation with Sugar: Jam and Jelly

Jams and jellies can be prepared with little technology. This made them popular among American pioneer women and common today with modern cooks who preserve only a few foods by canning. Historically, these very sweet items have been consumed as accompaniments to other foods or as desserts. They have not been a staple food for any known culture, but are favorites of many.

Food preservation in sugar probably occurred much later in history than salting since salt was much more plentiful. The process may have been discovered in the course of cooking when it was noted that "boiled down" fruits—sugar from fruit was concentrated as water evaporated—did not spoil as rapidly as those cooked for shorter periods of time. Romans preserved whole fruits by dipping them in honey, and Apicius, who lived in the first century, was believed to have had a tasty recipe for sweetened pickled peaches. In the sixteenth century when Spain began commercial harvest of sugarcane in the West Indies, sugar became plentiful enough for widespread use as a preservative. By the 1800s sugar was affordable for the middle classes and preserves were frequently made in home kitchens.

Increasing the sugar content preserves food by dehydrating bacteria, yeasts, and molds. Sugar has a very strong attraction (high affinity) for water (Fig. 11.1). When sugar is added to fruit, water moves from inside fruit cells to the surrounding sugar. This movement is apparent when juice rapidly surrounds the previously dry fruit and sugar. The bacterial cells also lose water that they need to grow. Water that is bound to sugar is unavailable for microorganisms to use because sugar binds it tightly. When sugar is added to a food, the available water, measured as water activity ($a_w$), decreases. Cane and beet sugar, corn syrup,

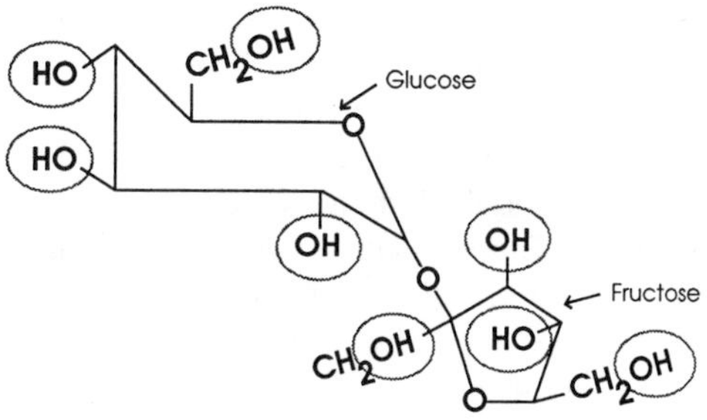


**11.1.** Sucrose has a high affinity for water due to its many bonding sites (circled).

honey, and concentrated fruit juices can be used in recipes to decrease the $a_w$.

When spoilage occurs in jam and jelly, it is usually due to mold. Yeasts and molds are much more resistant to low $a_w$ and the dehydrating effect than bacteria. Some molds can multiply in the presence of 60% sugar. However, since molds need air, spoilage can be prevented by heating jars of jam and jelly in a boiling-water canner so that they are vacuum sealed. Mold growth can also be prevented by freezing for long-term storage or refrigerating for short-term storage.

Householders preserve a variety of fruit products with sugar. Recipes and the fruits used in preservation change with each generation but the following has been standard terminology for most areas of the United States during the twentieth century: Jelly is made from strained fruit juice and its clarity and sparkle largely determine quality. Its firm gel strength holds the shape of the container when it is turned out. Marmalade is a soft jelly with small fruits or pieces of fruits distributed evenly throughout. Marmalades often contain pieces of citrus fruits. Jam is a thick sweet spread made from crushed small fruits or fruit pieces; its gel is generally not firm enough to retain the shape of the container when removed. Conserves are jams made from more than one fruit and may contain nuts, raisins, or coconut. Preserves are small fruits or pieces of fruits in a thickened, heavy syrup. Butters, such as apple butter, are made by cooking fruit pulp and sugar until the mixture is thick.

With increasing interest in reducing caloric intakes, consumers have demanded products with reduced sugars. Low-methoxyl pectins (LMP) and gums make lower sugar jams and jellies at the household level feasible. Fruits themselves contain sugars so no-sugar is a misnomer and the term no added sugar is more accurate. These special ingredients will be discussed at the end of the chapter; the following discussion applies only to conventional pectin products.

Flavoring, pectin, acid, sugar, and liquid are needed to make a jelly or jam. Flavoring usually is provided by a fruit. The pectin and acid may come from the fruit itself or be added from another source. Some liquid may need to be added depending upon the juiciness of the fruit, and large amounts of sugar must be added. Concentrations of pectin, acid, and sugar may vary within a tolerated range; a higher amount of one may compensate for less of another.

Householders should select the fruit carefully for their recipe. As with other preservation methods, top quality fruit results in top quality finished products. Underripe fruit usually contains more pectin and acid than mature fruit but lacks the characteristic ripe flavor. Therefore a combination of less- and more-ripe fruit is used. Recipes may even specify the proportions of mature and underripe fruit to use. Fruit overripe for canning may make an excellent, pectin-added jam but does not contain sufficient pectin for conventional recipes.

## The Role of Pectin

The conventional method of making jam and jelly relies on pectin naturally present in the fruit. It is possible to extract pectin from fruits such as apples and use it to thicken jams and jellies.

In the plant—pectin is not found in animal tissues—pectin has a structural role in the middle lamella and cell wall, serves as an ion exchanger, and also binds large amounts of water. Pectin is concentrated in the structural and rapidly growing parts of plants such as root tips, shoots, sprouts, and fruits, especially the core and peel. The cell sap usually is void of pectin so fruit juices expressed from raw fruit do not contain pectin. A few fruit juices such as unfiltered blackberry are exceptions because they contain large amounts of cell wall and therefore pectin.

Pectin is a non-nutritive fiber. It cannot be broken down into simple units by human digestion. The pectin molecule is a polysaccharide; but, unlike starch and cellulose, it has more than one kind of smaller molecule composing the backbone structure. These smaller molecules are five or six carbon rings with an oxygen in the ring structure. These are primarily galacturonic acid units with and without methyl groups attached, but rhamnose, xylose, arabinose, and acetic acid are also present in small amounts (Fig. 11.2). An ordered pattern for the inclusion of these other molecules has not been discovered; they appear randomly spaced throughout the pectin backbone, yet the functional properties of pectin remain fairly constant. An α-(1-4) linkage connects galacturonic acids in the chain (Fig. 11.2). The angle of this alpha linkage produces a slight twist along the backbone. Where a rhamnose molecule is present in the chain, there is a kink, almost a 90° bend. These kinks are instrumental in keeping an open structure during gel formation. The presence of methyl esters (the -O-$CH_3$ compound) on some of the D-galacturonic acids in the pectin backbone is important in gel formation and discussed later.

Pectin gels form principally through hydrogen bonding between the -O and the -OH groups on different pectin molecules or different parts of the same pectin molecule. Hydrophobic (carbon-carbon) associations also strengthen the network. Jelly and jam may continue to become firmer even after cooling is complete because bonds continue to form. Heating a pectin gel breaks these hydrogen bonds which decreases the firm structure; the result is a

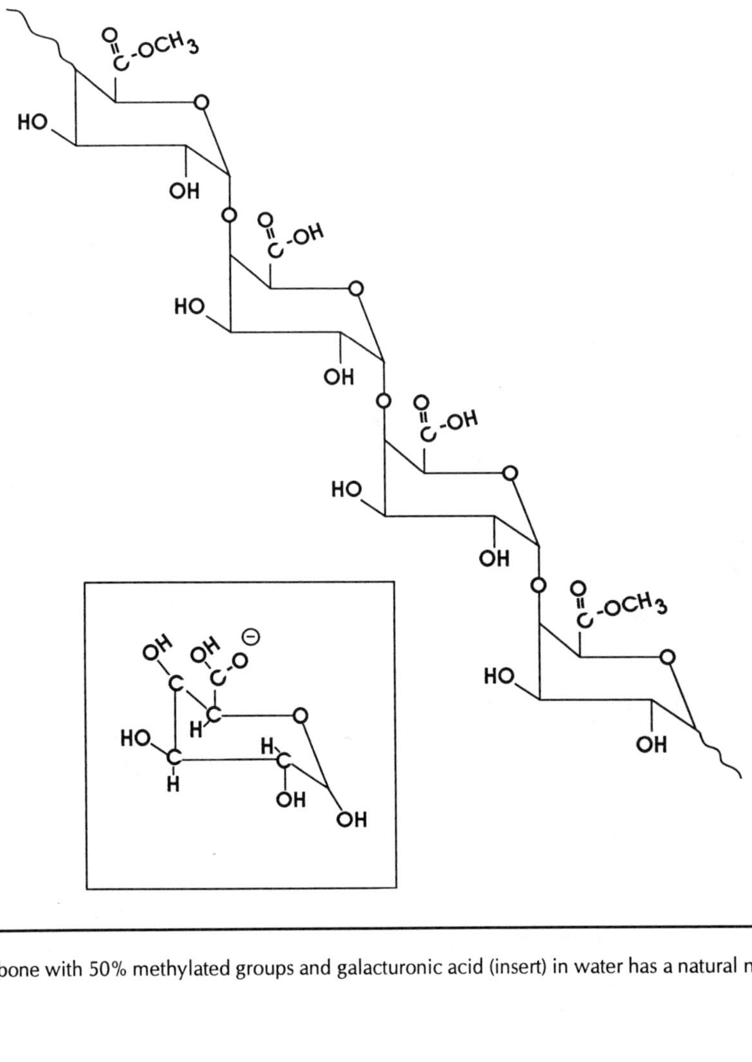


**11.2.** Pectin backbone with 50% methylated groups and galacturonic acid (insert) in water has a natural negative charge.

liquid called a sol. Heating jelly, then pouring the liquid over a cake or fruit is a common use of sols.

A brush-heap formation involving several pectin molecules results in the most desirable gels (Fig. 11.3). Water is held in the pectin gel by being (1) simply trapped in spaces among the molecules, (2) held by capillary action where it is drawn into narrow spaces, and (3) hydrogen bonded to the pectin molecule. The brush-heap formation has ample sites for all three of these methods of holding water.

## Gel Strength

The firmness and strength of pectin gels is dependent upon the pH, sugar, and concentration of the pectin as well as its percent esterification. A lower pH (higher acid) and increased proportion of methyl esters (esterification) increase gelation by reducing the number of negatively charged sites which repel each other (see Fig. 11.2).

Pectin molecules have an overall negative charge in water. Since fruits are mostly water (90+%), the pectin present in fruits added to the jelly pot initially is negatively charged. These like charges repel each other and prevent pectin-pectin bonds. In the presence of acid ($H^+$), the negative charges are neutralized (see Fig. 11.2). Positively charged hydrogen ions are attracted to the negatively charged -$COO^-$ groups along the pectin molecule. Acid could come from fruits or may be added in the form of lemon juice or vinegar. Citric acid is added in commercially packaged dry pectin to increase the likelihood of achieving a good gel.

Jelly and jam recipes are formulated to have a pH between 2.8 and 3.4 so the pectin molecule is charged for optimum gel strength. Variation in acid content among fruits (ripeness, cultivar) results in differences between batches, though usually the product is acceptable. Solely neutralizing the charge does not result in pectin-pectin bonds and gel formation. Even with the charge barrier reduced, pectin tends to bond to water molecules instead of to itself. Sugar is needed to bond with the water before a gel can form.

Being a hydrophilic—water loving— molecule, pectin has water bound along the chain at its many available sites. As the water leaves its binding sites on the pectin and binds to the sugar, the pectin is left with free sites which can participate in pectin-pectin bonds necessary for gel formation (Fig. 11.3). The pectin-pectin bonds give strength and the pectin-water bonds give juicy softness. Both types of bonds are necessary in correct proportions for a desirable texture in jellies and jams.

In addition to the well-known effect of sugar (sucrose) competing for the water, sugar may also enter into hydrogen bonding with pectin. When fructose is substituted for sucrose in recipes, even in amounts that result in the same osmotic pressure—a measure of solute concentration that takes into account type of sugar, a different quality of jelly results, perhaps due to the lack of sucrose-pectin bonds. Householders wishing to use honey or corn syrup

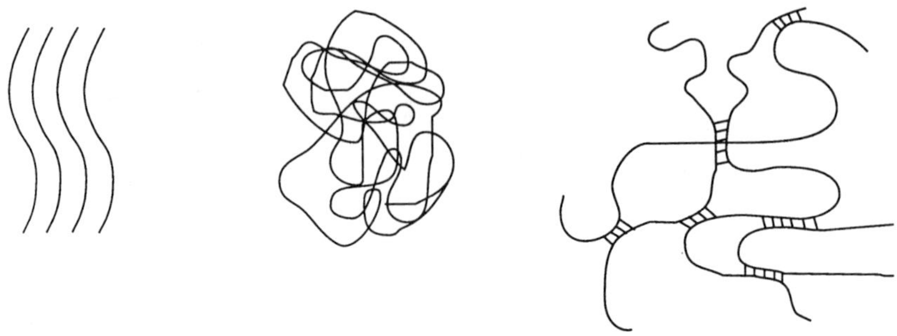

**11.3.** The gel structure of pectin holds large amounts of water. (Left to right: linear/capillary water, entrapped water, hydrogen-bonded brush-heap.)

instead of cane or beet sugar should expect a softer gel. Many fruit concentrates, except apples, also contain higher proportions of fructose than sucrose so substituting these as sweeteners may decrease gel strength also. Reducing the amount of sugar in the final product will result in a softer gel or even in failure to gel.

## Pectin Types

The degree of esterification (DE) is used to describe different types of pectin. DE can be calculated as the number of esterified D-galacturonic acid residues per total number of D-galacturonic acid residues multiplied by 100. Figure 11.2 has a DE of 50%. The DE of a particular pectin will influence gel formation. The charged $-COO^-$ repels, while the neutral $-OCH_3$ does not. Standard pectins have a DE of at least 55. Those with a DE over 70 have rapid gel formation, probably due to increased interaction between pectin molecules.

Various types of pectin have different gelation characteristics (Table 11.1). The rapid set high-methoxyl pectin contains fewer anions (negative charges), so can gel at a higher pH. Low-methoxyl pectins (LMP) require calcium ions, which form calcium bridges between the pectin molecules, to set instead of hydrogen ions so there is no pH nor sugar requirement for LMP. The availability of LMP to consumers dramatically changed household jelly making in the 1980s.

Pectin is available from chemical supply companies and some health food stores in different grades. The grading system is based on the parts of sugar (water also present) that one part of pectin will gel. One part of 100 grade pectin gels 100 parts of sugar. One part of 150 grade pectin gels 150 parts sugar. Powdered pectin commonly marketed for householder use contains fillers such as dextrose to provide easy recipe measurements and also to prevent the pectin from hydrating in slightly moist storage conditions, thus one part of this dry pectin product will not gel 100 parts of granulated sugar.

**Table 11.1. Hydrogen ion requirements of different types of pectin**

| Type | Esterification (%) | pH Required |
|---|---|---|
| High methoxyl | | |
| Rapid set | 70–90 | 3.0–3.4 |
| Slow set | 50–70 | 2.8–3.2 |
| Low methoxyl | 30–50 | wide range |

## Acid

Some fruits such as pears and peaches, and all overripe fruits, lack necessary amounts of both pectin and acid to make gels (Table 11.2). In most finished jellies the concentration

**Table 11.2. Acid and pectin levels in some fruits**

Usually contain adequate acid and pectin
- Apples (tart)
- Crab apples
- Blackberries
- Cranberries
- Currents
- Gooseberries
- Grapes (Eastern Concord and wild)
- Guavas
- Lemons
- Plums (sour)

The raspberry-blackberry crosses (Marion, Logan, Tay, Boysen) are most often adequate but some may lack acid

Usually lack acid
- Apples (sweet)
- Italian plums (called prunes when dried)
- Plums (sweet cultivars)
- Oranges (some cultivars)
- Quinces (some cultivars)
- Raspberries

Low in good quality pectin
- Apricots
- Pomegranates
- Rhubarb
- Strawberries

Deficient in both acid and good quality pectin
- Peaches
- Pears
- Cherries
- Elderberries

Grapes (Western Concord and most table cultivars, such as Thompson and Red Flame).
- Guavas
- Overripe fruit

of pectin is 0.5 to 1.0%, and the pH is between 2.8 and 3.4. Too much acid in jelly results in syneresis (water squeezed out) of the gel, and too little, in failure to form a gel. As fruit ripens, enzymes which catalyze the conversion of acids to sugars and the breakdown of pectin are active. Insoluble protopectin in unripe fruit is converted by the plant to the more soluble pectin, and finally into pectinic acids—by demethylation and splitting into smaller molecules. The end result is a sweet, mushy fruit in which often the cell structures have broken down. Since the composition of the fruit is influenced by cultivar, growing conditions, and maturity at harvest, Table 11.2 is only a guideline and each juice should be tested during jelly making.

## Methods for Making Jam and Jelly

There are four main ways householders make jams and jellies: conventional, added pectin, uncooked freezer, and low-methoxy pectin (LMP). All of these methods produce satisfactory products. Some require more skill than others and some can only be used with certain fruits.

### Conventional

Apple jelly will be the focus of this discussion since apples with their natural amounts of pectin and acid lend themselves to conventional jellies and are the most frequent choice of householders. Tart, firm apples should be selected. As fruit ripens, flavor-developing enzymes increase, pectin-degrading (tissue-softening) enzymes increase, and starch-degrading (sugar-forming) enzymes increase. Choosing 1/4 of the fruit underripe assures a good gel without noticeably decreasing the desirable apple flavor which develops as the fruit ripens. Mature fruit should make up the remaining 3/4 of the lot and overripe fruit should be avoided. The apples are cut into pieces to enhance heat penetration but not pared or cored since these portions of the fruit contain the largest concentrations of pectin and must be present during the pectin-extraction step.

If the fruit used lacks sufficient hydrogen ion concentration, acid—lemon juice, slices of citrus fruit, vinegar—is added at this step because it enhances the extraction of pectin. Some of the flavonoid compounds in the citrus peel become bitter with heating so recipes using citrus in this manner may suggest partial peel removal.

The fruit must be heated to extract the pectin. Commercially bottled and frozen juices have not gone through a heat extraction so pectin must be added to gel them. Water (maximum of 1 c per lb of apples) is usually added to the kettle to decrease the chance of scorching. Fruits with less structure lose water easily from the cells when heated so may need little or no added water. Excess water should be avoided as it dilutes the pectin, flavor, color, and odor compounds and results in a longer concentration step.

The fruit and water mixture is brought to a boil, then the heat is lowered and the apples simmered for 15 to 25 minutes. During this simmering period, the naturally occurring pectin-degrading enzymes, pectin methyl esterase (PME) and pectin polygalacturonase (PG), are inactivated which is necessary to avoid loss of gel strength. The linkage between compounds in the pectin backbone is susceptible to acid hydrolysis and overcooking will result in decreased gel structure as acids from the fruit catalyze the breakdown of pectin molecules. The shorter pectin chains would result in less firm gels.

If this heated mixture contains large chunks, as it would if using apples, it is strained before being put into a jelly bag. Jelly bags are made of woven material that allows only certain size particles to pass through. Pectin molecules pass, but larger particles, which can result in cloudiness, do not unless pressure is applied. Squeezing the bag is a common householder method of applying pressure. Unripe apples contain higher proportions of starch than the ripe—during ripening much of the starch is converted to sugars. Starch is a small enough molecule to pass through the jelly bag without pressure and produces a cloudy jelly. To avoid this problem, the recommendation is made in some sources to not include underripe apples for jelly. There are a variety of bag holders on the market that work well but bags may simply be tied to kitchen sink faucets or other available supports. Four thicknesses of wet cheesecloth lining a colander and draped over the sides is a good substitute for a jelly bag.

Some recipes include a second pectin extraction for apples since they contain such large amounts of pectin. For the second extraction, the pulp from the strainer and jelly bag is combined with additional water and reheated. The majority of compounds that produce flavor and color and the salts are no longer present in this pulp, but the second extraction may double the amount of pectin from the fruit.

One to 1 1/3 cups of juice can be expected from 1 pound of fruit. The amount of pectin in the juice should be tested as it will determine the optimum amount of sugar required for a good gel. If the amount of pectin is unknown, the old recommendation was to add 1 cup sugar per each cup juice. The current recommendation of 3/4 cup sugar per cup of juice is more reliable since that will provide greater concentration of pectin by the end of cooking.

An estimate of the amount of pectin can be made with the alcohol test or with a Jelmeter. Jelmeters were once available in retail markets for several dollars so some householders own one, and the alcohol test is conducted with ingredients most households have on hand.

The results of the alcohol test can give a reliable indication of the amount of pectin in a juice, particularly when it is done by one who has experience. To make this estimate, one teaspoon of the cooked juice is added to 1 tablespoon of 70% rubbing alcohol in a saucer. It is stirred and observed (but not tasted). Juices with adequate pectin will form a gel that holds its shape enough to be lifted with a fork. Low-pectin juices will form a few pieces of weak gel. If the juice is found to be low in pectin, more can be added in the form of a commercial pectin product or by the addition of a high-pectin juice such as apple, if there are apples on hand to make some by hot extraction. To further complicate conventional jelly making, the different types of pectin which are found in various fruits form optimum gels at a pectin concentration specific for each, so experience is an important factor.

Jelmeters have recently become difficult to locate. One nationwide preservation supply distributor explained that he ceased carrying them when he was informed that he would be liable should this glass instrument break and any pieces which fell into the jelly pot were later consumed. Since some Jelmeters are already in kitchens and because they are an asset in making conventional jelly, an explanation of their use is provided here. The Jelmeter provides an estimate of the concentration of long pectin molecules in the juice based upon juice viscosity. Since the viscosity will be temperature dependent (colder, thicker; warmer, thinner), room temperature (72F, 20C) is used. The greater the gel-forming capacity of the juice, the slower the flow from the Jelmeter. By timing this flow, the amount of sugar needed to form a good gel can be determined. Larger flows indicate decreased amounts of good quality pectin which require smaller amounts of sugar initially so that a greater amount of water is evaporated. Smaller flows indicate increased amounts of good quality pectin which support the addition of more sugar per cup of juice.

A successful jelly or jam will also depend upon the hydrogen ion concentration—amount of acid. Human senses can test for adequate acidity: combine in a saucer 3 tablespoons water and 1/2 teaspoon sugar, then add 1 teaspoon lemon juice. This mixture is then tasted and the tartness compared with the extracted juice. Juice at least this tart contains adequate acid to gel. To juice that is not as tart, 1 tablespoon lemon juice per cup of juice should be added and the taste test repeated. As people age, the sense of taste for tartness (acid) decreases. If children are present, this is an excellent phase of jelly making to involve them in.

Juice extracted from the fruit or crushed fruit is combined with an appropriate amount of sugar—determined by the juice's gelling power or 3/4 c of sugar per cup of juice—and brought to a boil to dissolve the sugar and evaporate water. During heating, sugar protects the pectin molecules from being split (depolymerized) which would lessen the strength of the gel. The long pectin molecules enhance foaming so this evaporation step should be done in a large container. An addition of 1/4 teaspoon butter, margarine, or oil reduces surface tension and foaming. The amount of juice or fruit for one batch should not be greater than 8 cups. This requires a large pan so that the juice will not boil over and evaporation will be rapid. Too long a cooking period causes hydrolysis of pectin and a weaker gel.

Determining when the evaporation step is completed is best done by observing the thickness of the solution. The sheet test is the best indicator of the gelling capacity of the syrup because it takes into account all the factors that affect a gel. However, householder experience and judgment are required. The sheet test is conducted after the syrup has boiled. First remove the pan from heat to slow cooking—since the syrup and pan themselves are hot, some evaporation continues. A shallow spoonful of syrup is removed in a metal spoon. It is cooled slightly and then the spoon is tilted to the side so the syrup drips off. Drops that join together and fall slowly at the edge of the spoon (sheet from the spoon) indicate sufficient acid pectin and sugar concentrations (Fig. 11.4). The first time that these conditions occur is optimum, so the test needs to be performed at several intervals to prevent overcooking. Overcooked syrups give fairly good sheet tests, but do not produce top quality finished products. There is variation in the way syrups with different pectins drop from the spoon edge, so even householders experienced in the sheet test may be confused by the way a new fruit syrup appears. An alternative method which is slower but which avoids this problem is to pour about 1/4 teaspoon onto a chilled plate; if a gel forms in a few minutes the product is done. If the sample still is liquid, additional boiling and retesting should be done. Conserves and butters are cooked until thick, with recognition that they will have increased thickness when cold.

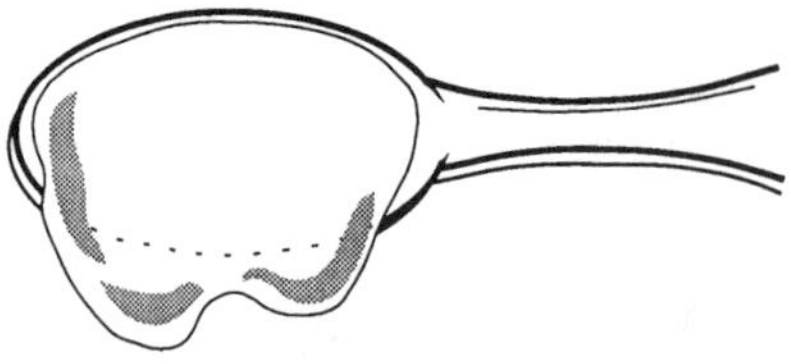

**11.4.** Sheeting test indicates cooking endpoint when making jelly.

It is also possible to successfully determine the doneness by measuring the boiling point of the solution. If the pectin, acid, and sugar concentrations were all in the necessary balance in the original solution, the required amount of water loss through evaporation can be determined by the sugar concentration in the solution. The boiling point of water is determined by the number of solute particles with only very small molecules such as sugar having a large effect. The optimum endpoint temperature will be 217–221F (103–105C) at sea level. Endpoints for other elevations are 5–9F (3–5C) degrees above the boiling point of water. A source of error in this method results from a long cooking time because the formation of invert sugar is enhanced by acid, heat, and time. If the sucrose (1 molecule) originally added to the fruit becomes invert sugar (2 molecules), the boiling point will be increased twice as much but the dehydration effect on pectin does not double so insufficient water will have been evaporated.

Success in using temperature as a cooking endpoint occurs only if the amount of sugar initially added was appropriate for the amount of pectin in the juice. Therefore, the sheeting test is usually used along with boiling temperature to give a check on the endpoint.

The concentration of sugar in finished jellies and jams varies with the type but is usually in the 40–70% range; averaging 65%. By definition, commercial jellies have 55% soluble solids by weight. At room temperature, 67% is the saturation level for sucrose. If all the sugar in jelly were sucrose, precipitation during storage and an accompanying gritty texture would be problems. Invert sugar—hydrolysis of a sucrose molecule to one glucose and one fructose molecule—is more soluble and therefore less likely to crystallize, and sucrose is also more soluble with fructose present. When cane or beet sugar is used in jelly making, inversion occurs during the cooking process. This chemical change is favored by slow cooking so boiling on a high setting for a very short time results in less invert sugar and later a problem of crystallization.

Error in judging the endpoint during preparation of a conventional recipe produces noticeable changes in the finished product. Understanding the endpoint tests and careful monitoring during cooking are critical, but the balance of ingredients also determines gel strength. Adding too little sugar initially in a conventional recipe may result in a tough, rubbery gel structure as the pectin becomes extremely concentrated by the time the sugar is concentrated enough through evaporation of water to raise the boiling point sufficiently or to give a positive sheeting test. Sugar concentration which is too high in the finished jelly results in a product that is too sticky and moist since the sugar draws moisture from the air into the food. In excessive cases, the product will not be able to form a firm gel and a sweet fruit syrup results. Too much sugar can also overpower the fruit flavor. The starting proportions of ingredients in conventional recipes varies widely for strawberry jam made by the conventional method (Table 11.3). The products differ in consistency but not everyone prefers the same thickness in jam, so none of the recipes are considered inaccurate.

**Table 11.3. Proportions of ingredients vary in jam recipes**

| | Joy[a] Cooking | Sunset[b] | Keeping[c] Harvest | Keeping[c] Harvest |
|---|---|---|---|---|
| Berries, crushed | 2c | 4c | 4c | 6c |
| Sugar | 4c | 4c | 3c | 4.5c |
| Lemon juice | 2Tb | none | 0–4Tb | none |

[a]*Joy of Cooking,* Rombauer, I.S. and Becker, M.R. 1975. New York. Bobbs-Merrill. p. 835.

[b]*Home Canning,* 1993. Menlo Park, CA. Sunset Publ. Co. p. 37.

[c]*Keeping the Harvest,* Chioffi, N. and Mead, G. 1991. Rev. ed. Pownal, VT. Storey Commun., Inc. p. 67, 68.

A very weak gel may result if not enough bonding sites on the pectin molecules are dehydrated by sugar such as when the sugar amount is decreased in an added-pectin recipe and then the mixture is boiled for a standardized time. Householders with this problem end up with an ice cream or pancake topping instead of jelly or jam.

After cooking is completed, the hot syrup is poured into clean jars. Any foam present may be skimmed from the top of the finished syrup before it is added to jars for processing. This step was particularly important when jelly was sealed with paraffin since the foam structure traps air which favors mold growth. The foam is edible as a fresh product.

In the last decade, paraffin sealing for jellies has ceased to be recommended as an option. Jams have uneven surfaces so paraffin sealing was never recommended for them. Heat processing in boiling water is now recommended because it prevents spoilage from molds and yeasts during storage. Though bacteria are inhibited by the concentration of sugar in jelly, yeasts and molds are much more resistant to low $a_w$ and the dehydrating effect. Some molds can multiply in the presence of 60% sugar; therefore, jars are heat processed to insure a vacuum seal and remove the oxygen required for mold growth. Low temperature also prevents the multiplication of yeasts and molds so products may be frozen for long-term storage or refrigerated for short-term storage.

Screw-capped jars to be vacuum sealed are filled within 1/4 inch of the top; leaving little air since they are not processed for long times. The jars are then covered with 2-piece lids and placed in boiling water to cover. Processing time is 5 minutes at altitudes of 1,000 feet or below. One minute should be added for each 1,000 feet above sea level (example: at 3,100 ft above sea level, process for 8 min).

If paraffin were used, a 1/8-inch thick layer was optimum to allow the paraffin to expand and contract with temperature changes and still maintain an airtight seal. Thick paraffin layers were especially prone to becoming concave as they hardened, pulling away from the jar sides and leaving access for molds from the air. Even though the cooking step kills mold spores originally present in the juice, spores may enter as the jars are filled.

Our emphasis has been on potential problems with the conventional method and reasons for their occurrence, but the conventional method was used by many householders in the past to make economical and tasty products. By following good procedures with an understanding of the steps, even inexperienced cooks can have success with conventional recipes. When using homegrown or inexpensive fruit, the conventional method produces low-cost jams and jellies which may not otherwise be allowable in some household budgets.

### Pectin-Added Gels

Powdered pectin has been available to householders for many years and liquid pectins became popular when they were introduced several decades ago. Both are made by extracting pectin from fruits. In the United States, most of the pectin comes from the white (albedo) layer of citrus peels but some also is from apples.

Both the liquid and powdered types produce fine gels. The liquid pectin is often more expensive and has a limited refrigerator storage time but many householders find it easier to use. The products are added at different times in jelly recipes. When using powdered pectin, the fruit juice and pectin are boiled to dissolve the pectin, then sugar is added and all is boiled again to dissolve the sugar. In the case of liquid pectin, the fruit and sugar are boiled to dissolve the sugar then the pan is removed from the heat source and the pectin added. Since heat increases the hydrolysis or breakdown of pectin and the commercial liquid pectin is already dispersed, heat is not necessary and would only reduce the gel strength.

Fruit cooked in a sugar syrup remains firmer than fruit cooked in water because sugar molecules interact with the cell wall hemicelluloses and pectins of the fruit, resulting in a more rigid wall. When making preserves with whole, soft fruits such as strawberries, the shape is best retained when the cooking step is done in the presence of sugar. Some householders feel the extra cost of liquid pectin is warranted for these jams.

The pectin-added homemade products require less householder skill, judgment, and time, thus are advisable when expensive, purchased fruit is being used or in the case of low-pectin fruits. Many householders today consider themselves experienced jelly makers but have never attempted a conventional jelly recipe.

### Uncooked Jams

These gained great popularity in the early 1970s because of their fresh-fruit color and flavor. They are made with commercial pectin. About twice as much pectin is added initially as in cooked jams and the proportion of sugar is greater in the final product since there is little evaporation. They are stored in the freezer or refrigerator to prevent yeast and mold growth. Many consumers find that their flavor is worth the expensive storage required. Preparing uncooked jams is usually an uncomplicated and fairly quick procedure. The ingredients are simply measured and combined in a bowl, stirred to speed dissolving of the sugar, and then the jam is packaged for frozen storage. A small amount of corn syrup is added to uncooked fruit freezer spreads to prevent sucrose hydrate, a hard mass that develops on the surface during frozen storage. Typically, the jam is stored at room temperature for a short time to increase gel formation. If fruit is very ripe, it may ferment during this time. If this occurs in a batch, the rest should be held at refrigerator rather than room temperature. The high sugar content of these products prevents them from freezing solid so they thaw rapidly for use.

There is an assortment of freezer jam recipes available. Table 11.4 shows the amounts of ingredients used in recipes for one package (1 3/4 oz) pectin. With the different proportions of ingredients, different products result. The microwave version of frozen jam in Table 11.4 contains no added pectin and is closer to a concentrated, sweetened fruit puree than to a traditional jam.

**Table 11.4. Strawberry freezer jam recipes are varied**

| | Sunset[a] | Ortho[b] | Ball[c] | PFB[d] | Sunset[a] Micwav. |
|---|---|---|---|---|---|
| Berries, crushed | 4c | 2c | 2c | 2c | 2c |
| Sugar | 6c | 4c | 4c | 4c | 1 1/2c |
| Pectin | 1 pkg | 1 pkg | 1 pkg | 1 pkg | none |
| Water | none | 1c | 1c | 3/4c | none |
| Lemon j. | 2 Tb | none | none | none | 1 1/2 Tb |

[a]*Home Canning*, 1993. Menlo Park, CA. Sunset Publ. Co. p. 37, 40.
[b]*12 Months Harvest,* Ortho series, 1979. San Francisco, Ortho Book Division. p. 43.
[c]*Ball Blue Book*, 1990. 32 ed. Muncie, IN. Ball Corp. p. 81.
[d]*Putting Food By,* Greene, M., Hertzberg, R., Vaughan, B. 1988. 4 ed. Lexington, MA. Stephen Greene Press. p. 274.

## Low-Methoxyl Pectin Gels

Improvements in the flavor of artificial sweeteners and consumer acceptance of less sweet fruit products has led to the popularity of special commercial pectin products that may be used with less sugar or no sugar. The low-methoxyl pectin (LMP) jams and jellies rely upon calcium bridges between pectin molecules to form the gel.

The LMP molecule differs from traditional pectin in that it has more galacturonic acid groups in the acid form. Approximately 30% of the groups are methylated. This structure eliminates the need to have sugar for a dehydrating effect and for acid to neutralize repelling charges. LMP molecules bond to each other via cross bonding with a divalent ion, usually calcium ($Ca^{++}$) (Fig. 11.5). It is usually added to commercial LMP preparations in the form of calcium phosphate crystals ($CaPO_4$). Calcium is naturally present in fruits as part of the cell wall and membrane systems and also is found as an ion in other parts of the cells. This calcium content varies with the type and cultivar of fruit, and also with maturity and growing conditions. These are variables, along with the hardness (calcium content) of water supplies across the country, that the LMP manufacturers cannot predict. Therefore, recipe adjustments at the household level commonly need to be made. Many home preservers make a very small test batch each season—when they change fruits, and when they switch LMP brands.

The calcium necessary to form good LMP bonds for gelled foods is usually included in the LMP powder mixture, but it is possible to buy brands with calcium packaged separately. This calcium needs to be dispersed in liquid before adding to fruit. Combining 1/2 teaspoon calcium (monocalcium phosphate) and 1/2 cup water is a common proportion.

LMP jams and jellies require minimum cooking and sugar is added only to taste or artificial sweeteners may be used. They may be preserved by frozen or refrigerated storage, or vacuum sealed in a boiling water process. Directions with the product should be followed.

## Low-Methoxyl Pectin Recipes

LMP recipes differ from regular pectin-added recipes and the two pectin types cannot be used interchangeably. Most LMP products sold at the retail level are added to the fruit or juice, boiled, optional sweetener stirred in, and the mixture put into jars for processing. A few LMP powders on the market must be combined thoroughly with water, juice, or sweetener

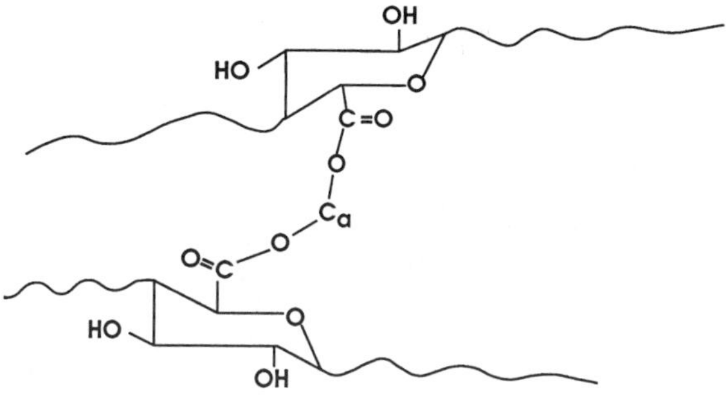


**11.5.** Low-methoxyl pectin bonds with divalent calcium ion to cross-link 2 pectin molecules through carboxyl groups.

before adding to the fruit to prevent lumping. Combining this type of pectin powder with boiling water and stirring in a blender 1–2 minutes may be helpful in dispersing it. If it was dispersed in hot water, this LMP will gel as it cools. Either the hot or cooled, gelled pectin can be used for cooked recipes, but freezer jam recipes must have this pectin added before it forms a gel. When adding pectin—in any form—to the fruit, vigorous stirring is necessary to disperse it for a smooth product.

A typical uncooked LMP freezer strawberry jam recipe uses 4 cups mashed, room temperature berries, to which 1 package LMP pectin or 1/2 cup hot pectin sol is added while stirring vigorously, then adding sweetener if desired. Two teaspoons calcium solution is then stirred in if calcium was not included with the pectin, then the mixture is poured into freezer containers. LMP freezer jams should remain at room temperature for 1–2 hours before freezing to facilitate gel formation.

Spoilage by yeasts or molds or fermentative bacteria can occur since sugar is not present as a preservative. Microorganisms will be killed if there is a boiling step but contamination occurs later. The acidity of the fruit (below pH 4.7) prevents the growth of pathogenic bacteria. Processing these jams and jellies is somewhat different from the traditional. Directions on the product package should be followed. Some artificial sweeteners develop off-flavors when heated, so those jars are usually sealed by inverting 5 minutes then turning upright. As the gel cools, a vacuum is formed inside the jar. This seal is weaker than for boiling water processed jars. These products are generally stored frozen or refrigerated.

In addition to LMP, gums such as carrageenan or specialty starches may be used singly or in combination to provide the gel. As with LMP, sugar or noncaloric sweetener may be added.

Nonpectin jams and jellies appear similar to those made a generation ago, but with their low sugar content are a very different product. LMP pectin and gums are emerging as new ways to produce gelled spreads; though it is not preserving with sugar.

## Spoilage of Products Preserved with Sugar

The concentration of sugar in conventional jams and jellies prevents the growth of bacteria and most yeasts which may have contaminated the product after cooking. Mold growth will occur on the surface if the jar is not sealed. Some molds which have been found on jams and jellies are known to be able to produce toxins (mycotoxins); therefore, moldy jams or jellies should be discarded. Doing research to identify more specifically the risk from such spoilage has had low priority so the level of risk has not been identified.

## Summary

Jams and jellies

**SAFE IF:**

No spoilage

**TOP QUALITY IF:**

**a.** Reliable directions are followed.

**b.** Jellies or jams canned in jars with 2-piece lids and given short boiling water processing to give airtight seal to prevent molding.

**c.** Most low-sugar type products or uncooked jams and jellies are frozen for long-term storage or refrigerated for short term.

## References

Cox, R.E. and R.H. Highby. 1944. A better way to determine the jellying power of pectins. Food Ind. 16:441.

Glicksman, J. 1982. Food carbohydrates. ed. D.R. Lineback and G.E. Inglett. Westport: Avi Publ. Co.

Lopez, A. and L.H. Li. 1968. Low-methyl pectin apple gels. Food Technol 22:1023.

Woodroof, J.G. 1986. Other products and processes. In Commercial Fruit Processing, 2d ed. Woodroof, J.G. and Suh, B.S. ed. Westport: Avi Publ. Co.

## CONSUMER QUESTIONS

**Q.** *Which types of fruit contain enough pectin and acid to make jelly without adding a commercial pectin?*
**A.** Usually tart apples, blackberries, and all the blackberry crosses such as Tay, Logan, and Marian berries, crabapples, cranberries, gooseberries, lemons, quinces, and many plum cultivars are good candidates for the conventional method. Some blackberry crosses may lack acid so taste-test for it as described in the chapter and have lemon juice on hand during preparation.

Often there are special concerns in making conventional jelly with sweet apples, overripe blackberries, cherries, elderberries, grapes, loquats, and citrus fruits, as they lack either acid or pectin. If using one of these fruits, test pectin content of the cherries, elderberries, grapes, and loquats with a Jelmeter and have liquid pectin and lemon juice on hand to add if needed. Obtaining a tried and true recipe from a friend that uses the same cultivar of the fruit is a good starting point.

Poor candidates for the conventional method are apricots, figs, Western concord grapes, guavas, peaches, pears, pomegranates, raspberries, and strawberries.

**Q.** *Jelly syrup doesn't sheet off spoon by the time it has reached 221F (105C). What should be done?*
**A.** Remove it from heat. Review process for sheet test. Any errors in your methods?

Cool a small portion and taste it. If it does not taste sufficiently tart, lack of acid may be the cause. Add lemon juice to taste and try sheet test again. If the flavor would be too tart with additional acid, continue to boil to 224F (107C) which concentrates the sucrose to a 70% solution. Try sheet test again. If it still does not pass, add 1 teaspoon liquid pectin per cup of fruit juice originally used and boil 1 minute. Test again. It should now gel. If so, too much sugar may have been added to the juice before boiling.

**Q.** *Why does jelly foam during boiling?*
**A.** This is normal. Pectin is one of the large compounds in the juice that traps air underneath it and holds bubbles. Phenolic compounds in juice may also foam. Lowering the surface tension of the juice by adding 1/4 to 1/2 teaspoon oil or butter is recommended in some recipes. It reduces foaming; however rancidity and its accompanying off-flavors in the finished product are possible during room temperature storage. Cooking in an oversized pan prevents boiling over and the foam is easily skimmed off at the end of the cooking period. Foam inside a sealed jar is unattractive, but does not increase the probability of mold growth that foam under paraffin did.

**Q.** *What are the crystals that are found in grape jelly?*
A. The glasslike potassium acid tartrate crystals are common in grape jelly as the compound that forms them comes from the grape itself. The problem can be minimized by chilling the extracted juice for 48 hours and then pouring off only the top liquid to use and discarding the sediment. These crystals form slowly with storage, so are rarely a problem when the jelly is consumed promptly. They are not a health hazard and are similar in composition to cream of tartar used in baking. Using grapes in mixed-fruit jellies will decrease the concentration of this compound.

**Q.** *Can jelly and jam recipes be doubled?*
**A.** This is not likely to result in good quality products. The rate of evaporation will be slower if there is less surface in relationship to the total quantity of product. Even and rapid heat penetration are also a problem with doubling. Freezing an abundance of berries for use later is a better practice than increasing recipe size.

**Q.** *What if the finished jelly is too soft?*
**A.** Common causes: too little acid, too little sugar in final product, increasing recipe size. If conventional method, reboil and test for doneness. If made with powdered pectin, correct by making a pectin sol by combining 4 teaspoons pectin, 2 tablespoons lemon juice, and 1/2 cup water, then bring it to a boil, stirring constantly. Add 1 quart of the soft jelly and 1/4 cup sugar. Quickly bring all to a boil and boil for 30

seconds. Remove from heat, do sheet test, skim foam and seal. To correct liquid pectin product, bring 1 quart jelly to boil, add 3/4 cup sugar, 2 tablespoons lemon juice, and 2 tablespoons liquid pectin. Boil hard (rolling boil) for 1 minute. Remove from heat, do sheet test, skim off foam and seal. Note: Syrupy jelly can occur if too much sugar is used. Recooking will not correct this problem. Find another use for the sweetened fruit syrup.

**Q.** *What if jelly or jam is too firm?*
**A.** Too much pectin (excessive amounts can come from the fruit itself or from added) or overcooking. Make note on recipe and next time add up to 1/2 cup more juice or soft fruit to pectin-added recipes. Using a tried and true recipe from a neighbor who uses the same cultivars is another approach. If the problem occurred with a conventional recipe, review methods for determining the amount of sugar to be added and for testing doneness. It is important to test frequently before an overcooked stage is reached.

**Q.** *What causes weeping jelly?*
**A.** Too much acid causes excessive bonding of pectin molecules which squeeze out the water that should be held in the spaces between them (syneresis). Make note on the recipe to decrease added acid next time. Overcooking also concentrates the acid. Some pectins such as those in cranberries, are more prone to this problem.

**Q.** *What if paraffin was used on jelly?*
**A.** Jelly and jam may weep around paraffin and break the seal if it is too soft or stored in a warm place. If this weeping has broken the seal and no mold is present, the product is still safe to consume. In the future, sealing in a boiling water process will prevent this problem. The lack of heat processing in paraffin-sealed jelly makes spoilage by yeasts (fermentation) or molds much more likely.

**Q.** *How should liquid pectin be stored?*
**A.** Unopened bottles should be stored in a cool dry place. Liquid pectin should be used the same season it was purchased for consistent, good results. Partial bottles may be stored in the refrigerator for up to a month. Liquid pectin thickens at refrigerator temperatures and should warm at room temperature for at least 1 hour before being added to a recipe. After reaching room temperature, if the pectin is still very thick or is thin and watery, it will not make a good gel and should be discarded.

**Q.** *How is powdered pectin stored?*
**A.** A cool dry place such as a pantry or cupboard is ideal for storage but good results are most consistent when pectin is used the same season as purchased. Powdered pectin that has been stored since last season is generally usable if it has not browned or caked.

**Q.** *Is sterilizing or scalding used for jars before filling with jelly and jams?*
**A.** Jars should be thoroughly washed before filling. Growth of pathogenic bacteria is not possible in high-sugar or -acid foods such as fruit jams and jellies. Sealing by processing in a boiling water process removes air needed for mold growth. Since a sterilized jar would be recontaminated by air during filling, the sterilization step is not relevant.

# BOOK II

# Practice

Hands-on experiences with food preservation are valuable for bringing up questions about procedures that may seem clear when reading descriptions in a text. Practical experiences also enhance long-term memory of the principles. The authors encourage students of food preservation, whether in a classroom setting or by yourself in a household kitchen, to perform as many of these practical experiences as your facilities allow. Most of the experiments can be conducted in a household kitchen and many of them are excellent learning experiences even when seen only as demonstrations.

We recommend that you read through each experiment before starting; some of them require several days to monitor. If you are studying food preservation by yourself, or in a small group, it is possible to eliminate some of the different treatments and still see the effect of altering recipes.

The experiments are designed to be low-cost to readers nationwide. If you have locally available specialty foods, they should be substituted to make the experiments more directly applicable to your region. Substitute entire recipes and procedures for products. For example, if practicing with fresh pressure processed crab in Alaska, use USDA recommended processing times and methods for crab. Experiment with cactus jelly in New Mexico, huckleberry jelly in Idaho, and flower petal jelly in Florida.

Remember to have fun in the kitchen and be creative, but put safety first. Before changing preservation recipes, take time to read about the basis for safety for each method and do not alter those ingredients or procedures. The consumer question and answer section for each method addresses preservation mechanisms that should not be altered. For example, do not change processing times or temperatures for canned foods, do not change the amount of salt in country hams, and do not increase the refrigerator's temperature. We encourage you to experiment canning fruits with artificial sweeteners, freezing casseroles with "imitation" salt, and reducing the amount of sugar in jelly by varying the type of pectin. This is the way to explore food preservation principles. Always remember to relate each change to your household's preferences, because whether or not the food is actually consumed is a critical evaluation of food preservation methods.

Each experiment section is preceded by questions that are commonly asked by household food preservers. If you need help in answering these, or wish to check the accuracy of your solutions, brief answers are provided in Appendix B.

NOTE: If you are completing the review of literature sections as a home-study student, the *Journal of Food Science* is a research journal that most large public libraries and community colleges carry. Many of the articles in it are understandable with a limited background in chemistry. We cited it whenever possible because of its availability. *Food Technology* and the *Journal of Food Protection* are other journals that are often carried by large community libraries, and may also be of interest to you in your studies. Consult your librarian for other suggestions. Sometimes it is possible to order photocopies of specific journal articles from another library in your state and have them sent to your local library.

The Cooperative Extension Service is mentioned as a source of many USDA recommendations. We encourage you to visit your local office and familiarize yourself with the wealth of information it has to offer. Usually there is an office located in each county seat across the United States. Sometimes their telephone number is listed under the name of the state's land grant university. For example, in Portland, Oregon, it is listed as Oregon State University Cooperative Extension Service, though the University is not located in Portland.

Experiments are rated to describe the level of laboratory equipment required.

HOUSEHOLD indicates the experiment uses utensils normally found in the kitchen of someone who has been preserving food by the particular method emphasized in that chapter, such as a pressure canner for canning experiences, a food dryer for studying dehydration, and jelly bags for preservation by sugar.

HOUSEHOLD+ experiences require items readily available from a drugstore such as litmus paper or calcium carbonate.

MODERATE indicates some laboratory or unusual household equipment is needed. A moderate experiment could call for use of a brine-injecting pump, a Jelmeter, a pipet, or petri dishes. Sources for this equipment are listed in Appendix A.

LABORATORY rating indicates that equipment such as highly accurate scales, centrifuges, or microscopes is required. Laboratory experiments are not practical for kitchen settings.

## Experiment Ratings

**1**

Identifying Quality Changes .... Laboratory
Subjectively Evaluating Foods ... Household
Quality Standards of Foods ..... Laboratory
Objective Quality Evaluations ... Laboratory

**2**

HACCP .................. Household

Consumer Education . . . . . . . . . Household
Representative Samples . . . . . . . . Moderate
Experience with Microbiological
Equipment . . . . . . . . . . . . . . . Laboratory
Safe Holding Temperatures
for Food . . . . . . . . . . . . . . . . Household
Adapting Old Recipes for Safety . Household
Foodborne Illness Scenarios . . . . Household
Dehydrating Proteins Delays
Coagulation . . . . . . . . . . . . . . Household

**3**

Pretreatment with Antibrowning
Dips . . . . . . . . . . . . . . . . . . . Household
Blanching . . . . . . . . . . . . . . . . . Household
Fruit Leather . . . . . . . . . . . . . . . Household
Blanched Vegetables . . . . . . . . . Household
Herbs . . . . . . . . . . . . . . . . . . . . Household
Jerky . . . . . . . . . . . . . . . . . . . . Household
Plant Pigments . . . . . . . . . . . . . . Household

**4**

Types of Acidifiers and Their
Characteristics . . . . . . . . . . . . Household
Different Vinegars in Recipes . . . Household
Types of Sweeteners and Their
Characteristics . . . . . . . . . . . . Household
Types of Salts and Their
Characteristics . . . . . . . . . . . . Household
Effect of Hard Water
upon Pickles . . . . . . . . . . . . . Household
Crisping Agents . . . . . . . . . . . Household+
Brine Strength . . . . . . . . . . . . . . Household
Anthocyanins and Acid . . . . . Household+

**5**

Measuring and Changing
Salt Solutions . . . . . . . . . . . . . . Moderate
Sauerkraut . . . . . . . Household to Laboratory
Fermented
Cucumbers . . . . Household to Laboratory
Kimchi . . . . . . . . . . . . . . . . . . . Household

**6**

Sweet Pickle Curing
Brine . . . . . . . . . . Household to Moderate
Dry Cured, Unaged
Ham . . . . . . . . . Household to Laboratory
Combination Cure,
Unaged Ham . . . . Moderate to Laboratory
Corned Beef Pickle . . . . . . . . . . Household
Survey of Sausage Flavors . . . . . . Household
Homemade Cured
Sausage-type Product . . . . . Household+
Salt Content of Cured Meat . . . . Laboratory
Cured-Meat Pigments in Sausages . . Laboratory

**7**

Foods of Different Composition . . Household
Storage Temperature and
Speed of Freezing . . . . . . . . . Laboratory
Packaging . . . . . . . . . . . . . . . . . Household
Retrogradation Study . . . . . . . . . . Moderate

**8**

Survey of Refrigerated
Food Spoilers . . . . . . . . . . . . . Household
Time in Growth Zone . . . . . . . . Household
Short-term Refrigeration . . . . . . . Laboratory

**9**

Rate of Heat Penetration . . . . . . . Moderate
Factors Affecting
Quality of Tomatoes . . . . . . Household+
Type of Tomato Pack
and Processing Times . . . . . . . Household
Quality of Tomato Juice . . . . . . . Household
Quality of Canned Pears . . . . . . . Household
Quality of Canned Snap Beans . . Household
Effect of pH on Processing
Time for Beets . . . . . . . . . . . . Household
Effect of Container on
Canning Asparagus . . . . . . . . . . Moderate
Canning Meat, Fish, and Poultry . Household
Thickened Pie Fillings . . . . . . . . . Moderate

**10**

Carrot Storage . . . . . . . . . . . . . . Household
Sugar Accumulation in Potatoes . Household
Relative Humidity . . . . . . . . . . . . Moderate
Condensation . . . . . . . . . . . . . . Household

**11**

Gelling Experiment . . . . . . . . . . Household
Conventional Apple Jelly
Experiment . . . . . . . . . . . . . . Household
Jelmeter . . . . . . . . . . . . . . . . . . . Moderate
Lower-Sugar Jam and Jelly . . . . . Household

# 1 Quality Changes in Foods during Storage

## Typical Consumer Questions

Decide for the following situations if a quality change and/or a safety change is involved. In each case, should the food be discarded to prevent foodborne illness? If you need more information to make decisions, state which facts are lacking.

**1.** *Refrigerated cottage cheese acquires a thin liquid on top after a week of household storage.*

**2.** *A large block of cheddar cheese which has been frozen to prevent mold growth is crumbly when thawed.*

**3.** *Frozen ground beef has excessive drip when thawed.*

**4.** *The top layer of canned peaches has floated above the syrup and browned.*

**5.** *Dried apples were properly packaged and stored in the cupboard. When first removed from the dryer, they were slightly brown. After 6 months storage, they are a medium brown.*

**6.** *A box of snack crackers was purchased in November. One of the individually sealed packages was opened and served for Thanksgiving hors d'oeuvres. When the other package was opened for New Year's Day, the crackers were crisp, but had an off-odor.*

**7.** *A poppy seed cake with custard filling is stored in the refrigerator. By the second day of storage, the cake has a coarser texture.*

**8.** *Whipped cream was served on Sunday and the leftovers were refrigerated. On Tuesday, it is no longer fluffy.*

**9.** *I prepared beef pot roast on Sunday and served the leftovers in lunch sandwiches. On Wednesday, family members complain of an off-taste.*

**10.** *Pies for the family reunion are made a day in advance. The crust is perfect when just out of the oven. However, the next day when served, the crust is not crisp.*

**11.** *Salad and dressing are tossed in the afternoon before the dinner party. When served, the lettuce appears wilted.*

**12.** *A starch-thickened cheese and potato soup becomes very thick in the refrigerator after only 1-day storage, and has a clear liquid layer on top.*

**13.** *Dry rice is stored in the cupboard. After cooking, the rice is stored on the counter for 1 day.*

**14.** *Grandma moved from her house into an apartment. Extra food from the pantry was given to neighbors. The blackberry syrup she canned 5 years ago is as thick as jelly.*

**15.** *Commercially manufactured, shredded, dried coconut becomes a tan color in the cupboard.*

**16.** *An opened package of graham crackers loses its crispness after only a few days.*

### ■ Food Quality Literature

Read a scientific research journal article that contains both subjective and objective testing. Photocopy the abstract and tape it to a blank piece of paper. Below the abstract, list the subjective and objective tests used in this article. Briefly (5 min max.) describe your article to the other students.

## PRACTICAL EXPERIENCES

### ■ Identifying Quality Changes

The laboratory guide will display an assortment of foods which exhibit quality changes. Storage history and product identification will be provided. Individually, examine each food and write down the changes you notice. If any foods also contain microbiological spoilage and are potentially unsafe to consume, they will be placed in a separate area and labeled as such; do not evaluate these by tasting. Food with only quality changes should be evaluated by flavor, odor, color, and texture. Unsalted soda crackers and water will be provided between samples that are tasted.

Each student should then take a turn explaining the changes they found in one of the foods. If the changes could have been prevented, explain how. If foods with several degrees of rancidity or Maillard browning are presented, at what level do you feel the food is objectionable? At what level would you discard it? Does this vary among the students in your class?

### ■ Subjectively Evaluating Foods

Success of the preservation method is partially determined by a subjective evaluation of quality. Flavor, odor, texture, and color can be assessed by people using their senses (subjective evaluation), or by laboratory instruments (objective evaluation). Human evaluation is usually preferred, even though instruments give a numerical rating of a particular characteristic, because appeal to human senses is the most important factor. In the case of human evaluation, several common formats are used.

#### Consumer Preferences

Hedonic methods are used by commercial food manufacturers to sample consumers. The consumers are presented with the food and then asked "Do you like it?" Hedonic scales ask the respondent to rate the overall quality of the food—from poor to excellent, from 0 to 5, from 1 to 10, etc. Hedonic sampling is quick, but the consumer's responses can be affected by variables other than the food quality.

Monitoring sales in a test market is another form of consumer preference sampling. The question "Do they buy it again?" is of utmost importance in commercial foods. At the household level, hedonic evaluations are made informally by household members and the major food preparer considers them when the same food is served again.

#### Trained Sensory Panels

Panel members are trained to detect low levels of a particular characteristic, such as rancidity in refrigerated cooked chicken. Coded samples are presented to panelists with score cards for evaluating the degree of each characteristic that is being tested. This is a time-consuming method, but it produces very accurate assessments of food quality.

Since trained panelists are very sensitive to minute differences in odors, it is very important that cooking odors not be present in the evaluation room. Lighting is also critical since it can greatly affect the appearance of the food or create a bias. Usually incandescent lightbulbs are used, and the evaluation booths are coated with special paint or coverings. In one study of the flavor of eggs from chickens fed different rations, the color of the yolk varied from almost white to bright red. The trained panel evaluated all samples under red lights which gave the eggs a similar appearance; thus only flavors were different among the samples.

### Sensory Difference-Testing

Either trained or untrained evaluators can participate in sensory difference-testing; although trained panels are usually used. There are several styles of this testing.

The triangle form involves three samples of food; two are identical. Each sample is coded with a three or four digit random number. The A, B, C designation is not used because A may be associated with excellence to some people. For example, two samples of frozen applesauce are presented along with one of canned applesauce. Evaluators are then asked to note only one characteristic at a time, such as "Which sample differs from the other two in tartness?" This type of test is also called the odd-sample test.

In a paired test, the evaluator is presented with two samples of apple sauce, and asked "Which sample has the smoothest texture?" Other questions such as "Which sample exhibits the greatest separation?" may follow.

Ranking involves the presentation of several samples, often five, and the evaluator is asked to rank them in order for one characteristic. "Rank the five samples of applesauce in descending order for darkness of color," is an example of instructions on a ranking scorecard. Below the instructions, five lines, positioned one above the other, are drawn for the evaluator to write the code numbers of samples.

In scoring, the evaluator is presented with a reference sample and then asked to make a comparison between the coded sample and the reference. A typical scoring question would be "Score the coded sample from 1 to 10 in tartness with 5 being the tartness of the reference sample, 1 being least tart, and 10 being very tart." It is important for sensory testing that evaluators judge only one food or set of foods at a time. If they are required to make several judgments, the previous food is removed before presenting the second test.

**1.** What are the advantages and disadvantages of the various types of sensory testing?

**2.** Which do you feel would be the most accurate method for evaluating the overall quality of 15 different cultivars (varieties) of canned tomatoes?

**3.** Look through a volume of the *Journal of Food Science*. Is sensory testing used in any of the published articles? Did you notice any experiments in which it would have been appropriate?

## ■ Quality Standards of Foods

In a group discussion format, examine samples of several foods. Some suggestions are cottage cheese, a canned pasta and meat combination, prepared frozen fish in batter, uncooked potatoes, canned peaches, and fruit leather. What characteristics are of primary importance? Do these vary among the foods?

Have a food researcher visit your class. What quality characteristics does this person routinely evaluate? How?

### Evaluating Food Quality

Divide the class into half. One group will design a test to assess the quality of two different brands of American cheese. Unsalted soda crackers and water may be provided to the evaluators between samples. The other group will design a test to assess the quality of two brands of frozen french fries prepared according to package directions. Each group will be panelists for the other's test. First determine the important characteristics to assess, then decide on a test method and design the scorecard. It is important that each evaluator be presented with samples that are the same size and shape, and prepared to the same degree of doneness.

**1.** Did the test you designed accurately assess the important quality characteristics? What problems arose?

**2.** How may statistical tests be used to assist in identifying differences?

## Objective Quality Evaluations

Visit a food research laboratory in a university or industrial setting. Ask the researcher to demonstrate some of the equipment.

**1.** What instruments are used to assess quality?

**2.** Do they accurately reflect average human food preferences?

**3.** Ask the researcher the length of time each piece of equipment has remained state of the art. How has objective food quality evaluation changed over time? Which instruments of 30 years ago are still considered to accurately assess a particular quality characteristic? Were there any instruments used in the past which this researcher considers to have been so inaccurate that the research was invalid?

# 2 Foodborne Illness

## Typical Consumer Questions

If you are studying this section with a group, divide the following questions among class members, then share the answers. Give a consumer answer plus a technical answer. If possible, provide a reference(s) for the latter. In some cases, you may need additional information. Add the questions you need to ask and possible consumer responses.

**1.** *The incidence of Salmonellosis is greatest during warm months of the year. Consider the usual modes of transmission and suggest some reasons for this.*

**2.** *Several members of a household have botulism. They had consumed green salad, Italian dressing, milk, apple juice, home canned tomatoes, leftover pot pie, leftover birthday cake with chocolate icing, and ice cream. Which food is the most likely candidate to have been the vehicle? What is a possible scenario for introduction of the organism, toxin production, and the poisoning to occur?*

**3.** *Why is* S. aureus *such a common form of foodborne illness in the United States today? What do you feel is the most practical and effective approach to this problem? State reasons for approach and implementation.*

**4.** *The milk in a home refrigerator does not seem to be as cold as usual, and the ham and ice cream in the top portion is not solidly frozen. The householder wonders if the food is being held safely. How would you respond?*

**5.** *A mother is concerned when her 9-year-old child is ill shortly after coming home from school. The child ate a leftover portion of a turkey-ham and mayonnaise sandwich from his lunch while walking home. The mother has heard recent news stories about* Salmonella *and fears the turkey ham was infected. State your belief of the most likely foodborne illness organism and food vehicle, the illness scenario, and future means of prevention.*

**6.** *On a guided bus tour of Mexico, you notice that many of your fellow passengers are taking bran and vitamin C supplements throughout the day to avoid Montezuma's revenge. You did not bring any. Are you therefore at greater risk of missing some of the sights? How can you improve your chances of not becoming ill?*

**7.** *I head the food planning committee for a club picnic. We have to do all preparation in* *the morning for this evening picnic and we don't have any ice chests. What do we do?*

**8.** *I've canned pickled beets by the USDA recipe which has 1 cup of vinegar to 1 cup of water for the liquid but my family says they are too sour. What can I try next time?*

**9.** *Is macaroni salad, which I made yesterday but then forgot to serve, safe to eat today?*

**10.** *On the way home from a field trip which included lunch as part of a farm tour, several children became ill. How can we determine if this was due to foodborne illness?*

**11.** *I'm planning to demonstrate cutting up a chicken as one way to save money on food. Are there any points about food safety that I should include?*

**12.** *A dinner guest who has become interested in wild foods has brought me mushrooms that he collected to be served for dinner. Should I cook them?*

**13.** *My family has a question. We were on a trip to the coast in August and planned to dig clams to cook for lunch. When we got to the beach where we always dig them, there was a sign saying that shellfish cannot be eaten from this beach. What's happened?*

**14.** *A call from a local company asks your help in solving a problem. They purchased 60 large frozen turkeys to give to employees before Christmas. The day before they were ready to distribute them, they checked and found that the freezer was at 60F (16C) and the turkeys were thawed. Should they be discarded or distributed?*

**15.** *I've recently moved to a state that approves the sale of certified raw milk. Is this a safe product to drink?*

**16.** *I usually like hamburger cooked well done. However, a local fast-food restaurant serves very juicy, rare hamburger in their sandwiches. Is it safe to eat a rare hamburger or should I send it back for further cooking?*

**17.** *I've been told that since I'm pregnant I should be especially careful to avoid* Listeria *in foods. What do you advise?*

## Survey of Food Safety Literature

Read a current journal article on any aspect of food safety. Write a 1 page abstract and post the abstract and citation in the classroom for all to share.

## Hazard Analysis Critical Control Point (HACCP) Approach

Frank L. Bryan has published several journal articles on hazard analysis and control of food in foodservice establishments. His HACCP approach is a good way to focus on the various steps necessary in food preparation and the microbiological implications of each.

Write out your own hazard analysis of one of the following topics or a topic of your own suggestion approved by the class leader. First, you may find reading some of Bryan's articles helpful.

Bryan, F.L. 1981a. Hazard analysis critical control point approach: Epidemiologic rationale and application of foodservice operations. J. Env. Health, 44:7.

Bryan, F.L. 1981b. Hazard analysis of food service operations. Food Tech. 35(2):78.

Bryan, F.L. and C.A. Bartleson. 1985. Mexican-style food service operations: Hazard analysis, critical control points and monitoring. J. Food Prot. 48:509.

Bryan, F.L. and J.B. Lyon. 1984. Critical control points of hospital foodservice operations. J. Food Prot. 47:950.

Bryan, F.L., S.C. Michanie, P. Alvarez and A. Paniagua. 1988. Critical control points of street-vended foods in the Dominican Republic. J. Food Prot. 51:373.

### Topics

1. Deli potato salad.
2. Take-out pizza.
3. Catered cold-cut platter and potato salad for company picnic.
4. Tuna salad sandwich in 7-year-old's lunch.
5. Poached red snapper at senior center dinner.

6. Baked beans at grange hall potluck.
7. Oysters in sauce at four-star restaurant.
8. Mother makes large batch of macaroni and cheese at beginning of week and serves at each dinner, along with other foods, so her children will eat something.
9. Home canned green beans donated by 12 households for the fire station benefit dinner and afternoon open house. A bean/cream of mushroom soup casserole will be served.
10. Deli broasted chicken.

### Consumer Education

Select a perishable commercial product. Compose a "care label" for the package which takes into account both consumer safety and marketing strategies. Write a short (1 page max.) explanation of the reasoning behind your care label. Does your advice also address physical and chemical changes in this food? If you have access to food packages, bring some with care labels to class to share. How do you rate the overall food industry labeling practices?

## PRACTICAL EXPERIENCES

### Representative Samples

The analysis will be meaningful only if the analyzed sample is typical of the product it was taken from. Microorganisms are not usually uniformly distributed throughout the food so blending of a sample which contains both surface and interior portions may be necessary as in the case of a macaroni and cheese casserole. Blending a food also puts the bacteria into a pipetable homogenous suspension. A weighed representative sample can be blended with sterile diluent, or surface swabs can be placed in the sterile diluent and shaken to suspend the bacteria.

#### Making the Dilution

It is important that the dilution bottles, any utensils used, blender container, and blender blades all be sterile. To sterilize utensils, wrap each in paper using the drugstore fold, cap dilution bottles loosely, package blender jar caps in paper, and cover the glass blender jars with foil. The blender jars made only of glass and metal will withstand repeated sterilizations. Those with plastic portions should not be sterilized. Autoclave these prepared items, or place in a pressure canner and process as you would a canned food at 15 pounds for 30 minutes. There may be some moisture in the bottom of the blender container after sterilizing in the pressure canner, but the small amount that cannot be shaken out should not greatly affect your dilution. Blender blades and rings should be sterilized by immersing in a 72% ethanol (alcohol) solution (1516 mL of 95% ethanol and 484 mL of water) for 30 seconds and then rinsing by three successive dips in large beakers of sterile water.

From the blended or swab sample, make dilutions based upon expected numbers of microorganisms in the sample. One mL when plated should contain between 30 and 300 bacteria or molds.

A 1:10 dilution is also referred to as a 1/10 or $10^{-1}$ dilution. In such a dilution, there is 1 g of food in every 10 g of the total mixture; in 1 g or mL of the total mixture there is 0.1 g of food. To make a 1:10 dilution, 25 g of representative food are added to 225 mL of diluent, or 50 g of food added to 450 mL of diluent, or 1 g of food to 9 mL of diluent. A mL of diluent is equal to 1 g of diluent. If enough food sample is available, starting with 50 g is more likely to be representative than starting with 10 g.

Figure 2.1 is a schematic representation of the dilution procedure. Blending 2 minutes at low speed is necessary for the first $10^{-1}$ dilution, but shaking the dilution bottle is acceptable for the others. Follow the shaking demonstration from your laboratory leader (shake 25 times through a 30 cm arc in 7 sec). Notice (Fig. 2.1) that after the second dilution has been made the example accelerates the process by

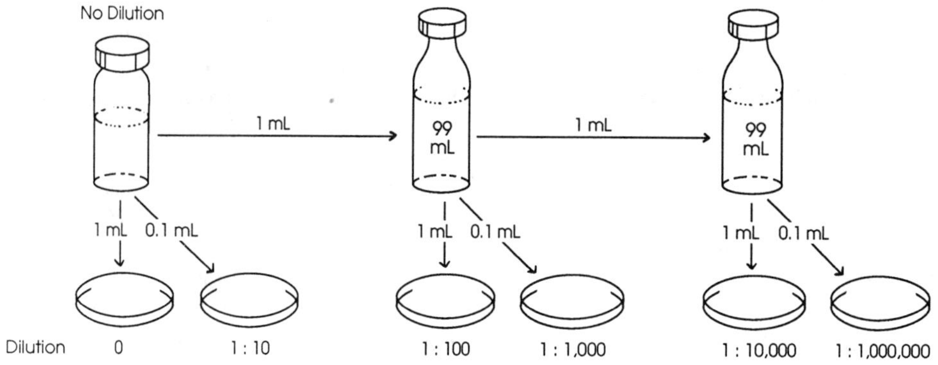


**2.1.** Dilutions to make countable plates from a food sample.

diluting 1:100 at each step. This will decrease the number of dilution bottles used and the time required. Foods likely to have large numbers of microorganisms, such as fermented products, require high dilutions to achieve countable plates (between 30 and 300 colonies/plate). Plating 1 mL from several of the dilutions made increases the chance of having countable plates.

After the plates have been incubated and counted, the microorganism load in the food must be calculated. If 1 mL of the 1:10 dilution was plated, the colonies are from 0.1 g of food. To report the microorganism load as number or as colony forming units (CFU) per g or mL of food, multiply the plate count by 10 (1/10 of a g of food × 10 = 1 g of food). Similarly for each dilution, multiply the actual count by the dilution factor.

Not all bacteria are dispersed into single-cell units during the dilution process, the clump of cells may produce only one visible colony, not all viable cells will produce a visible colony, and environmental conditions may determine if a visible colony grows. With these factors affecting every pour plate, the actual plate count represents between 10 and 90% of the real numbers of bacteria present. This often makes comparisons between labs difficult, but comparisons within one lab for the same foodstuffs over time—as in the refrigerated storage experiments—can be made. Peptone water is an easy diluent to prepare and it protects the microorganisms during the dilution process. However the counts will be increased if the dilution and plating process takes longer than 20 minutes as microorganisms can grow in peptone water.

## Plating

Pour plates are usually used for estimating the living organisms in a food. After the dilutions are made, a measured aliquot (amount) is put into a sterile petri plate. Disposable sterile plates may be purchased from microbiological supply companies. Sterilized agar medium that has been remelted and cooled to 108 to 113F (42 to 45C) is then added. The agar is mixed with the inoculum by gently swirling the plates on the table so the colonies will be dispersed across the plate. After the agar solidifies, the plates are turned upside down so moisture does not condense and fall to the colonies below causing them to spread. The medium may be either nonselective—supports the growth of most bacteria—or selective—formulated to suppress the growth of certain bacteria and favor the growth of specific type(s) of interest. Dehydrated media or sterile prepared media can be purchased from biological supply companies. The plates are then incubated aerobically for 2 to 10 days depending upon the incubation temperature. For anaerobic bacteria or those with other special requirements such as $CO_2$, special incubation conditions (anaerobic jars and $CO_2$ packets) are provided.

Streaked plates (Fig. 2.2) are made by drawing the loop across the surface of solidified agar in the petri dish in the pattern shown. The objective is to obtain separate colonies in at least one portion of the plate. Streaked agar slants are made by drawing the loop across the surface of solidified agar in a test tube—slanted after pouring—in a zigzag fashion. Stabbed slants are made with only one stab per tube. They favor anaerobic organisms below the agar surface.

For standard total counts in foods, the plates are incubated at 35F (1.7C) for 48 hours (+ or − 2 hr). For comparative purposes if no incubator is available, room temperature can be substituted—incubate at 72F (22C) for 5 days. For counts of psychrotrophic bacteria related to the spoilage of refrigerated foods, incubation is generally at 45F (7C) for 10 days.

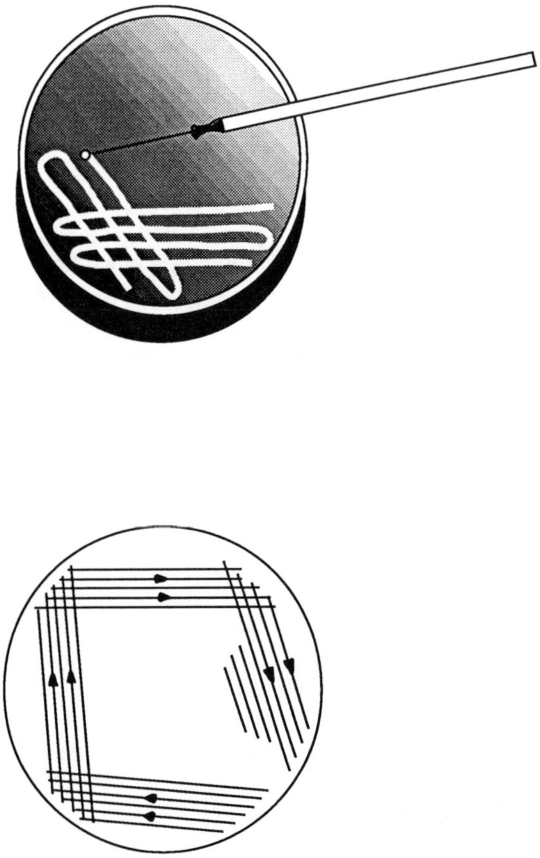

2.2. Streak plate for isolating bacterial or mold colonies.

## Experience with Microbiological Equipment

1. Observe demonstrations of the pour plate method, a streaked plate, streaks and stabs on agar slants, the use of a counter, the incubators, a pH meter, and electronic and manual balances. The laboratory leader may also wish to demonstrate mixing and autoclaving media using both professional lab equipment and also equipment readily available to those teaching outside of a science laboratory setting, such as pressure canners, household jars, and an alcohol burner. Some pharmacies sell large and small screw-cap bottles for those demonstrating in rural areas.

2. Look at the display of dry media samples, a *Difco Manual*, and catalogues for ordering microbiological supplies. You may want to note the addresses of these companies for later use. Also familiarize yourself with the *Compendium of Methods for the Microbiological Examination of Foods*, by M.L. Speck. Note that incubation conditions are suggested in this source. Use these suggestions for incubating your samples prepared below.

### Sources for Microbiological Media

Becton Dickinson Microbiological Systems, Customer Service, P.O. Box 243, Cockeysville, MD 21030-9986

Difco Laboratories (media, no other supplies), P.O. Box 331058, Detroit, MI 48232-7058

### References on Methodology

Food and Drug Administration. 1992. Bacteriological Analytical Manual. 7th ed. Assoc. Official Anal. Chemists, Arlington, Va.

Vanderzant, C. and Splittstoesser, D. ed. 1992. Compendium of Methods for the Microbiological Examination of Foods. 3d ed. American Public Health Assoc. Washington, DC.

**3.** Make duplicate pour plates of 1 mL samples from $10^{-1}$, $10^{-4}$, $10^{-6}$, and $10^{-8}$ dilutions of an available food. Record the time you have the sample in peptone water. Use plate count agar and choose your incubation temperature depending upon the food and its storage history. What was the number of visible colonies for each dilution? Calculate the number of organisms per gram from the countable plates.

Streak a plate with a sample from your $10^{-1}$ dilution. After incubation, compare this plate to the pour plates made above. Is one type of plate more likely to give a better overall assessment of the microbiological flora of a food?

Streak an agar plate with a calibrated loop. After incubation, were there isolated colonies to study? If not, what were your errors?

Practice using the pH meter and balances. If your laboratory has the "portable" type of pH meters that are made for field work, find out their price and frequency of repair. Are they mailed to the manufacturer for repair or is an expert needed locally? What is the cost and accuracy of a balance and a set of weights?

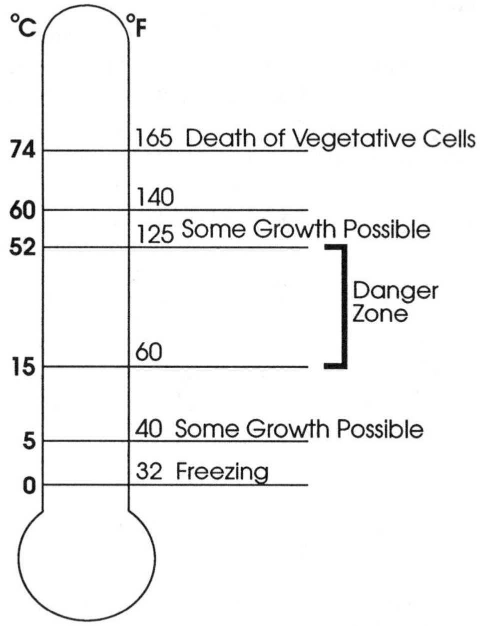


**2.3.** Critical temperatures for food safety.

## Safe Holding Temperatures for Food

Using commercially canned refried beans, divide them into several lots and heat (or chill) them to temperatures below the danger zone, barely in the danger zone, in the middle of the zone, and barely out of the zone (Fig. 2.3). Display these lots with a thermometer in each for classmates to touch. Remember that your body temperature is usually 98-99F (about 37C), when you feel the warmth of these lots. Are you able to accurately judge temperatures by touch only?

How long does it take a serving-temperature sample (bake four cans of refried beans in a casserole topped with grated cheese, reheat for each test) to cool to safe holding temperatures when sitting on the counter? when placed in the refrigerator? when surrounded by cold water in the sink? when placed in the freezer? What do you feel is the safest way to handle a 2-quart leftover casserole (noodles, white sauce, meatballs, cheese) in the household? Is this also the most practical way? How would you recommend cooling a 2-gallon pot of soup? How do you currently cool such a food at home?

## Adapting Old Recipes for Safety

Salmonellae are a concern in raw eggs. Householders wishing to make traditional recipes safe can eliminate the salmonellae threat by acid or heat treatments of the raw egg.

Do this experiment as class demonstration so all can observe the appearance and taste of safely cooked eggs. As eggs cool, the yolk becomes firmer so display of these eggs for later evaluation is not applicable.

Check the thermometer for accuracy by putting it in rapidly boiling water. Water boils at 212F (100C) at sea level, 208.4F (98C) at 2,000 feet, and 203F (95C) at 5,000 feet.

### Fried Eggs

Preheat skillet (oiled if necessary) to 280–300F (137–148C). Break five refrigerated eggs into the skillet and cook turning once. Start testing one egg at a time when the yolks begin to firm around the edges. Test temperature by putting thermometer in the center since the center has the slowest heat penetration. A temperature of 140F (60C) is typical when yolks are still semiliquid in the center. Hold at this temperature for 3.5 minutes. Cook one egg to a 160F (71C) endpoint without holding. Both of these methods render the egg safe. Repeat with sunny-side up eggs.

### Scrambled Eggs

Prepare two batches of scrambled eggs using four eggs each. Heat one set of scrambled eggs to 140F (60C) and hold for 3.5 minutes.

Heat the other to 160F (71C) without a holding time. Sample each.

### Soft Cooked

Put five large eggs (at refrigerator temp.) in saucepan and add cool tap water to cover by 2 inches. Quickly bring water to simmering (200F, 93C), remove from heat and let eggs stand covered with hot water for 1 minute. Remove one egg and test temperature of yolk. Hold egg in pot holder and insert thermometer through a small opening in the shell and penetrate center of yolk. Keep testing at 1 minute intervals until 140F (60C) is reached. Hold an egg at 140F (60C) for 3.5 minutes and then observe its appearance and taste.

### Soft Meringue on Pie

Put meringue on top of hot pie filling, then bake at 325F (163C) for 25 to 30 minutes. This longer, slower cooking is needed for heat penetration. A nicely browned crust is obtainable by following the old instructions of a few minutes at high temperature, and that recommendation is still printed on product packages and in recipe books although it does not reliably give a safe temperature throughout.

### Hollandaise Sauce

Follow your usual recipe with the exception of heating the butter to 220F (104C) and testing the temperature after combining it with the eggs. The mixture should be at least 160F (71C). If it is between 140 and 160F (60 and 71C), hold it for 3.5 minutes before proceeding to other steps.

### Egg Nog

Original egg nog recipes were made with both raw milk and raw eggs. Safe recipes now recommend both of these ingredients go through a heat treatment.

EGG NOG

6 eggs
½ c sugar
¼ tsp salt (optional)
1 qt milk, divided
1 tsp vanilla

In large saucepan, beat together eggs, sugar and salt. Stir in 2 cups very warm milk. Cook over low heat, stirring constantly until mixture is thick enough to coat a metal spoon. Remove from heat and stir in remaining milk and vanilla. Cover and refrigerate until thoroughly chilled before serving. This recipe is based on one from the American Egg Board, 1460 Renaissance Dr., Park Ridge, IL 60068.

### Pork Roast in Microwave

If available, use a microwave range that has a temperature probe and place it in the center of a pork roast. If using a model without the temperature probe, check temperature of the roast by removing it from the oven, making a hole with a skewer and then inserting a thermometer into the hole.

Cook the roast following directions for your particular range. At the end of the cooking time and stand-by time (if any), test the internal roast temperature. Make temperature readings in the center portion and also in any area shielded from the rays by curved bones.

**1.** If you did not use a thermometer would you be able to assess the doneness accurately by the color of the meat? If your answer is yes, do you feel demonstrations similar to the experiment above are a good way to educate the general public? What are some other ways?

**2.** What organisms are of concern in undercooked pork in the United States?

### Roast Beef

If available, display pictures of beef at various stages of doneness.

Cook your own beef samples to various stages of doneness if photos are not available in your laboratory. Display these with labels of endpoint temperature and common terminology. Generally, rare is cooked to 140–150F (60–65C), medium is 150–160F (65–71C), and well done is 160–170F (71–76C), however local standards vary widely.

**1.** Does the medium-rare picture resemble the medium-rare stage of doneness in your local restaurants?

**2.** When you order beef, what is a good way to describe it so your food is cooked to a safe temperature?

**3.** What organisms are of concern in undercooked beef?

**4.** Have one member of your class contact someone who is teaching prenatal classes or preschool nutrition to the general public. Do they mention undercooked meat consumption in their classes? How accurate is their nutritional information? Why is it especially important to reach these groups with understandable and correct facts? What are some ways of reaching these educators with updated information?

## Foodborne Illness Scenarios

If there are class members that feel they may have had a foodborne illness incident and care to share it, have them describe the foods eaten, the time elapsed before symptoms appeared, the symptoms, and how the illness was handled. What does the class believe are the most likely organisms involved? Could non-foodborne illness also be possible in this case? What was the most likely contaminated food? How could the illness have been prevented? Has the afflicted person taken additional precautions since the illness and if so were they appropriate precautions?

Considering families in the United States, what are some of our lifestyle patterns that make foodborne illness common? How would you educate consumer groups to address these problem areas? If members of your class have had experience working with consumers, have them share some techniques they have found particularly useful. Are some groups especially difficult to reach? Why?

## Dehydrating Proteins Delays Coagulation

The membrane surrounding microorganisms is largely protein. This membrane allows selective entry and exit of gases, water, and nutrients between the organism and the environment. This protein membrane is vital for microorganism survival. Enzymes are also proteins. Heat destroys protein by first denaturing and then coagulating it. In a dehydrated (dried) state, proteins are more resistant to heat stress. Since even slightly dried bacteria are more heat resistant, they require longer processing times.

Look in a canning resource. What is the difference in the heat treatments used for canning fresh salmon and canning smoked salmon? Why? Is the recipe formulated to be safe only when the smoked salmon is dried to a specific endpoint, or is it a general recipe for any smoke or dry treatment?

Can you find recipes for canning other slightly dried foods? Are these safe, recommended procedures? For those that are unsafe, what are your concerns?

### Dehydration-Coagulation Relationship Experiment

Read an angel food cake recipe to familiarize yourself with the ingredients. Egg white proteins provide the main structure. The cakes are done when the protein of the egg whites coagulates. This sets the foam structure. Sugar dehydrates proteins by pulling water away from them. Proteins may also be dehydrated by hot air, as in making beef jerky in a food dryer. Use a cookbook angel food cake recipe or prepare an angel food cake mix according to package directions. Divide into 5 lots. Prepare the lots as follows:

1. One lot should be used as is (control).
2. Add 1 tsp sugar.
3. Add 1 Tb sugar.
4. Add 2 Tb sugar.
5. Add 3 Tb sugar.

Put equal amounts of each batter into custard cups or individual-serving size loaf pans. Bake using the oven temperature recommended on the package. If time is given for small cake, use this as a guide. Test the doneness of the control cake and remove all cakes when it is done. Invert cakes in the pans until cool. Note: Some cakes may not be done and cannot be inverted. An alternative procedure is to bake each cake until it is done and note the time required. Was more heat treatment required to change (coagulate) the protein when the proteins were dehydrated by sugar?

Relate the results of this experiment to changing heat treatments for processing home preserved foods. What would you conclude about changing recipes?

# 3 Drying

## Typical Consumer Questions

Answer the following questions. Give both consumer advice and a technical answer. In a classroom setting, students should take turns discussing the questions aloud.

**1.** *I bought a large wooden dryer at a garage sale. It heats with light bulbs at the bottom and has homemade screens. Since wood is a porous material, I'm wondering if this dryer is as safe to use as the plastic ones?*

**2.** *I made apricot leather in August. I want to serve it with cream cheese for a New Year's Eve party. It is discolored, but the flavor is okay. Will this be safe to eat?*

**3.** *Is it safe for meat jerky to be displayed at checkout counters unrefrigerated? What about all those hands reaching into jars of unwrapped jerky?*

**4.** *My neighbor makes all of her own baby food. Most of it is vegetable combinations which are dried then pulverized in a blender. She claims this is better for her infant than the commercially canned vegetables I feed my baby. Is it?*

**5.** *Our garden produced a lot of excess corn this year. We don't have a freezer, but can borrow a neighbor's food dryer. How do I proceed?*

**6.** *I grew horseradish this summer. In December we will grind it for gourmet gifts. Should it be dried to keep from September to December?*

**7.** *The farmer's market had basil on sale and I bought several bunches. The drying instructions all recommend hanging it, but the times are much longer than in a food dryer. Can I put it in the food dryer?*

**8.** *I received fish jerky as a gift. It looks and smells awful. What do I do with it?*

**9.** *I dried zucchini chips hoping my family would eat them as a low-fat snack. They won't. What do I do with 20 sacks of zucchini chips?*

**10.** *I did not blanch my green beans before drying. What will happen?*

**11.** *I cook for one. Onions often spoil when I try to keep them fresh. Do you recommend chopping and drying them?*

**12.** *I am very concerned about additives in foods my family eats. Can I simply omit the pretreatment steps in drying?*

**13.** *There are wire screens on my food*

*dryer. Is there an easy way to clean them? Are they safe as they age?*

**14.** *My 6-year-old ate about a cup of dried apples. She now has a stomachache. Is it food poisoning?*

**15.** *I don't like my children to eat a lot of sugars, so I try to always have fruit leather available for them to snack on. Their dentist just told me to treat fruit leather as a dessert. Why?*

**16.** *Will a solar food dryer actually work?*

**17.** *I always peel peaches before pureeing for leather, but do I need to peel nectarines?*

## Review of Drying Literature and Principles

**1.** Read the following *Journal of Food Protection* article and be prepared to discuss it in class.

Chung, D,S. and Chang, D.I. 1982. Principles of food dehydration. J. Food Prot. 45: 475.

**2.** Read a journal article of your choice on dehydrated foods; either home or commercial methods. Write a 1 to 2 page report on this article and prepare a 5 minute presentation for class. Turn in a photocopy of the article with your report.

**3.** Photocopy a jerky recipe. Write the citation on the top of the page. Evaluate its safety. Write your comments directly on the recipe and post it in the classroom for all to share.

**4.** For one of the foods in laboratory, calculate the percent moisture lost in the dehydration process. Weigh a sample amount before and after drying. Use a composition-of-foods book to find an estimate of the initial water levels in the food. Post these results in class for all to share. If use of a hygrometer is available, make some $a_w$ measurements.

# PRACTICAL EXPERIENCES

## Pretreatment with Antibrowning Dips

For this experiment, light-colored fruits—green grapes, bananas, and apples—will be used. Prepare and label the various dip treatments.

| Treatments | Time in dip |
|---|---|
| 1. Ascorbic acid 8.9 g/L | 5 min |
| 2. Ascorbic acid 44.9 g/L[a] | 5 min |
| 3. Fruit Fresh[b] (conc. rec. on pkg.) | 5 min |
| 4. NaCl 15.3 g/L | 10 min |
| 5. NaCl 7.6 g/L | 10 min |
| 6. $Na_2S_2O_5$ 7.6 g/L | 15 min |
| 7. $Na_2S_2O_5$ 15.2 g/L | 15 min |
| 8. Untreated (no water immersion) | 0 min |
| 9. Water | 10 min |

[a]This is the concentration recommended for apples.
[b]Or other ascorbic acid commercial dip. Calculate the concentration of ascorbic acid from package information if possible.

**GRAPES.** Divide Thompson seedless green grapes into eight even amounts. Slice one of the eight groups into halves using a sharp knife. Immediately put them into a dip and start timing. Repeat for the remaining grapes. At the end of the specified time period, remove the grapes and place on labeled drying trays. When all the grapes are prepared, place in dryer and start timing length of the drying period. Do not rinse after dipping for this experiment.

**BANANAS.** Use one whole banana for each of the eight treatments as the blossom end contains more sugars than the stem end which may affect browning. Peel one banana at a time, slice evenly using a slicer set to approximately 0.8 cm, quickly immerse slices in a dip, and place on drying tray. When all the bananas are ready, put tray into dryer and start timing the drying period. Do not rinse the slices after dipping for this experiment.

**APPLES.** Select eight apples of the best quality available. Gravenstein, Granny Smith, Newton, and Jonathans are good varieties for drying. During summer months, if the commercially stored apples are exhibiting storage disorders (internal browning), then fresh Yellow Transparents may need to be used. Because of their unusual taste and texture, sample the undried apple before evaluating dried Yellow Transparents. If the Cortland variety is available, run a small sample size of it along with the experiment apples. Cortland apples have very little, if any, enzymatic browning due to a lack of, or low levels of, these enzymes. Red-fleshed apples, such as the cultivars "Pink Pearl" and "Airle Red," are available to backyard orchardists. If possible, dry a small sample of this type too.

Work quickly and slice each peeled, cored apple into eight vertical wedges. Do not slice into rings for this experiment as the sugar concentration will vary in different parts of the apple. Combine one wedge from each apple for each treatment. Mixing of the slices will give more accurate results in this experiment as the sugar content of apples will also vary with the position of the fruit on the tree and among trees. Dip each lot of slices as directed, place on trays, and time the drying period. Do not rinse slices after dipping for this experiment.

Evaluate the dried foods by their odor, color, texture, and unless you are allergic to sulfites, by their flavor. Color is especially emphasized here due to the significance of Maillard browning in dried foods.

**1.** Which apple treatment do you believe consumers will prefer overall after 4 months storage?

**2.** If you were giving directions over the telephone for pretreatment, would you be able to accurately convey all methods to someone inexperienced in the kitchen?

**3.** How would fresh fruit of varying sugar content affect this experiment? Do you feel representative samples were used for all treatments?

**4.** Compare the stored, dried samples made in class to their purchased counterparts. List positive and negative points for each. Try to include dried banana slices sold in the bulk section of grocery stores; they are often sprayed with coconut oil to prevent sticking and then stored for various times. Longer times may result in rancidity.

**5.** Why were the grapes cut in half before drying?

**6.** Rehydrate the apple slices and use in a pie or cobbler. Evaluate the product. NOTE: include cost comparisons for in-season and off-season apples in your evaluation.

## Blanching

Temperatures near 212F (100C) inactivate enzymes within 1 minute of exposure under moist conditions. Enzymes are fairly heat stable to dry heat. They can withstand exposures near 400F (204C) if the heating and the food are dry. Steam and water blanching are often used in households to inactivate enzymes before a food is dried. The enzymes, catalase and peroxidase, are generally used as indicators of residual enzyme activity in foods. Catalase is less resistant to heat than is peroxidase.

Select good quality apricots. Divide into 3 lots for the blanching experiments.

### Syrup Blanching

Combine and bring to a boil 1 cup white sugar, 1 cup white corn syrup, and 2 cups water. Cut apricots in half and remove pits. Hold most of the product in a weak sulfite solution (1 Tb sodium bisulfite per gal of water) until all are sliced and pitted. Hold a sample in water without sulfite. Keep the treated and untreated lots separate. Add prepared apricots to the boiling syrup. Simmer 10 minutes then remove pan from burner and leave the fruit in the hot syrup for 30 minutes. Pour all into a colander to drain the syrup from the fruit, then rinse the fruit lightly with cold tap water from the sink's sprayer attachment. Place the apricot halves on dryer trays and dry at 160F (70C) for 2 hours, then at 130F (55C) until dry. Apricots will be pliable when dry.

### Steam Blanching

Bring water in the lower part of a steamer pan to a boil. Cut apricots in half and remove pits. Hold most of them in a weak sulfite

Pretreatment with Antibrowning Dips

| Treatment | Color | Drying Time | Comments | Color after Storage |
|---|---|---|---|---|
| Grapes | | | | |
| 1. | | | | |
| 2. | | | | |
| 3. | | | | |
| 4. | | | | |
| 5. | | | | |
| 6. | | | | |
| 7. | | | | |
| 8. | | | | |
| Bananas | | | | |
| 1. | | | | |
| 2. | | | | |
| 3. | | | | |
| 4. | | | | |
| 5. | | | | |
| 6. | | | | |
| 7. | | | | |
| 8. | | | | |
| Apples | | | | |
| 1. | | | | |
| 2. | | | | |
| 3. | | | | |
| 4. | | | | |
| 5. | | | | |
| 6. | | | | |
| 7. | | | | |
| 8. | | | | |

solution until all are sliced and pitted. Hold a sample in water without sulfite. Keep the treated and untreated lots separate. Place the prepared apricots in the top section of the steamer and put the lid on. Steam for approximately 3 minutes or until heated through—remove an apricot and cut it to test the center. Place the steamed fruit on drying trays and dry at 160F (70C) for 2 hours, then at 130F (55C) until dry.

### Control

Cut the apricots in half and remove pit. Hold most of them in a weak sulfite solution until all are prepared. Hold a sample in plain water. Keep the lots separate. Then place on a drying tray and dry at 160F (70C) for 2 hours, then at 130F (55C) until dry.

Evaluate all lots after storage. Note flavor, color, odor, and texture. Which method would you recommend to a householder drying in July for December gifts-from-the-kitchen?

## Fruit Leather

### Peach Leather

Scald peaches in boiling water 2 minutes and then dip in cold water to remove skins. Cut the fruits in half to remove the pits and puree in a blender. If fresh peaches are unavailable, canned or frozen fruits may be used but with poorer quality. Heat in saucepan to almost boiling. If you wish to add spices; do so sparingly. As the leather dries, the spice will be more concentrated because it does not evaporate. A recommended amount is 1/4 teaspoon per quart of puree. Cool in a cold waterbath to room temperature and place on drying trays which have been covered with plastic or place on specially designed trays for leather. Make sure the leather is spread evenly on the tray for uniform drying. Thicknesses of 1/8 inch to 3/8 inch are common. If a leather is poured thin (1/8 in.), the fruit must be very pulpy or when the water evaporates, holes will appear. Leave at least 1-inch border to allow for spreading. At this point, garnishes such as chopped nuts or sunflower seeds may be added if storage will be short (rancidity). Dry leathers at temperatures under 140F (60C) until they feel leatherlike and pliable without sticky spots. Remove the leather from the tray while it is still warm, roll it, and wrap in plastic wrap. Place the wrapped leathers in an airtight container, label, and store in a cool dry place.

**1.** Why, when making blackberry or raspberry leather, are the seeds removed?

**2.** If you wish to serve the peach leather spread with cream cheese and then sprinkled with nuts, when is the best time to add these toppings? Why?

**3.** A householder with preschool children finds the leather is too tart for the family. What advice would you give this person?

**4.** Explain in scientific terms any quality defects in your peach leather.

## Blanched Vegetables

Divide fresh green beans into 3 equal lots. Slice them into 1-inch segments with diagonal cuts to increase the surface area for more rapid dehydration.

### Steam Blanched with Sulfite

Put a measured amount of water into the bottom of a steamer. Dissolve 4 teaspoons of sodium metabisulfite per cup of water and bring to a boil. Place one-third of the beans in the top part of the steamer and steam with the lid on for 5 minutes or until heated through—slice a sample bean to check internal temperature. Place on drying tray and dry at 140F (60C) for 2 hours then at 130F (55C) until dry.

### Steam Blanched

Repeat as for the steam blanched with sulfite except use plain tap water in the lower part of the steamer.

### Untreated

Place one-third of the beans directly on the drying trays. Dry at 140F (60C) for 2 hours then

| Treatment | Color: Immed. | Color: After Storage | Rehyd. after Storage: Flavor | Rehyd. after Storage: Color | Rehyd. after Storage: Texture |
|---|---|---|---|---|---|
| Steam, sulfite | | | | | |
| Steam | | | | | |
| Untreated | | | | | |

at 130F (55C) until dry.

After storage, rehydrate the beans. Cook them in the simmering rehydration water, keeping the 3 lots separate. Combine each of the cooked bean lots with a seasoned white sauce or condensed cream of mushroom soup and evaluate and record in the chart above.

A new product for vacuum packing dried foods (and other items) in plastic, marketed as Foodsaver, is now available to householders. If a householder asked you as a consumer education specialist for helpful hints on storing home dried green beans and sunflower seeds purchased from a bulk foods store by this process, what advice would you give?

## ■ Herbs

Some of the volatile odor and flavor compounds may be lost during drying, but drying generally tends to concentrate the flavors as the fresh herbs lose approximately 80% of their weight.

Select a leafy herb with long stems such as sweet basil, parsley, cilantro, or mint. Tie the stem end of a small bunch together. Place the bunch upside down in a small paper bag. Tie another string around the opening of the bag, securing the herb stems also. Cut 1/2-inch holes in the bag for air circulation. The bag is optional, but it catches the leaves that fall during drying. Hang the bag from the ceiling in a warm dry area. Most herbs will dry in a week. They are very brittle and crumble easily when thoroughly dry. Rolling the bag gently between your hands and a table will remove most of the leaves from the main stems which can then be removed. The small pieces of stem should then be picked out as they contain only a fraction of the flavor found in the leafy parts. Package in plastic bags and store in a cool dark place.

## ■ Jerky

Use lean cuts of beef or large game—venison, elk, antelope, mountain sheep, moose. Remove tendons and muscle sheaths leaving only tender muscle tissue. All visible fat should also be removed. Game should be frozen for at least 1 month before drying to eliminate any parasites. Higher myoglobin levels in game meat than in beef result in a darker color jerky.

Slice the meat with the grain 1/4 inch thick, 1 inch wide, and in strips up to 12 inches long. A butcher can do this uniformly with an electric slicer. Slicing with a knife is easier if the meat is partially frozen. There are many recipes for flavoring the jerky. Brine curing, also called pickle cured, involves soaking the meat in a salt, water, and seasoning brine for approximately 12 hours before drying or smoking. In dry curing, a mixture of salt and seasonings is applied to the meat surface and then refrigerated for 24 hours before drying or smoking.

### Dry Cured, Oven Dried, Unsmoked Jerky

Prepare 2 pounds of meat slices. Sprinkle with 4 teaspoons salt, 3/4 teaspoon pepper, 3 teaspoons sugar, and other spices if desired. One teaspoon of chili powder, garlic powder, onion powder, or coriander are some commonly used. Refrigerate in a glass dish for 24 hours. Spread the meat out on cake drying racks. Place these racks on top of oven racks for drying. Raw meats should first go through a precooking period at 145F (63C) to kill bacteria—salmonellae—early in the drying process. Place a thermometer on the rack as oven controls often are not accurate below 250F (121C). When the meat tests "medium" doneness then reduce heat to 120F (49C) for the remaining drying process. Lining the bottom of the oven with foil aids in cleaning. Leave the oven door ajar for moisture to escape as the meat dries. Dry 48 hours or until done, occasionally blotting the meat with paper towels to absorb beads of oil on the surface. Test for doneness with a piece that has cooled, as jerky is more pliable when warm. A well-dried jerky cracks when bent, but does not break completely through. Place the cooled, dried jerky in plastic bags and store it at refrigerator or freezer temperatures.

## Plant Pigment Experiment

### Anthocyanin (red)

Puree fresh or thawed frozen blackberries, raspberries, loganberries, or marionberries. Sieve to remove seeds. Add spices if desired. Spread evenly on a leather tray and dry at a maximum temperature of 140F (60C) until dry.

### Anthocyanin (blue)

Wash fresh blueberries in tap water and remove stems. Place on drying tray and dry at 150F (65C) for 2 hours and then at 130F (55C) until dry. If fresh blueberries are unavailable, use the frozen in a leather.

### Betalain

Select small beets. Detroit is a good drying cultivar that is readily available. Wash and remove tops. Place in top of steamer above boiling tap water and steam until just tender. Peel and slice crosswise into 1/8-inch slices. Cut each slice into uniform strips a maximum of 1/2 inch wide. Place on drying trays. Dry at 160F (70C) for 1 hour and then at 130F (55C) until dry.

### Chlorophyll

Select good-quality green peppers. Slice into 1/4-inch strips or 1-inch dice. Dry at 140F (60C) for 2 hours then at 130F (55C) until dry.

### Lycopene

Slice tomatoes crosswise 1/4 inch thick and place on drying trays or prepare a tomato puree by blending fresh tomatoes or using canned tomato sauce or paste. Season with oregano or basil. Spread evenly on leather trays or plastic film, and dry at a maximum temperature of 140F (60C) until dry. Dry sliced tomatoes at 150F (65C) for 2 hours then at 130F (55C) until dry.

### Carotene

Wash and peel carrots. Slice 1/4 inch thick. Divide carrot slices into three groups. Place one group directly on drying trays. Steam blanch one of the other groups for 3 minutes over a sodium bisulfite solution (1 tsp $Na_2S_2O_5$/1 c water), then place on drying trays. Steam blanch the third group for 3 minutes over tap water before placing on drying tray. Dry at 140F (60C) for 2 hours, then at 130F (55C) until dry.

Prepare a recipe using each of the dried foods in this experiment. Present the prepared dish alongside the dried food for comparison. Some recipe suggestions are tomato leather spread with cream cheese, berry leather spread with marshmallow cream and nuts, blueberries in a quick bread, beet soup, bean dip or scrambled eggs with peppers, and carrots in a soup.

Plant Pigment Experiment

| Pigment | Dried Color | Product Acceptability | |
|---|---|---|---|
| | | As Is | In Recipe |
| Anthocyanin (red) | | | |
| Anthocyanin (blue) | | | |
| Betalain | | | |
| Chlorophyll | | | |
| Lycopene | | | |
| Carotene (untreated) | | | |
| Carotene (w/sulfite) | | | |
| Carotene (water blanch) | | | |

# 4 Quick Pickles

## Typical Consumer Questions

Give both a practical householder answer and a technical answer to the following questions.

**1.** *It is not possible for any bacteria to grow in my dill pickles, but the lady running the local food drive refused to accept any home canned food. Which one of us is wrong?*

**2.** *This year's batch of dill cucumber pickles is shriveled and some have large hollow spaces inside. What happened?*

**3.** *The USDA recipe for dilled green beans is too spicy for my family. Can I leave out the peppers?*

**4.** *For generations my family has made top quality pickles without using a boiling-water canner. Why have recommendations changed?*

**5.** *I want to make pickled cocktail onions and cannot find a recipe for them. Can I use a recipe I have for pickled peppers?*

**6.** *Pickles give me heartburn. How much vinegar can I leave out of the recipe and still preserve a safe product?*

**7.** *My neighbor said I should serve my 3-year-old only sweet pickles because the acid deteriorates the stomach lining in small children. I haven't noticed any problems and my child likes dills. Should I eliminate them from her diet?*

**8.** *Pickling cucumbers are a good price in the supermarket now, but my recipe advises use of only freshly picked cucumbers. What will happen if I pickle the cucumbers in the store?*

**9.** *How do I make the crispest cucumber dills possible?*

**10.** *Can I use my favorite brine to pickle jars of mixed vegetables?*

**11.** *I pickled ears of baby corn last summer. Now, in December, when I opened them, they do not smell sour at all and I'm afraid to taste them. What do I do?*

**12.** *My aunt adds a few carrot slices to her pickled green beans for color. Is this safe?*

**13.** *I have a jar collection of unusual shapes and I would like to use some of these for pickled gifts. How do I proceed?*

**14.** *I followed the recipe from the Extension Service for pickled onion slices. After 3-months storage, the onions are a tan color. What happened? Are they safe?*

**15.** *My husband was given a jar of pickled apples. How do I know if they are safe to serve my family?*

**16.** *In the gourmet section of our supermarket, a mixture of pickled vegetables is for sale. The jar has a label that does not look professionally drawn. How do I know these are safe?*

## Survey of Pickling Literature

**1.** Read the most recent pamphlet on pickling available from your local Extension Service. Is this a USDA publication, or written by the local office staff?

**2.** Skim through a book on pickling, or a section of a home food preservation book on pickling, or the pickling chapter of a cookbook. What is the author's background? Is this book published by a product manufacturer? What is the copyright date? How do the recipes, clarity of instructions, processing methods, and cost compare with those distributed by the Extension Service?

**3.** Read the following *Journal of Food Science* reference. Be prepared to discuss this article in class.

Heaton, E.K., Shewfelt, A.L. and Henderson, L. 1978. Effects of varying levels of sucrose, corn syrup solids and vinegar on the quality of sweet pickled peaches. J. Food Sci. 43:1015.

**4.** Talk with householders of various ages about pickling. What products are popular? How do they determine which recipe to use? Has their household's pickling changed in the last 15 years? How do they rate their home pickled products with those available commercially? Do they feel confident they are preparing a safe-to-eat food?

**5.** After talking with householders, how do you rate the safety of home prepared quick pickles?

**6.** Select a quick pickle recipe from a popular press source or an old family favorite. Photocopy the recipe, then briefly (one–two phrases) write the scientific principle for each recommended procedure directly on the photocopy. Revise if it is not already a safe recipe. Post the recipe and your reasonings in the laboratory room for all to share.

# PRACTICAL EXPERIENCES

## Types of Acidifiers and Their Characteristics

Sample the following solutions by dipping a piece of lettuce into the bowl and then eating it. Standardize the pH meter before testing each different sample. Fill out the chart below.

## Different Vinegars in Recipes

Prepare the Dilled Onion Slices recipe two separate times, using white vinegar and cider vinegar. Evaluate the color immediately after processing and again after a minimum of 3 weeks storage.

### Dilled Onion Slices

½ c sugar
2 tsp salt
½ tsp dill weed
½ c white vinegar
¼ c water
2 large white onions, sliced thinly

Boil all ingredients except the onions for 1 minute. Cool this solution by placing the saucepan in a sink or pan of cold water. Pack the onions in a canning jar. Pour room temperature pickling solution over the onions leaving ½-inch headspace. Process in boiling-water canner for 10 minutes.

**1.** What recommendation would you give the consumer who wished to make this recipe so it tasted less sour for her children's preferences?

**2.** What is the scientific principle behind cooling the hot brine surrounded by cold water instead of room temperature or cold air?

| Acidifier | pH | Flavor | Odor | Color | Cost | Best Appl. |
|---|---|---|---|---|---|---|
| White vinegar | | | | | | |
| Cider vinegar | | | | | | |
| Tarragon vinegar | | | | | | |
| Rice vinegar | | | | | | |
| Red wine vinegar | | | | | | |
| White wine vinegar | | | | | | |
| Lemon juice (fresh) | | | | | | |
| Lemon juice (bottled) | | | | | | |
| Citric acid soln. (5%)[a] | | | | | | |

[a]Make a 5% citric acid solution by combining 1 Tb crushed citric acid with 1 c water.

## Types of Sweeteners and Their Characteristics

Mix the following sweetening agents with 1/4 cup water and 1/2 cup white vinegar. Bring to a boil then simmer 5 minutes. Cool by placing each labeled container in cold water. Sample by dipping a piece of lettuce in each solution and then eating it. Use handheld refractometer to test degrees Brix. Fill out the chart. Use the amount of artificial sweetener that is equal to 1/2 cup sugar according to the manufacturer's package directions.

**1.** A popular trend in some communities is substituting honey for refined sugar. What recommendation would you give a person wanting this change in their pickles?

**2.** How would you advise the home pickler who ran out of white sugar while preparing Bread and Butter pickles and lives 30 minutes from the nearest grocery store?

| | | After 3 Weeks | | | |
|---|---|---|---|---|---|
| Vinegar | Immediate Color | Color | Flavor | Odor | Comments |
| White | | | | | |
| Cider | | | | | |

| Sweetener Applications | Brix | Characteristics | | | Cost |
|---|---|---|---|---|---|
| | | Flavor | Odor | Color | |
| ½ c (3.42 oz) white sugar | | | | | |
| ½ c (3.42 oz) lt. brown sugar | | | | | |
| ½ c (3.42 oz) dk. brown sugar | | | | | |
| ¾ c + 1 Tb (3.42 oz) powdered sugar | | | | | |
| ¼ c honey + ¼ c white sugar | | | | | |
| ½ c honey | | | | | |
| 2 Tb corn syrup + 6 Tb white sugar | | | | | |
| Artificial sweetener Brand: Active ingredient: | | | | | |
| Artificial sweetener Brand: Active ingredient: | | | | | |

## Types of Salts and Their Characteristics

Prepare Dilled Onion Slices 3 ways; with iodized table salt, with noniodized table salt, and with pickling salt that does not contain anticaking additives such as silicoaluminate. Evaluate the color and clarity of liquid immediately after processing and the other characteristics after at least 3 weeks.

## Effect of Hard Water upon Pickles

Many local water supplies contain large enough amounts of iron to discolor pickled products. Prepare dilled onion slices using tap water from a variety of sources. Evaluate color immediately after processing and the other characteristics after at least 3 weeks.

List any information you have about the type of water from the various locations. How do the local people use it for pickling?

Do the local householders use this water for drinking and laundry?

## Crisping Agents

Three separate batches will be prepared for the 5-hour soaking process (2 Tb salt and 2 c water) to compare differences. Then one batch of seasoned brine will be prepared to pour over all three saucepans of brined, rinsed cucumbers.

*Treatment One.* Using the Short-Brine Chunk Pickle recipe, substitute 2 teaspoons food-grade calcium carbonate plus 4 teaspoons pickling salt for the 2 tablespoons salt in the soaking brine recipe.

*Treatment Two.* Using the Short-Brine Chunk Pickle recipe, substitute 2 teaspoons food-grade calcium chloride plus 4 teaspoons pickling salt for the 2 tablespoons salt in the soaking brine recipe.

*Treatment Three, Control.* Prepare the Short-Brine Chunk Pickle recipe as written, using 2 tablespoons pickling salt and 2 cups water for the soaking brine.

Evaluate the texture. What pickle recipes would benefit from calcium addition?

**SHORT-BRINE CHUNK PICKLE**

*Five-Hour Soak*
6 cucumbers, about 3 in. long
2 Tb pickling salt
2 c water
*Seasoned Brine*
½ c + 2 Tb sugar
12 c vinegar
1 Tb mustard seed
¾ tsp celery seed
¼ tsp curry powder

Wash cucumbers. Remove any remaining blossoms as the enzymes they contain may affect pickle texture. Cut cucumbers into 1-inch chunks. Combine the salt and water, stirring to dissolve. Pour this brine over the cucumbers in a nonmetallic container. Cover and let stand at room temperature for 5 hours. Drain cucumbers and rinse thoroughly, then place in a saucepan.

| | | After 3 Weeks | | | |
|---|---|---|---|---|---|
| Salt Type | Immediate Appearance | Color | Flavor | Odor | Texture |
| Iodized | | | | | |
| Noniodized | | | | | |
| Pickling | | | | | |

Comments:

| Water Type, Texture | Immediate Color | After 3 Weeks | | |
|---|---|---|---|---|
| | | Color | Flavor | Odor |
| Distilled water | | | | |
| Lab tap water | | | | |
| Local well water | | | | |
| Other water | | | | |
| Other water | | | | |

In a separate saucepan, bring remaining ingredients to a boil. Stir and heat until sugar is dissolved. Divide this seasoned brine evenly and pour over the cucumber chunks, then heat each lot to a boil again. Fill two pint jars for each treatment to within ¼ inch of the top. Process in boiling-water canner for 10 minutes (altitude 0–1,000 ft). When jars are cool, attach a label to the side of each jar.

**WINTER SQUASH PICKLE**

(To be substituted for cucumber recipe in crisping agent experiment if top quality cucumbers are not available. If neither cucumbers nor winter squash are available, substitute zucchini in the pickled squash recipe.)

1 c prepared winter squash
⅔ c vinegar
⅔ c sugar
1 small cinnamon stick

Remove skin and seeds from squash. Cut the flesh into 1-inch cubes. Put the cubes in the top of a steamer and steam until just tender. Allow to drain. Simmer vinegar, sugar, and cinnamon for 15 minutes. Add the squash and continue to simmer for 3 minutes. Set aside for 24 hours, covered. Reheat and simmer for 5 more minutes. Remove cinnamon. Pack hot squash into a half pint jar and then pour in liquid leaving ½-inch head space. Process in a boiling-water canner for 10 minutes (altitude 0–1,000 ft). Label jars when cool.

## Brine Strength

A popular trend now is to decrease the salt content of foods. In pickling, this can have a great affect on the final product. Prepare the Short-Brine Dill Pickle recipe using the 5% soaking brine as written and also prepare it with the increased and decreased amounts of salt. Change only one brine at a time. As written, nine different batches would be prepared. For smaller groups of students, omit one of the low-salt and the no-salt batches. This results in only five short-brine pickling experiences.

**1.** Which treatments do you think would be acceptable to those not on a physician recommended salt-restricted diet? Note: the texture of the cucumbers is directly related to salt concentrations, however the acceptability

| Crisping Agent | Texture | Comments |
|---|---|---|
| Calicum carbonate | | |
| Calcium chloride | | |
| Control | | |

| Treatment | Texture | Comments |
|---|---|---|
| Control | | |
| Add 2 tsp calcium carbonate | | |
| Add 2 tsp calcium chloride | | |

of flavor with different salt concentrations will vary depending upon the seasonings in the recipe.

**2.** What conclusions would you draw from this experiment about the role of sodium chloride in quick pickles?

**3.** Survey your local market shelves. Are low-salt pickles made simply without sodium chloride or with the addition of salt substitutes?

**SHORT-BRINE DILL PICKLE**

*Overnight brine*

4½ lb cucumbers, 3-5 in. long
2 qt water
3 Tb salt

*Pickling brine*

3 c vinegar
3 Tb salt
1 Tb sugar
2½ c water

| Salt Amount | Salometer Reading | Texture | Appearance | Comments |
|---|---|---|---|---|
| Overnight brine | | | | |
| 6 Tb | | | | |
| 3 Tb (5% control solution) | | | | |
| 2 Tb | | | | |
| 1 Tb | | | | |
| no salt | | | | |
| Processing brine | | | | |
| 6 Tb | | | | |
| 2 Tb | | | | |
| 1 Tb | | | | |
| no salt | | | | |

1½ tsp mixed pickling spice, tied in cheesecloth
½ tsp whole mustard seed
1 clove garlic
3 heads dill

Wash cucumbers, scrubbing with a vegetable brush. Be sure to remove any blossoms. Combine the 2 quarts of water and the 3 tablespoons of salt. Stir to dissolve salt. Cover whole cucumbers with the brine in a nonmetallic container. Let stand overnight at room temperature then drain and discard the overnight brine.

To make the pickling brine, combine the vinegar, salt, sugar, water, and bag of pickling spices. Heat to boiling. Put cucumbers into a quart jar. Add mustard seed, whole garlic clove, and the dill heads. Cover the cucumbers with the boiling hot pickling brine. Leave ½-inch headspace. Process in boiling-water canner for 20 minutes. When cool, attach label to side of glass.

## Anthocyanins and Acid

A frequently heard complaint about pickled pears and cauliflower is a product with a pink tinge. The degree of pinkness depends upon the concentration of anthocyanins in the fruit or vegetable, the pH, and heat. Prepare the Pickled Pears recipe increasing the acidity and also prepare one lot by wrapping the jars in a towel so they cool slowly. This is a total of three separately prepared recipes. If fresh small pears are unavailable, use previously canned pears. The anthocyanin reaction will follow the same trends.

**PICKLED PEARS**
*(makes approx. 2 pints)*

2 c sugar
1 c white vinegar
½ c water
2 cinnamon sticks
1½ tsp cloves
1½ tsp allspice
2 lb Seckel* pears

Tie the spices in cheesecloth. Combine all ingredients except the pears. Bring to a boil and simmer 30 minutes. Peel the pears and immediately put them in an antioxidant solution—make according to package directions. When all are peeled, put the pears into the simmering syrup and continue simmering 20 to 25 minutes. Pack the pears into two pint jars. Put a piece of cinnamon stick in each jar. Remove bag of cloves and allspice. Pour syrup over pears leaving ½-inch headspace. Process in boiling-water canner for 20 minutes. When cool, label.

*Any pear will show pinkness, but Seckel cultivar is the best pickling pear.

| Treatment | pH | Color |
|---|---|---|
| 1 c vinegar (control) | | |
| 1½ c vinegar | | |
| 1 c vinegar cooled slowly | | |

# 5 Fermentation

## Typical Consumer Questions

Give both a scientific explanation and instructions consumers can follow. If you need more information to make a decision, state the facts you assume.

**1.** *I want to purchase fermented pickles that taste like the ones my mother used to make. What do I look for on the label?*

**2.** *There is a booth at the farmer's market that sells fermented pickles. Are these safe?*

**3.** *I want to make sauerkraut for a person on a low-salt diet. How do I proceed?*

**4.** *Why does kimchi smell so different from sauerkraut? They both seem to be a cabbage product.*

**5.** *I decreased the salt in a 7-day pickle recipe. What will happen?*

**6.** *There is not much brine in my finished sauerkraut. Is it safe to can anyway?*

**7.** *The sauerkraut I made 6 months ago has turned brown. Why?*

**8.** *How can I make dill pickles that are as crisp as the commercial types?*

**9.** *The little gherkin cucumbers in my garden this year aren't so little. Can I use them in the USDA sweet gherkin recipe anyway?*

**10.** *How do I know the plastic pail I ferment in does not contain contaminants from recycled plastics?*

**11.** *Can I use my enameled boiling-water canner for a pickle crock?*

**12.** *My pickle brine is cloudy and the pickle jars in the store all contain clear liquid. Must my pickles be discarded?*

**13.** *Why do my pickles taste bland?*

**14.** *Can I ferment lemon cucumbers? They are the only kind I am able to grow at home.*

**15.** *I want to make fermented, black olives and cannot find a USDA recipe. How do I proceed?*

**16.** *Can I add some garlic and hot peppers to a USDA fermentation recipe?*

**17.** *What should I do with 3 quarts of sauerkraut that didn't seal?*

**18.** *I haven't been able to grow cucumbers so that they all ripen at the same time. Some of the ones I am fermenting were past optimum ripeness and some were a little underripe. How will this batch of pickles turn out?*

**19.** *I would like to put something pretty such as a flower or sprig of herbs in the pickle jar before I seal it. What do you suggest?*

**20.** *I found an old pickle crock in a friend's barn. It is very dirty. Can it ever be used for food again?*

## Survey of Fermentation Literature

Read one of the following articles, and be prepared to discuss it in class.

Etchells, J.L., Bell, T.A., Fleming, H.P., Kelling, R.E. and Thompson, R.L. 1973. Suggested procedure for the controlled fermentation of commercially brined pickling cucumbers—the use of starter cultures and reduction of carbon dioxide accumulation. Pickle Pak Sci. 3:4.

Lampi, R.A., Esselen, W.B., Thomson, C.L. and Anderson, E.E. 1958. Changes in pectic substances of four varieties of pickling cucumbers during fermentation and softening. Food Res. 23:351.

Read a scientific journal article of your choice on brined vegetables. If you are having trouble finding one, check chapter references. Write a 1–2 page report to be turned in and share a summary orally with the class.

Read the section on brined pickles in a consumer publication. Select a recipe designed for household brining of vegetables and break it down into the various steps. What is the scientific basis behind each of these procedures? A text on basic food microbiology may be a helpful resource for this assignment.

# PRACTICAL EXPERIENCES

## Measuring and Changing Salt Solutions

A salometer is used to measure brine density. It has a scale of 0 to 100. Each SAL represents 0.26% salt by weight. A fully saturated brine is 26.4% salt.

To measure brine strength, pour a sample into a 500-mL graduated cylinder. Put the salometer into the cylinder with the wide end down and the narrow end with numbers up. The salometer will float. The brine level will correspond to a number on the narrow end. This number is the brine strength.

Salometers are designed to be read at 60F (15.5C). For every 10F (4.7C) above 60F (15.5C), one degree salometer should be added as a correction factor. For every 10F (4.7C) below 60F (15.5C), one degree salometer should be subtracted as a correction factor.

| SAL | % NaCl by Weight | # NaCl Pounds per Gal of Water |
|---|---|---|
| 4 | 1.1 | .089 |
| 6 | 1.6 | .134 |
| 8 | 2.1 | .179 |
| 10 | 2.6 | .226 |
| 12 | 3.2 | .273 |
| 14 | 3.7 | .320 |
| 16 | 4.2 | .367 |
| 18 | 4.8 | .415 |
| 20 | 5.3 | .468 |
| 22 | 5.8 | .512 |
| 38 | 10.0 | .982 |
| 58 | 15.3 | 1.505 |
| 76 | 20.1 | 2.019 |

**EXAMPLE:** Cucumber fermentation was started in a low-salt brine (2 1/2%). The householder then believed the product was going to be of inferior quality and wished to raise the brine strength to 5%. The following procedure is recommended.

**1.** Estimate the amount of brine. This is most accurately done by measuring it. We will state that it is 3 quarts (3/4 gal) for this example.

**2.** The brine currently contains .226 pounds NaCl/gallon or .170 pounds NaCl/3 quarts.

$$3(.226 \div 4) = .170$$

The brine will contain .464 pounds NaCl/gallon or .348 pounds/3 quarts.

$$3(.464 \div 4) = .348$$

**3.** The difference is .178 pounds.

$$.348 - .170 = .178$$

This is the amount of salt that needs to be added to the brine to change it from 2 1/2 % to 5%.

Practice using a salometer and calculating changes in brine strength.

## Sauerkraut

Approximately 3.5 pounds of cabbage will make 1 quart of kraut. Ideally, two jars of each treatment should be prepared to compensate for product failures. Prepare four jars of the 2.5% salt lot and store two jars in the dark and two jars in sunlight. This can be easily done by wrapping a heavy paper around one set of jars.

Remove the outer leaves from the cabbage if it has not already been cleaned, also remove any bruised leaves. Rinse off any visible sand or dirt. Cut the head into quarters and core it. Shred the leaves finely—about the thickness of a dime, using a food processor or sharp knife. For this experiment, more accurate results will be obtained if all the cabbage is shredded mechanically and then mixed so variation among cabbage heads will not be a factor.

Each lot of cabbage should be weighed and the specified percentage of salt be added (for 50 lb of cabbage, 2.5 lb of added salt will be a 5% treatment; for 2 lb, 1.5 oz salt is 5%). Mix the salt among the shreds well, being careful not to bruise them as that releases softening enzymes. Let the salted cabbage shreds stand for several minutes. Some liquid will be drawn out and the shreds will wilt. This reduces bruising when they are packed into jars. Pack the shreds and liquid firmly into a glass jar or other suitable container using your hand or a wooden spoon. Press it into the jar until liquid rises to the surface to cover the shreds. Enough liquid may not appear in the low-salt types to cover. Check the following day and press so that liquid covers the cabbage.

Place a brine- or water-filled plastic food storage bag on top to enhance anaerobic conditions. You may need to smooth out creases in the plastic with a table knife. Store the jars at room temperature. Sample as the chart directs using sterile pipets. The 5% jars should complete fermentation in about 3 weeks.

To process sauerkraut, slowly heat it just to boiling, then pack the shreds into a hot clean jar. Cover with the hot juice. Leave 1/2-inch headspace. Process by boiling in a boiling-water canner for 10 minutes for pints, 15 minutes for quarts.

### Assignment

Prepare a graph of the pH of the product at 1, 2, 3, 5, 7, 14, and 21 days of fermentation. Post the graph on the wall for other students to share. Write a brief description of the microscopic examination at 7, 14, and 21 days. Prepare gram stains and do total counts with pour plate agar. Plates are incubated at 72F (22C) to be evaluated at 24 and 48 hours. Note: *Leuconostoc mesenteroides* is a gram positive coccus, *Lactobacillus* is a gram positive rod. Figure 5.2 in Book 1 is an example of flora that are common with varying salt concentrations. Use this graph as a guide when observing changes in the different lots of kraut.

If possible, analyze a sample of sauerkraut for vitamin C content. How does this content compare with that of other foods? In June of 1863 when Southern troops captured the town of Chambersburg, Pennsylvania, during the Civil War, 25 barrels of sauerkraut were demanded from the community. General Harmon, leader of these Confederate troops, wanted the kraut to combat the scurvy his men were suffering from. How would you evaluate his request?

**1.** Why should the cabbage and salt be weighed?

**2.** What did you observe as soon as the salt was mixed with the cabbage? Explain.

**3.** Why is the shredded cabbage packed to exclude air?

**4.** What are the functions of the salt? Under what conditions might 2% salt be used? When is 3% appropriate?

**5.** What substance in the cabbage is fermented?

**6.** Name the microorganisms which participate in the fermentation and the functions of each.

| Types Prepared | pH Readings during Fermentation (days) 1 | 2 | 3 | 5 | 7 | 14 | 21 | Final Appearance and Taste[a] |
|---|---|---|---|---|---|---|---|---|
| w/o salt | | | | | | | | |
| 1% salt | | | | | | | | |
| 2.5% salt | | | | | | | | |
| 3% salt | | | | | | | | |
| 5% salt | | | | | | | | |
| 10% salt | | | | | | | | |
| 2.5% salt in dark | | | | | | | | |

[a]At 21 days.

**7.** Why is the room temperature when the cabbage is packed important?

**8.** What factors influence the fermentation sequence? How?

**9.** What contributes to the sourness of the kraut? The aroma?

**10.** Which lot of kraut fermented fastest? Why? Which was the slowest? Why?

**11.** What happens to kraut exposed to sunlight? Explain.

**12.** Which lot of kraut was the most tart? Why? Which had the best aroma? Why?

**13.** What happens when the salt is not evenly distributed in the shredded cabbage?

**14.** Why is the kraut frozen or heat processed after fermentation is complete? Under what conditions could it continue to be stored?

**15.** Account for the short processing time and the low processing temperature for canned kraut. What other canned food is processed in a similar manner?

**16.** Give the causes of the following defects in kraut:

Soft:

Slimy:

Pink:

Dark spots:

Poor flavor:

White scum:

## Fermented Cucumbers

### Laboratory Preparation

Ten experimental products are compared. The type 10 pickle in the experiment will take 5 to 6 weeks. Store the other treatments at room temperature until all are ready to eat, or start type 10 two weeks before the others. The fermentation temperature for this experiment is high to shorten the process. The temperature of all the lots may be lowered if the class schedule permits; a firmer pickle should result.

To make the 2 1/2% brine, combine 1/3 cup salt per gallon of water. To make the 5% brine, combine 3/4 cup salt per gallon of water.

If more brine is needed to cover the cucumbers, prepare it the strength specified in each recipe. One part brine for each 2 parts vegetables, by volume, is a good starting amount to make. The shape of the container will largely determine if this ratio is adequate. A nonmetal container about 1 gallon (3.8L) is best for these experiments.

The inoculation step is optional in types 2–9. Ideally, these types should be done both with and without inoculation if the lab situation permits. To omit the inoculation step, simply do not add bacteria but follow the other procedures, and omit blanched types since the natural fermentative bacteria will have been killed. Householders make pickles without inoculation, but inoculation is suggested here so the finished product is achieved more quickly and the effects of other variables on pickle texture are seen without as many spoilage problems. However for students wanting more experience with household pickle problems, uninoculated lots would be more appropriate.

A recommended inoculation is $10^8$ cells of *Lactobacillus plantarum* WSO per 3.8 L of brined cucumbers. The culture can be grown in cucumber juice broth containing 5% NaCl at 86F (30C) until the broth reaches O.D. 2–4.

For detailed instructions on preparing the bacteria cultures, see:

Etchells, J.L., Costilow, R.N., Anderson, T.E. and Bell, T.A. 1964. Pure culture fermentation of brined cucumbers. Appl. Microbiol. 12: 523.

For detailed instructions on adding the cultured bacteria to the pickles, see:

Fleming, H.P., Thompson, R.L., Bell, T.A. and Hentz, L.H. 1978. Controlled fermentation of sliced cucumbers. J. Food Sci. 43:888.

A source is Quest International, 1-800-237-6831.

### Cucumber Experiment

For all lots, select unwaxed cucumbers 1.5–2.25 inches (3.8–5.7 cm) in diameter. For lots 2 through 9, slice the cucumbers 1/4 inch (0.5 cm) thick with a stainless steel blade. For consistency, all lots may be sliced, but types 1 and 10 are written as typical household recipes to give experience with whole cucumbers. Use 5 pounds of cucumbers for each type.

Place the cucumbers in the bottom of a crock or jar, with seasonings if recipe directs, then pour the liquid over the top. Cover with a heavy plate or glass lid as a weight to hold the cucumbers under the brine. If you are using a jar to hold the cucumbers, a food-grade plastic bag (doubled) filled with water or brine and tied with a wire twist makes an acceptable weight. However, scum removal is difficult with this system. A piece of food-grade, perforated plastic weighted down with small, washed rocks (Fig. 5.1) works well. Householders adapt an assortment of plastic items (and broken plastic utensils) already present in their kitchens for this. Some of the more appropriate include a piece of a plastic colander, and a plastic grid for making "fake" lattice pies. Cover all loosely with a clean cloth.

For all types: Store at 80F (27C). Remove scum daily as it forms. If necessary, make up more brine of the same strength and add to keep the cucumbers covered. Determine the pH on the same schedule for each type; suggested interval is day 1, 2, 3, 5, 7, 14, and 21. Other time intervals may be selected but should be closer together in the first week and then at longer intervals until fermentation is complete. If a salometer is available, do a salt determination on the brine at weekly intervals.

At the end of 3 weeks, the cucumbers should have a desirable flavor and be an even olive green color. When fermentation is com-

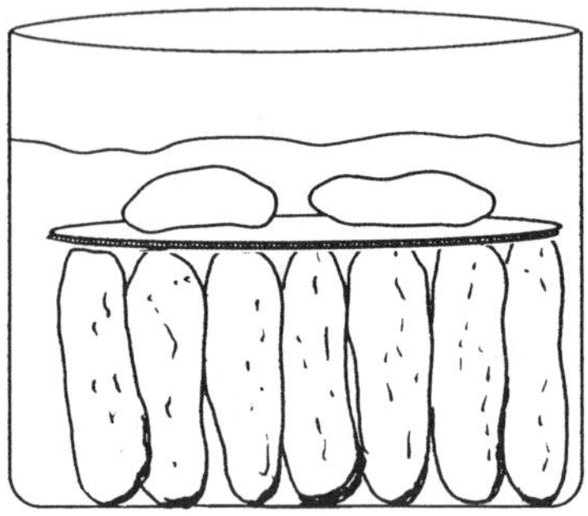

**5.1.** Fermenting jar.

plete, remove a pint of the cucumbers and enough brine to cover, leaving 1/2-inch head-space. Process in boiling-water canner for 15 minutes. Compare the processed to the unprocessed.

## Dill Pickle Experiment

### Type 1 (control)—*A Homestyle Brined Dill Pickle with Acetic Acid Addition*

5 lb cucumbers
3½ Tb whole mixed pickling spice
1 bunch fresh dill (may substitute dried)
⅔ c vinegar
6 Tb salt (= ¼ c + 2 Tb)
⅔ c water

Wash cucumbers then drain or wipe dry. Place half the spices and half the dill in the bottom of a crock or jar. Add the cucumbers. Place the remaining dill and spices on top. Combine the vinegar, salt, and water, then pour over the cucumbers.

### Type 2—*Unblanched with 2½% Salt*

Wash, slice, and put into 64F (18C) water for 5 min. Pack into a container with 2½% salt brine. Inoculate with *Lactobacillus plantarum.*

### Type 3—*Unblanched with 5% Salt*

Wash, slice, and put into 64F (18C) water for 5 min. Pack into a container with 5% salt brine. Inoculate with *Lactobacillus plantarum.*

### Type 4—*Blanched, 2½% salt*

Wash and slice. Put in a wire basket and immerse in a 171F (77C) water bath for 3.5 minutes. Keep moving the basket for uniform heating. Cool in a 64F (18C) water bath for 5 minutes. Pack into a container with 2½% salt brine. Inoculate with *Lactobacillus plantarum.* Omit this type if not doing inoculated.

### Type 5—*Blanched, 5% Salt*

Wash and slice. Put in a wire basket and immerse in a 171F (77C) water bath for 3.5 minutes. Keep moving the basket for uniform heating. Cool in a 64F (18C) water bath for 5 min. Pack into a container with 5% salt brine. Inoculate with *Lactobacillus plantarum.* Omit this type if not inoculating.

### Type 6—*Blanched, Lime, 2½% Salt*

Wash and slice. Put in a wire basket and immerse in a 171F (77C) water bath for 3.5 minutes. Keep moving the basket for uniform heating. Cool in a 64F (18C) water bath for 5 min in which the water has been saturated with lime—0.15% $Ca(OH)_2$ pH = 12.4. Rinse well through four changes of water. Pack into a container with 2½ % salt brine. Inoculate with *Lactobacillus plantarum.*

### Type 7—*Blanched, Lime, 5% Salt*

Wash and slice. Put in a wire basket and immerse in 171F (77C) water bath for 3.5 minutes. Keep moving the basket for uniform heating. Cool in a 64F (18C) water bath for 5 minutes in which the water has been saturated with lime—0.15% $Ca(OH)_2$ pH = 12.4. Rinse well, through four changes of water. Pack into a container with 5% salt brine. Inoculate with *Lactobacillus plantarum.*

### Type 8—*Unblanched, $CaCl_2$, 5% Salt, Added Acetic Acid, Not Inoculated*

Wash and slice. Put in a wire basket and immerse in a 120F (49C) water bath that contains 0.2% water/volume $CaCl_2$ for 3.5 minutes. Keep moving the basket for uniform heating. Cool in a 64F (18C) water bath for 5 minutes (tap water). Pack into a container and fill with a 5% brine (½ lb salt, and 1 c vinegar per gal of water).

### Type 9—*Blanched, $CaCl_2$, 5% Salt, Added Acetic Acid, Inoculated*

Prepare as for type 8 but immerse in water at 171F (77C). Add the *Lactobacillus plantarum* inoculation at the time of brine addition. Omit this type if not inoculating.

### Type 10—*A Typical Household Recipe for Salt Stock Pickles*

Select 5 pounds of cucumbers. If they are very dirty, wash them; otherwise only wipe with a damp cloth. Pack them whole into a clean container. Cover with a cold brine strong enough to float an egg. This usually specifies a 10% brine made by combining 1 pound of salt per gallon of water. Put a weight on the vegetables to keep them under the brine. On day 2, add ½ pound of salt. Place the salt in a mound on top of the weight; often a dinner plate is used for the weight in a crock. At the end of the first week, add ⅛ pound salt. Continue to add ⅛ pound salt each week for the next 4 weeks (a total of 5, ⅛ pound salt additions). Daily, remove any scum that forms during the fermentation period. Add more 10% brine if necessary to keep cucumbers covered. Store at 80F (27C) during the entire fermentation period. Fermentation will continue 4 to 8 weeks, indicated by a few bubbles rising to the surface.

The cucumbers are ready when they are a consistent olive green color, translucent throughout and without white spots. Before consuming, de-salt the pickles by soaking for several hours in large quantities of fresh water which is changed several times, or in equal parts of water and vinegar. Note: If you are accustomed to lower-salt foods, an overnight soak may be preferred.

### Fermented Cucumber Laboratory Report

Each group needs to keep accurate records of the pH changes for the type of pickle process it was assigned. Prepare a graph of these results to post for all to share. At the end of the fermentation period, share your results orally with the class and complete a summary chart of the 3-week monitoring highlights of all the types prepared.

Make objective texture readings using instrumentation available to your class on the finished products. Testing of commercial cucumbers is done using a USDA Pressure Tester with a 5/16-inch tip. Eighteen pounds and above is considered very firm, 14 to 17 pounds firm, 11 to 13 pounds inferior, 5 to 10 pounds soft, and 4 pounds and below is considered mushy.

Sample each finished lot with unsalted soda crackers. Note the odor, color, texture, and flavor of each lot.

Compare the type 1 home recipe dills—the processed jar—with purchased genuine dills and with purchased dill pickles. Which flavor do you prefer? Which texture? What is the cost difference between the genuine and regular, purchased dills?

## Kimchi

Kimchi (or kimchee) is a popular fermented food in Korea and similar fermented-vegetable mixtures are made throughout much of Asia. Many Korean households prepare their own to have vegetables during the winter months. With populations of Asian descent increasing throughout the United States, kimchi is available in grocery stores and is being made frequently at the household level. Napa or Chinese cabbage is the main ingredient. Other ingredients may include radishes, red peppers, onions, garlic, ginger, meats, anchovies, shellfish, nuts, sesame seeds, and salt. Kimchi recipes vary from mild to hot. Some are true fermentations with salt added to the raw mixture and others marinate vegetables overnight in a salt brine then refrigerate for further storage. The odor of fermented sulfur-containing vegetables such as cabbage and garlic with fermenting radishes is often objectionable to Westerners; however, they may find the taste pleasant.

Try preparing kimchi with a local recipe or one of those provided here. Sample the products and evaluate them.

**1.** What is the purpose of the salt water soak?

**2.** Why are the seeds removed from the peppers?

**3.** Purchase a jar of fermented kimchi. Open it only after all other sensory evaluation is done for the day. Sample it and note the complex blend of flavors and odors. Remember that the fondness for kimchi is acquired. What is the pH of this commercial product?

**4.** Have pairs of class members interview householders of different ethnic backgrounds that prepare their own fermented vegetables. Give a short (5 min) oral report to the class.

FERMENTED KIMCHI (*yield 3 cups*)

2 lb Napa cabbage
2½ Tb salt
¼ lb daikon radish
2 green onions
3 cloves garlic
2 tsp. finely chopped red pepper
2 tsp. sugar

Rinse and then cut cabbage into 1-inch pieces. Let stand in a bowl with 2 tablespoons of the salt—toss well—for several hours or until the cabbage is reduced to about half of its original volume. Put cabbage into a colander and rinse thoroughly, then drain without pressing. Cut peeled daikon into matchstick pieces for authenticity or shred coarsely in food processor. Cut both the green and white portions of the onions into thin slivers. Mince the garlic and red pepper. Combine these vegetables with the wilted cabbage and add the remaining 1½ teaspoons of salt and the sugar; toss well. Pack loosely into a 1-quart jar and cover. Let stand at room temperature for 1 to 4 days—the fermentation proceeds more rapidly during warm weather and the flavor varies considerably with the length of fermentation, then store in the refrigerator.

**QUICK KIMCHI,** *a mild version*

2 lb Napa cabbage (sometimes marketed as Chinese cabbage)
½ c salt
1 qt water
2 chili peppers, remove seeds
1 Tb sugar
1 tsp salt
½ tsp ginger root, chopped
1 clove garlic, chopped

Rinse the Napa cabbage, then cut it into 1-inch lengths. Soak it in salt water (½ c salt and 1 qt water) for 2 hours or more. Rinse it and drain thoroughly. Combine the cabbage with the remaining ingredients and mix well. Press the mixture tightly into a jar, cover, and refrigerate for 3 days before examining. Determine pH.

# 6 Curing

## Typical Consumer Questions

Each student should select one or two of the following consumer questions and provide both a consumer answer and a technical answer. If possible, give a reference(s) for the latter. Share and discuss the answers with other class members.

**1.** *I heard that Ralph Nader called bacon a "public enemy." Should we stop eating it? I have eaten bacon all my life; am I at higher risk for cancer now?*

**2.** *My great-grandparents butchered their own hogs and cured their own hams. These were hung in a shed and used until late spring. Now I read that even the whole hams I buy spoil easily and should be refrigerated. Is this really necessary?*

**3.** *Why is it that ham is sometimes the food that causes staph food poisoning but sausage, pork chops, or spareribs do not seem to be involved?*

**4.** *We have a smoker and frequently smoke fillets of salmon we catch. How should these be stored?*

**5.** *Why doesn't the Extension Service publish a booklet on homemade cured meats? I'd especially like recipes for smoked bratwurst and pickled pigs' feet.*

**6.** *We have an additive-free kitchen. I want a recipe for homemade hams that can be cured without chemicals. What happens if I leave the saltpeter out of an old recipe?*

**7.** *I am concerned about the frequency of salmonella in poultry. I certainly do not want to expose my children to salmonella, but my 6-year-old wants to serve hot dogs at her birthday party and some of the guests do not consume beef or pork. Are turkey franks safe for children?*

**8.** *I have saltpeter leftover from when we bought our sausage stuffer in the early 1970s. Can I still use it and the recipes that came with the stuffer?*

**9.** *My neighbor brought over some bacon immediately after she took it out of their little tabletop smoker. I refrigerated it while still warm. If we cook it thoroughly, is it safe to eat?*

## Review of Curing Literature

**1.** Read one of the following references and be prepared to discuss it in class.

Hansen, L.J. 1960. Emulsion formation in finely comminuted sausage. Food Tech. 14:565.

Pearson, A.M. and Wolzak, A.M. 1982. Salt—Its use in animal products—A human health dilemma. J. Anim. Sci. 54:1263.

Yarbrough, J.M., Blake, J.B. and Eagan, R.G. 1980. Bacterial inhibitory effects of nitrite. Appl. Env. Micro. 39:831.

**2.** Read a scientific journal article of your choice on nitrites in the human diet, cured meats, or sausage products. Write an abstract (max. length two paragraphs) with the reference at the top of the paper to be turned in before the curing laboratory. Post a photocopy of your abstract in the laboratory room for all to share.

**3.** Have one member of your class contact the local county extension office and ask about the types of consumer questions they receive on home curing of meats. Are many householders in your area curing and smoking their own pork? Does the agent feel their office has adequate information to give these people?

**4.** Take a label from a cured product—some will be available at class for those without access to them—and list all of the ingredients, in order, in one column. Briefly list the functions of these ingredients in a second column. Post these lists in the laboratory for all to share. Be sure to attach the label with brand name to the paper. How do home cured products differ from these?

# PRACTICAL EXPERIENCES

Though sodium and potassium nitrate are not used commercially (nitrite is added instead), nitrate in the form of saltpeter is frequently included in home cures. It is listed as an ingredient in some of the following recipes to provide experience working with it. There are premixed curing salts available at grocery stores that many consumers find acceptable; however, in these experiments, we will prepare the curing mixture to better understand the components.

## Sweet Pickle Curing Brine

Repeat this experiment twice. Once with saltpeter and once with ascorbic acid. Label the crocks as to type of treatment.

Put 2.5 pounds of pork loin, beef brisket, or fatback pieces in a clean crock or other pickling container (see pickling lab for alternatives).

Mix together the following brine:
3 c pickling salt
1 c sugar
1 Tb saltpeter or 2½ tsp ascorbic acid
6 qt water

Pour the brine over the meat and weight it down with a plate and clean rocks so all portions of the meat are submerged. Use smooth surfaced rocks which have been scrubbed with a brush and washed in the dishwasher.

Store the crock at 36–38F (2 to 3C). At the end of a week, remove the meat, stir the brine, and replace the meat again. Do this each week for 2 to 3 weeks.

Cook and slice the meat; then evaluate its odor, flavor, texture, and color. Which cure do you prefer? Why?

The 1990 Code of Federal Regulations (CFR, 1990) requires commercial dry cures, not containing nitrite or nitrate, to begin with sufficient salt so the concentration of the finished brine is at least 10%. Use a salinometer to measure the concentration of your brine at the end of the curing process. It may be necessary to pour the brine into a graduated cylinder for measurement.

NOTE: A hydrometer with specific gravity (.999 to 1.1170) and Brix scales (−2 to 39) is available by mail order for $7.00 from a mail order source—see Appendix A if one cannot be obtained locally.

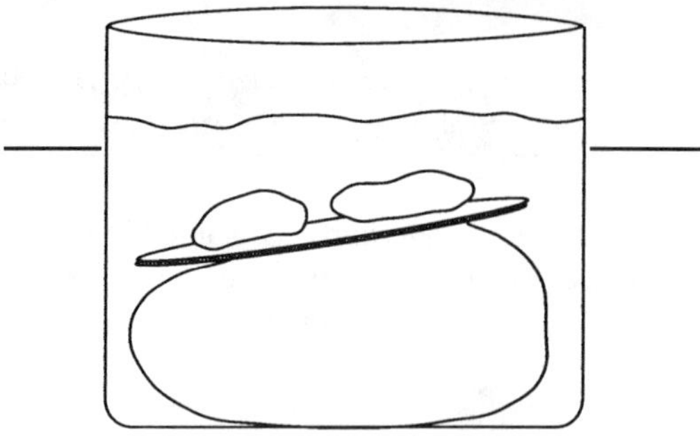

**6.1.** Ham submerged in brine for curing.

| Salinometer Degrees | Salt Conc. (oz/gal) | Salt Conc. (%) |
|---|---|---|
| 3.8 | 1.3 | 2 |
| 7.6 | 2.6 | 4 |
| 11.3 | 4.0 | 6 |
| 15.1 | 5.3 | 8 |
| 18.9 | 6.7 | 10 |
| 22.6 | 8.1 | 13 |
| 26.4 | 9.6 | 15 |
| 30.2 | 11.1 | 17 |
| 34.0 | 12.7 | 20 |
| 37.7 | 14.2 | 22 |
| 40.5 | 15.8 | 25 |
| 45.3 | 17.5 | 27 |
| 49.1 | 19.1 | 30 |
| 52.8 | 20.8 | 33 |
| 56.6 | 22.6 | 35 |
| 60.4 | 24.4 | 38 |
| 64.2 | 26.2 | 41 |
| 68.0 | 28.1 | 44 |
| 71.7 | 30.0 | 47 |
| 75.5 | 32.0 | 50 |
| 79.2 | 34.0 | 53 |
| 83.0 | 36.1 | 56 |
| 86.8 | 38.2 | 60 |
| 90.6 | 40.4 | 63 |
| 94.3 | 42.7 | 67 |
| 100.0 | 46.1 | 72 |

## Dry Cured, Unaged Ham

This experiment requires two small fresh hams or pork shoulders. They are cured identically, the duplicate is necessary only for the second core sampling. Different students should prepare each ham to provide a dry curing experience for more people. The following directions are for one ham.

Weigh the ham and prepare or purchase approximately 3/4 ounce (1 1/2 Tb) of a salt-nitrite mixture for each pound of meat. Spices can be added in addition to this amount. A recipe is provided for a simple dry cure mix if commercial mixes are not available in your area. Since the following recipe contains sugar in addition to the salt-nitrate mixture, use 3 tablespoons per pound of meat. If the commercial dry mix you choose already contains spices, use the amount of mix recommended on the package for the weight of your ham.

### Dry Cure Mix

3 c pickling salt
1 c sugar
1 Tb saltpeter

Rub half (divide the weighed cure in half by weight) of the cure mixture on the meat, making sure all of the surface is covered and also any exposed bones are well covered. If all of the measured curing salts do not stick to the meat, simply pile any surplus on the flesh side of the ham. Place each ham in a heavy plastic freezer bag and refrigerate at temperatures between 36 and 40F (2 to 4C). If the temperature is below 36F (2C), salt penetration is greatly slowed and temperatures above 40F (4C) increase the chance of bone-sour bacteria multiplying and spoiling the meat.

Keep the meat refrigerated for a total of 2 days per pound or 7 days per inch of the ham's diameter. Apply the second portion of cure after about ⅓ of the total curing time has elapsed. Return the meat to the refrigerator as before.

When curing is completed, take a core sample through the center of the ham for testing of the salt content. Repeat this procedure on the second ham after salt equalization. If a core instrument is not available, sample the various depths by removing a 1 1/2-inch thick slice with a knife from the center. Cut off the two fatty edges of the slice so the sample resembles the cores shown in the diagram below.

At the end of the curing period, place the meat in lukewarm water for 1 hour to dissolve the excess curing mix on the surface. Pat dry with clean paper towels and place in a new plastic bag and return to the refrigerator for 20 days. If you leave this bag partially open, slime bacteria are less likely to grow. If the surface does become slimy, scrape or wash it off at the end of equalization.

The ham with equalized salt content should not spoil even at temperatures up to 100F (38C). Salt does not penetrate the exterior fat well, so even after equalization these portions are less salty (Fig. 6.2).

You may wish to smoke your country ham after salt equalization. Follow directions on a commercial smoker.

Aging after salt equalization is also possible. At the household level, 6 months hanging at 70 to 85F (21 to 29C) achieves a good level of enzyme activity. Above 95F (35C), this enzyme is destroyed. Smoking after aging halts the enzyme too. Good air circulation during aging is essential to dry the surface of the ham. Evaporation will account for a 12 to 15% weight loss from the original fresh meat weight. This further increases the salt content of the meat. Remove any surface mold from the aged country ham with a cloth soaked in a 50/50 vinegar/water solution.

NOTE: Some dry cure ham recipes cure the meat for 2 days per pound instead of the 5 days per pound used here. They usually produce acceptable products too.

### Frying Your Country Ham

Remove the skin and most of the fat. Cut into generous 1/4-inch thick slices. Heat slowly in a heavy, uncovered skillet with a little water. Fry slowly turning pieces frequently. If grease is spattering turn heat down. When done, slice is thoroughly heated and light brown. To make Red Eye Gravy, remove ham from skillet and add 1 ounce of coffee for each slice that was cooked. Bring just to a boil and serve with the ham.

**1.** Estimate the total number of days from start to finish for your country ham. Approximately how many hours of your time was spent actually preparing it?

**2.** Was the quality uniform throughout?

**3.** Compare the flavor of your country ham to a purchased country ham and also to a purchased injection-cure ham. Does your cure mix meet the FDA standards for nitrite levels?

**4.** What is the cost per pound of the country ham you prepared? What is the cost per pound of the purchased country ham?

**5.** Consider a situation in which a pig is slaughtered in early October. The hams are dry cured in a 5-gallon plastic bucket outside, instead of in plastic bags in the refrigerator as in lab. Are these hams safe to consume? What do you base your decision on?

**6.** Weigh the ham at the end of the curing period. What percent of the fresh weight is the cured product? How does this compare with commercial country hams?

## Combination Cure, Unaged Ham

If you have access to a meat pump, you may wish to try injecting the curing solution in addition to rubbing the cure salts on the surface. The Morton Salt Company markets a 4-ounce, 12-hole pump for household use that is often available for sale at grocery stores with their premixed curing salts. Other pumps may be available where home food preservation

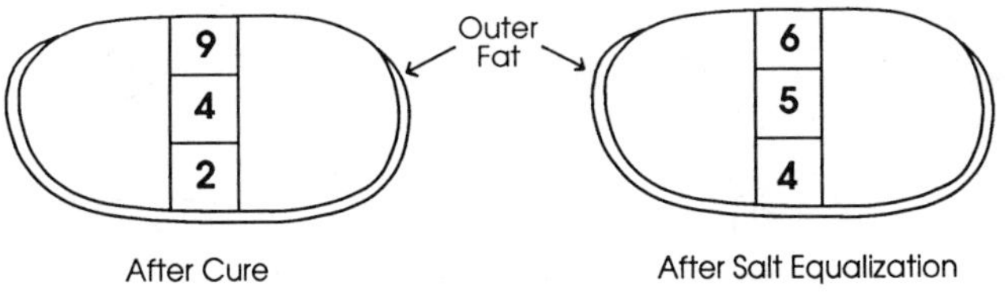


**6.2.** Salt content varies throughout the ham.

supplies are sold, or in urban areas, at butcher supply stores. See Appendix A for a mail order source.

### Combination Cure for Ham

Weigh the ham. Prepare 1 ounce of pickle for each pound of meat using a pickle recipe from this lab section or a recipe from another source. Fill the pump with pickle. This is called 1 stroke. A full pump reduces the air forced into the interior of the meat. Insert the needle completely and then after half of the stroke is completed, slowly withdraw it continuing the stroke until all the pickle is in the meat. Pinch the hole left by the needle with your fingers and hold for 10 seconds to decrease the amount of pickle that leaks out at this site. Start pumping at site number 1 (Fig. 6.3) and follow the numbers until running out of measured solution for the weight of ham. Sites 4 and 5 are optional and often not used for small pieces of meat. Injections parallel to the bone reduce the chance of bone sour by immediately increasing the salt content in this portion.

After pumping, rub dry cure on the surface of the meat as in the dry cured ham method. Use ½ ounce (1 Tb) of salt-nitrite mixture per pound of meat. Seasonings may be added in addition to this weight. Apply half of this dry cure initially and the remainder when ⅓ of the refrigerated curing time has elapsed. Finish the ham by following the dry cure instructions. Allow 5 days curing time per inch of meat thickness and 14 days for salt equalization in the ham for this combination method.

**1.** Do you think pumping is practical at the household level? What are the advantages and disadvantages?

**2.** What ingredients did you use in your pickle? How does this flavor compare to the commercial pumped hams?

**3.** Is this a high- or low-technology preservation method? How do you rate its safety compared to other ways of preserving pork?

**4.** Was curing more uniform in this ham compared to the dry cure?

## Corned Beef Pickle

This method was used by householders who butchered their own beef in the midwestern United States in the late 1800s. It is still practiced by those wishing to corn their own products.

Mix a pickle by combining 1 gallon of water, ½ ounce of saltpeter, 3 pounds of salt, and ½ ounce white sugar. Immerse a boneless 3-pound piece of beef brisket (the original cut used) or rump in the pickle, in a non-metal container. Add a cheesecloth bag of 2 stalks celery, 2 onions, 2 carrots, 6 bay leaves, 1 teaspoon mustard seed, ½ teaspoon allspice, and 3 cloves garlic—seasonings may be adjusted for personal preference. Refrigerate for 2 weeks. Remove the meat from the pickle and blot dry. It can then be smoked or simply wrapped for refrigerated or frozen storage until cooked for serving. Follow a cookbook for suggestions on preparing the corned beef. Many recipes freshen the meat to reduce the salt content.

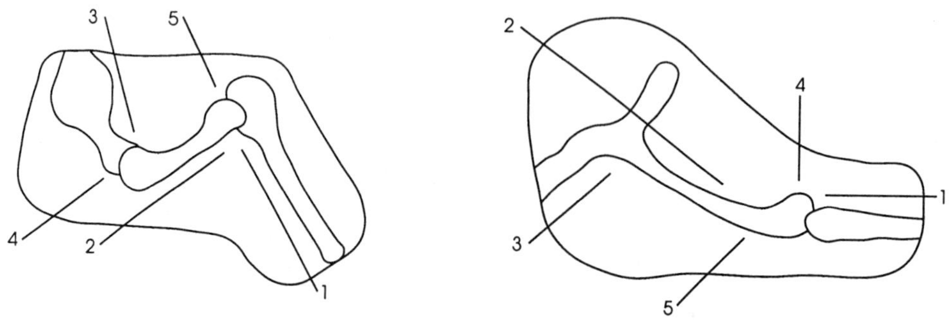


**6.3.** Pumping sites for hams and shoulders. Start pumping at site 1 and follow the numbers until running out of measured solution for the weight of ham. Sites 4 and 5 are optional and often not used for small pieces of meat.

**1.** What is the percent salt of this pickle?

**2.** Why is the curing done under refrigeration? Does this speed the process?

**3.** Why is curing done in a nonmetal container? If a householder does not have a pickling crock, what else could be used?

**4.** What is the chemical composition of saltpeter?

**5.** If a householder has purchased premixed curing salts and wishes to use them to corn beef, what advice would you give?

**6.** How does the home prepared corned beef compare to purchased? Consider cost, time, and quality.

## Survey of Sausage Flavors

If you are able to purchase a variety of sausages in your locale, heat them thoroughly and serve with bread or crackers (and the sauerkraut and mustard you prepared in earlier labs, for enjoyment after sensory evaluation). Display each prepared product with its wrapper. If your markets do not carry different types, the recipes at the end of this lab will provide some sampling experience.

**1.** Why do you think acidic foods are often served with sausages in the United States?

**2.** Did any of the fermented products have a "biting acid" flavor? What compound is responsible for this?

**3.** Which of these products appear to be emulsions and which are closer to mixtures?

**4.** Which of the labels recommended refrigeration? Do you agree with the labeling?

**5.** What advice would you give a householder who wished to make salami at home for their family pizzas?

### Varieties of Sausages

CHINESE SAUSAGE (*Bok Yu Chong type*)

| | |
|---|---|
| Pork trimmings (75% lean) | 3 lb |
| Pork back fat, diced | 1 lb |
| Salt | 1.6 oz |
| Sugar | .6 oz |
| Soy sauce (fermented type) | .15 oz |
| Cinnamon | .05 oz |
| Sodium nitrite | .01 oz |

Grind the chilled meat and fat through a ½-inch plate. Use mixer to combine the ground meats with the remaining ingredients. Store at refrigerator temperatures and cook thoroughly before serving.

| Type, Brand | Unusual Ingredients | Fresh, Smoked, or Fermented | Sensory Evaluation |
|---|---|---|---|
| 1. | | | |
| 2. | | | |
| 3. | | | |
| 4. | | | |
| 5. | | | |
| 6. | | | |
| 7. | | | |
| 8. | | | |

**SWEET ITALIAN SAUSAGE**

| | |
|---|---|
| Pork trimmings (65% lean) | 3 lb |
| Salt | 0.8 oz |
| White pepper | .13 oz |
| Fennel | .13 oz |
| Paprika | .07 oz |

Grind the chilled meat through a ¼-inch plate. Use a mixer to combine ground meat with the other ingredients. Store at refrigerator temperatures and cook thoroughly before serving.

**FRESH CHORIZOS** *(Mexican)*

| | |
|---|---|
| Pork trimmings | 3 lb |
| Neckbone trimmings | 1.2 lb |
| Salt | 1 oz |
| Sugar | .4 oz |
| Sweet red peppers | 3.03 oz |
| Chili powder | .03 oz |
| Red pepper, hot | .09 oz |
| Sodium nitrite | .01 oz |
| Garlic powder | .01 oz |

Grind chilled meat through a ¼-inch plate. Use mixer to combine with other ingredients. Store at refrigerator temperatures and cook thoroughly before serving.

## Homemade Cured Sausage-type Product

If you have access to a home-type sausage stuffer, each student ideally should make some links for the experience. The following recipe from Morton Salt (division of Morton Thiokol, Inc., Chicago, IL 60606) uses their premixed cure.

**COUNTRY STYLE BOLOGNA**

3 lb boneless pork trimmings
2 lb boneless beef trimmings
1 c ice cold water
¼ c Morton Tender Quick mix
1 Tb sugar
1½ tsp ground white pepper
1½ tsp ground coriander
1½ tsp ground mace
½ tsp onion powder
1 c NFD milk powder
1 Tb liquid smoke (optional)
1½- or 4-in. diameter casings (use of 1½-in. casings provides experience for more students)

Grind meat through a ¼-inch plate. Mix with other ingredients, cover, and store refrigerated overnight. Regrind through a ⅛-inch plate and stuff into casings. Simmer the large links until the innermost portions reach 160F (71C). If small links are made, they should also be thoroughly cooked. Cool in ice water for 15 minutes and store refrigerated.

## Salt Content of Cured Meat

If this class is held in a chemistry laboratory, students can use the following Volhard method which dates back to 1874 for calculating sodium chloride content. However students working in nonlaboratory settings may find it more feasible to only perform calculations based on the figures a lab technician has provided for their samples. The silver nitrate solution is expensive so choose your products to be sampled carefully. Read the entire experiment before beginning.

### Volhard Method

Put 3 g of a blended representative sample into a 300 mL Erlenmeyer flask. Add 25 mL of 0.100 N silver nitrate solution. Swirl contents to combine then add 15 mL concentrated nitric acid. Boil until the meat dissolves—approximately 10 minutes. Then add concentrated potassium permanganate solution in small portions, boiling after each addition, until solution becomes almost colorless. Add 25 mL water and boil again for 5 minutes. Cool and add distilled water to make a total volume of 150 mL. Add 25 mL diethyl ether and 2 mL saturated solution of ferric ammonium sulfate indicator, cover tightly and shake vigorously; the silver chloride ppt will coagulate. Titrate the excess silver nitrate with 0.1 N ammonium thiocyanate solution to a permanent light brown endpoint.

$$\%\ \text{salt} = \frac{(25.0\ \text{mL} - \text{mL}\ NH_4SCN)\ (0.1\ \text{N})\ (5.85)}{\text{Sample weight}}$$

$$\%\ \text{Sodium} = \frac{(22.997)\ (1.00)}{58.45}$$

NOTE: Atomic weight of Na is 22.997; atomic weight of NaCl is 58.45. The percent Na in salt is 39.34% or 39.34 g per 100 g salt. 0.100 N sodium chloride is 5.85 g/L.

**Reagents**

0.100 N Silver nitrate: Dry silver nitrate at 110C. Dissolve 17.04 g of this dried silver nitrate in distilled water to make a total of 1L solution.

0.100 N Potassium thiocyanate: Dissolve 0.72 g potassium thiocyanate in distilled water to make a total of 1 L.

The potassium permanganate is a 5% aqueous solution.

The ferric ammonium indicator is saturated aqueous solution $FeNH_4(SO_4)_2 \cdot 12H_2O$.

**1.** If the concentrated nitric acid were added to the flask before the silver nitrate, what would happen to the chloride and the final results?

**2.** Which products had the highest salt content? Do your human senses also detect this?

**3.** Was there a difference in salt content between the outer portion of a dry cure ham and the inner portion? Explain how this may occur.

If a specific ion (Na+ or Cl−) electrode is available, test representative samples of the same cured meats used in the Volhard experiment. Compare results.

Assoc. Official Anal. Chemists. 1990. Salt (Chlorine as sodium chloride) in meat, fish, and cheese. Volumetric method 935.47; Indicating Strip Method 971.19. Official Methods of Analysis, 15th ed. Arlington: Assoc. of Official Anal. Chemists, Inc.

## Cured-Meat Pigments in Sausages

The degree of meat curing can be expressed as a percentage from results of an acetone extraction of hematin. Pigment conversion to an 80 or 90% level of nitrosohemochromagen is considered acceptable. For all-meat sausages, 100 ppm is minimum and 140 ppm is very good. For all-beef sausages, 140 ppm is considered average.

The extraction must be conducted in subdued light to reduce pigment fading during the procedure.

**Equipment**

A spectrophotometer with 1 cm Beckman cells is used. For each sample tested, also have assembled the following:

| Sample | Sodium (%) | Salt (%) | Sensory Evaluation |
|---|---|---|---|
| 1. | | | |
| 2. | | | |
| 3. | | | |
| 4. | | | |
| 5. | | | |
| 6. | | | |

Comments:

2, 45 mL polypropylene centrifuge tubes with covers
2, 15 x 90 mm test tubes
2, 50 mm funnels
2, 2 in. diameter watch glasses
4, 9 cm diameter, No. 42 Whatman filters
2, 200 mL volumetric flasks
20 mL graduate
pipets and pipetting bulb
glass stirring rod with tapered tip that fits end of centrifuge tube

**Nitroso Pigment Extraction Solution**

Put 18 mL distilled water in a 200 mL volumetric flask. Add C.P. acetone to bring liquid up to the 200 mL level.

**Total Pigment Extraction Solution**

Put 4 mL concentrated hydrochloric acid in a graduated cylinder which contains 10 mL of distilled water; use extreme caution and pipet with bulb not mouth. Add distilled water to bring liquid up to the 20 mL level. Transfer this 20 mL of diluted HCl to a 200 mL volumetric flask. Add C.P. acetone to bring the total liquid up to the 200 mL level.

Take two representative 2.0-g samples from the center of a cured sausage-type product—often dyes are used on the outside surfaces of commercial sausages. Put each 2.0 g in separate, labeled centrifuge tubes. Pipet 9.0 mL of the nitroso pigment extraction solution into one tube and 9.0 mL of the total pigment extraction solution into the other tube. Mix the meat and the acetone solutions thoroughly using a special-tip stirring rod for 3 minutes. Put the cover on the tubes and swirl the contents gently. Let stand 10 minutes, then filter through two combined fine filter papers per funnel into two separate, labeled test tubes. Transfer the filtrates—they will not be pink—to 1-cm Beckman cells.

The OD should be read within 1 hour of the extraction. Take the nitroso pigment reading at 540 nm and the total pigment reading at 640 nm. Calculate both ppm and percent pigment conversion using the following formula.

$$\text{ppm nitroso pigment} = \text{OD at 540 nm} \times 290$$

$$\text{ppm total pigment} = \text{OD at 640 nm} \times 680$$

$$\%\ \text{conversion} = \frac{\text{ppm nitroso pigment}}{\text{ppm total pigment}} \times 100$$

Prepare the sausages by steaming them and display with the experiment results for tasting. How does the percent cure influence your purchasing decisions? Other than aesthetics, why is percent cure important?

| Product | Nitroso Pigment (ppm) | Total Pigment (ppm) | Percent Conversion |
| --- | --- | --- | --- |
| 1. National brand beef frankfurter | | | |
| 2. Generic brand beef frankfurter | | | |
| 3. National brand part-beef frank | | | |
| 4. Turkey frank | | | |
| 5. Homemade cured sausage | | | |
| 6. Commercial cured sausage | | | |
| 7. | | | |
| 8. | | | |

# 7 Freezing

## Typical Consumer Questions

For each of the following questions, give both a scientific explanation and recommendations for consumers to follow. If you need more information to accurately answer, state the assumptions you make.

**1.** *I froze a whole salmon about 6 months ago. It has an off-odor and some of the flesh is grey instead of pink. Is it spoiled?*

**2.** *Why can I freeze green peppers by simply cutting and putting them in a plastic bag, but when I do the same thing with my extra chopped onions, the thawed product is not good?*

**3.** *My boss gave me a 10-pound ham for Christmas, and then he presented me with another ham for New Year's Day. I am going to have to freeze the excess. What is the best way?*

**4.** *We received two large blocks of medium cheddar through a food program for low-income people. The last time this happened, I froze one of the blocks in the compartment above my refrigerator and it was awful. Why can't they give us a better quality of cheese that can be stored?*

**5.** *What happens if I keep food in my freezer for longer than recommended?*

**6.** *The electricity just went off. The electric company said it will be off for 4–6 hours while they fix a transformer. We just put a cut-and-wrapped side of beef in our freezer last week, can I make them pay for another side of beef?*

**7.** *My neighbor freezes pesto in ice cube trays and then packages it as cubes in plastic bags for use this winter. I can't find this procedure in any of my cookbooks. Is it okay?*

**8.** *Is it possible for frozen cakes to pick up odors from frozen fish? The cake I froze 2 months ago has an off-odor.*

**9.** *Can I freeze bacon?*

**10.** *How long can frozen orange juice keep in the freezer compartment above my refrigerator?*

**11.** *I would like to bake pies ahead of time for company dinner. Do they freeze well?*

**12.** *I bought 2 gallons of 2% milk because I was expecting the grandchildren to visit. They canceled. Can I freeze this milk?*

**13.** *The last time I purchased a box of frozen shrimp from the grocery store, it had an* *off-odor when thawed. I returned it, but would like to know how to evaluate commercially frozen foods before I purchase them.*

**14.** *I want to freeze green beans from my garden. What is the best way to blanch them?*

**15.** *How do I freeze corn on the cob?*

## Survey of Freezing Literature

**1.** Read any two of the following journal articles and be prepared to discuss them in class. Write a short (1/2-page max.) abstract of one of these articles, or of another freezing journal article, to turn in. Write the citation at the top of your abstract.

Bengtsson, B. & Bosund, I. 1966. Lipid hydrolysis in unblanched frozen peas (Pisum sativum). J. Food Sci. 31:474.

Brown, M.S. 1967. Texture of frozen vegetables; Effect of frozen water on green beans. J. Sci. Food & Agr. 18:77.

Sterling, C. 1968. Effect of low temperature on structure and firmness of apple tissue. J. Food Sci. 33:577.

Winter, J.D., Hustrulid, A. and Noble, I. 1952. The effect of fluctuating storage temperature on the quality of stored frozen foods. Food Tech. 6:311.

**2.** After reading the text chapter on freezing, prepare a list of the chemical, physical, and microbial changes that occur in slices of ham stored at 0F (−18C) for a period of 6 months. What is your opinion of using these ham slices in sandwiches at the end of the 6-month storage period? (Be brief, 2 page max.)

**3.** Have one member of the class contact the local county Extension agent—or Master Food Preserver if that program exists in your area—who receives most of the food preservation questions. What are the issues that consumers are most concerned about? Does this person feel there are "grey areas" in freezing that she does not know how to respond to? What are they?

**4.** Have one member of the class calculate the cost of freezing foods commonly preserved in your area opposed to the cost of canning this same food. The Extension Service may have a bulletin on the costs of energy use for these two methods.

**5.** List the ingredients from a commercially frozen food item. What is the function of each of the additives? Complete these functions as best you can before class and then finish with the help of class discussions of each product chosen.

# PRACTICAL EXPERIENCES

## Freezing Foods of Different Composition

If possible, students should prepare and wrap these foods and store frozen for evaluation by next term's class. Prepare freshly made samples for side-by-side comparisons with their thawed counterparts. If the class is offered only annually, many of the foods would be low quality so it is suggested in this situation that a lab assistant freeze the foods ahead of time or the class freeze them early in the term for later evaluation.

To conserve freezer space, small amounts may be used. Enough for all students to sample is appropriate. Freezing only 1 slice of pie, or preparing a 3-inch pie will show the effects of freezing.

Pretreat and wrap these foods according to the most recent USDA recommendations for that type of product. Label all with name of product, date, type of pretreatment (water blanched for 2 min) or additives if any, and prefrozen volume. Make a note on the container of any errors in preparation—example: creamed chicken was scorched—to help the next class in their evaluation of the effects of freezing.

**Suggested Foods to Freeze**

1 quart milk in original container
6 biscuits
1 cake layer
creamed chicken
raw fish

gravy
hard cooked egg
3 egg yolks w/ sugar or salt addition; 3 egg yolks w/o additions
3 egg whites w/ sugar or salt addition; 3 egg whites w/o additions
macaroni, tomato sauce, and hamburger dish
ham w/ water added; ham w/o water added
uncured pork
walnuts
baked pie shell
double crust, homemade, fruit pie
homemade pumpkin pie; commercial pumpkin pie
vegetable soup (green beans, celery, peas, carrots, potatoes)
peaches w/ antibrowning agent; peaches w/o browning agent
broccoli
bell pepper
cheese in a block; cheese grated
grapes (green, Thompson seedless)
1 cup water frozen in narrow container

Note the temperature and length of time in frozen storage. Prepare the foods for serving. Evaluate all products for flavor, color, odor, and texture. Was the blanch adequate? Which items were freezer burned? Why?

Prepare to discuss in class the scientific reasons for each product's quality. Which characteristics can be attributed to chemical changes and which to physical changes? After your group discussion, prepare a 1-paragraph written statement of these changes for each product.

**1.** Potatoes are not recommended for freezing. Are these thawed samples safe to eat?

**2.** If a food separates during thawing, is it safe to eat?

**3.** Were there any problems with container size and expansion of food? Why or why not?

## Storage Temperature and Speed of Freezing

Canned Vienna sausages and fresh fish—ling cod or salmon work well—are recommended foods for the following experiments, but local items may provide more applicable experiences. Divide the food into 6 representative lots, package all identically, and store under the following conditions. Longer storage times show the differences best.

32F (0C)
22F (−5.5C)
0F (−17.7C)
0F (−17.7C) freeze and thaw cycle several times
0F (−17.7C) frozen rapidly (use liquid N if available)
−10F (−23.3C)
−30F (−34.4C)

Poach or microwave the fish (unseasoned) and evaluate for quality (odor, flavor, color, texture) and also examine frozen samples microscopically for cross-sectional appearance.

## Packaging

Store samples of cooked and pureed winter squash in a variety of airtight and porous materials. Then evaluate for quality (odor, color, texture, flavor). Some suggestions of wrappings follow. Longer storage times show the differences best.

**1.** Rigid wall freezer container w/ 1/2-in. headspace.

**2.** Rigid wall freezer container partially filled with freezer plastic wrap flush on surface and crumpled wax paper on top.

**3.** Rigid wall freezer container only partially filled.

**4.** A clear freezer bag—expel air w/ straw or immerse in water.

**5.** A clear sandwich bag—expel air w/ straw or immerse in water.

**6.** Plastic-coated freezer paper seal drugstore fold w/ freezer tape.

**7.** Plastic-coated freezer paper seal drugstore fold w/ masking tape.

**8.** Freezer aluminum foil seal drugstore fold w/ freezer tape.

**9.** Light-weight aluminum foil seal drugstore fold w/ freezer tape.

**10.** Light-weight aluminum foil seal non-drugstore fold w/ masking tape.

## Retrogradation Study Using Modified Starches, Celluloses, and Gums

Examine the display of thickening agents used commercially and their literature. Select one for your product.

Decide on a food you would like to prepare in advance and remove from the freezer to serve guests easily. Use the conventional recipe for one batch and then alter this recipe only by thickening it with one of the modified products or a gum. You will need to follow the manufacturer's recommendations of amounts to use.

After frozen storage and final preparation, if any is required, evaluate both lots for quality—odor, color, texture, flavor. Display so all class members can sample all products. Label each with thickening agent and amount per cup liquid used.

**1.** Clear gel A became available to household canners in the late 1980s. Consumers as a whole were slow to adopt its use though some gave it glowing reports. What problems do you see in the marketing of such a phosphate-modified starch for home freezing?

**2.** Did the estimates of amounts to use based on the manufacturer's literature result in a desirable thickness?

**3.** Was the clarity and mouthfeel of all the products acceptable to all class members?

**4.** Record the mailing addresses from the labels and literature of these modified products for future reference in your career. Suggest a nutritional concern a consumer may voice for one of these products and address this concern using the information provided by the company and any other scientifically sound information you may have. Share these in a class discussion.

# 8 Refrigeration

## Typical Consumer Questions

**1.** *Your friend in Minnesota calls you on this 24th day of December to say happy holidays and ask a food safety question. Their family received a ham mid-November. It was one of many handed out to all employees at the day/swing shift change (3 P.M.) and left in the car during the 8-hour swing shift. There had been heavy snow on the road for several weeks so it took 30 minutes to drive home. The ham was immediately refrigerated at home. Is it safe to serve half of this ham to company on December 25 and keep the other half refrigerated for company on January 1? Write a short (1-page max.) explanation of the principles you base your decision on.*

**2.** *A 1-pint container of sour cream was purchased on sale Tuesday. A small portion was used Tuesday evening and the remainder refrigerated. Friday evening the container was opened to reveal a more concentrated white portion on the bottom and a translucent liquid on top. Is this sour cream safe to serve? Write your short (1/2-page max.) reasoning for this decision.*

**3.** *Visit a household and ask to test the temperature of their refrigerator—place thermometer in jar of pickles or similar item for 10 minutes. Is there a thermometer present in the refrigerator? What temperature does the householder believe the appliance should be at? Why? Share your findings with the class.*

**4.** *Have several class members visit retail grocery outlets. What is the temperature of their refrigerated cases? How do store personnel monitor this temperature? How long does it take the repair person to respond to their call when the temperature is too high? What is done with the food stored at the higher temperatures? Does the retail outlet have more discards from the refrigerated cases or from the frozen food cases? Share your findings with the class.*

**5.** *Divide the following refrigerated food categories among class members; dairy, prepared foods, desserts, fresh produce, miscellaneous. Survey local retail markets for open—consumer readable—dates. Share your findings with the class.*

**6.** *Read a journal article of your choice on either household or commercial refrigerated storage. Cite the reference at the top of the* *page and write a brief (1-page max.) abstract. Post this in the classroom for all to share.*

## PRACTICAL EXPERIENCES

### Survey of Refrigerated Food Spoilers

Students should clean out the discards from their own or friends' refrigerators, and collect each of the spoiled (both physical and microbiological) food items in 1-gallon plastic bags. Record the name of each item and the approximate length of storage directly on the bags. Put only one food item in each bag. Store these discards at refrigerator temperatures until laboratory time.

Display this food anonymously in the laboratory along with the temperature of the refrigerator. What is the most common reason for spoilage of the cooked foods?, dairy?, fresh produce? Was the spoilage due to miscalculated storage times, lost and forgotten items, or other reasons? How could the number of these discards be decreased?

### Time in Growth Zone

This experiment is to be done by one group only for all the class to observe.

Open four cans of refried beans into a large casserole dish so that depth is at least 5 inches, and top with a 1-inch layer of thick white sauce. Bake in a 350F (177C) oven for 1 hour. The top should become a golden crust. Immediately after removal from the oven, place a thermometer in the center portion and record the temperature. Leave the thermometer in place. Put the casserole on top of a trivet or potholder as you would for the serving table and hold it there for 45 minutes. Record the temperature again at the end of the 45 minutes. Then cover with plastic wrap—cut a small opening for the thermometer to protrude—and refrigerate. Check and record the temperature at 1/2-hour intervals until the center of the casserole reaches 40F.

**1.** How long is the casserole in the temperature danger zone? Discuss the safety of this procedure.

**2.** Thorough reheating of leftovers kills vegetative cells. Compare microwave range reheating to conventional oven reheating.

**3.** What suggestions do you have for speeding the cooling process? Make suggestions for preparing this recipe and storing in a safer manner.

### Short-term Refrigeration

The purpose of this study is to determine the sequence of organoleptic, pH, and microbiological changes during refrigerated storage of perishable foods. Each team of students will be responsible for monitoring one food. One student should record the refrigerator temperature for all groups before the door is repeatedly opened. Some suggestions for foods are: cultured cottage cheese, directly set cottage cheese, raw and pasteurized milk, yogurt, heavy cream, fresh burritos, sprouts, gelatin dessert, deli sliced beef, ground beef, chuck roast, and raw fish.

#### Day 1

Decide the organoleptic qualities to be determined for each product. Design a rating sheet and rate the product. Determine the pH of the product—directly if it is liquid or as a slurry. Do a total plate count (TPC) for microorganisms (method follows). Copy all information from the label, including open date. Package the product for storage as a householder commonly would and label with your names, day of laboratory, and refrigerate.

#### Second Lab Day

(And subsequent lab day intervals until the product is rated by you as spoiled.) Follow the procedures used on day one.

### After the Last Sampling Day

Summarize the changes that occurred, making a laboratory summary sheet with columns for storage time, pH, TPC, and organoleptic ratings. Draw a graph of pH and TPC (numbers converted to logarithms) plotted over time. Examples of such a graph are widely available in microbiology texts which will be provided in lab for you to observe before drawing the graph of your results. Post these results for all to share and have a group member summarize your findings for the class.

### Methods

**1.** Organoleptic rating: Do not use a hedonic scale (like/dislike) but a descriptive one of flavor (taste & odor) and appearance qualities.

**2.** pH: Follow directions with pH meter. Instrument must be turned off except when electrodes are in the sample. Product should be at room temperature. Setting a small beaker of sample in a pan of warm water will hasten its warming. Discard this portion of food.

**3.** Total plate count (TPC): Melt a bottle of agar (standard plate count agar: Difco, Detroit, Michigan, or other brand) by setting the loosely capped bottle in boiling water in the bottom of double boiler. This will take about 20 minutes.

Put bottle into warm water to cool to 117F (47C)—thermometer in water. This will take 15–20 minutes. Hold at this temperature.

Weigh 10 g of the food into a sterile blender jar—sterilize blender blades by 10 minutes in 70% alcohol followed by three rinses in sterile water—or dilution bottle using aseptic techniques. Add 90 g (mL) of sterile peptone water as a diluent. Blend for 1 minute or shake vigorously for 1 minute.

Pipette 1 mL into sterile 99mL dilution bottles to make desired dilutions. Check with instructor for some dilution suggestions for the various products. For cultured dairy products such as cottage cheese, you may need to plate dilutions from $10^{-8}$ to $10^{-14}$. For pasteurized milk, you may need to plate $10^{-1}$ to $10^{-4}$. Plate 1 mL or 0.1 mL dilution into sterile petri dishes. Add a small amount of agar and carefully mix by swirling dish on a flat surface. Invert cooled plates. Incubate at 30C for 3–5 days (nonstandard incubation time). Count colonies on those plates having 30–300 colonies, or a plate close to this, and determine total count. Base the dilutions to be used each day on the previous count plus the expected multiplication.

# 9 Canning: Preservation by Heat and Vacuum Seal

## Safety Decisions about Canned Food

Decide if the food will be safe to eat in each of the situations below. Why or why not? Suggest changes the home canner should make, that will result in safe procedures. If possible, take turns answering these questions in a group setting.

**1.** *Green beans hot packed in jars, sealed, and then processed in a boiling-water canner for 2 hours.*

**2.** *Whole kernel corn and lima beans canned as a mixture, using the time for lima beans.*

**3.** *Pints of salmon were processed for 90 minutes at 12 pounds pressure (altitude 3,500 ft). The food preserver later heard that the correct processing time was 100 minutes. Within 6 hours, the jars were reprocessed—without opening—for an additional 10 minutes. Is this salmon safe to serve chilled on crackers?*

**4.** *I collect and sell old glass jars. I have been preserving peaches in some of the less valuable ones from the 1940s. They have beautiful quilted designs imprinted on the glass and are interesting shapes. Is it safe to consume these peaches? Green beans?*

**5.** *My mother lives above 2,000 feet elevation. She cans her vegetables at 10 pounds pressure because that is the only weight on her pressure canner.*

**6.** *A jar of applesauce lost its seal 2 months after canning. There was a layer of mold on the surface.*

**7.** *A householder brings you a jar of asparagus with yellow crystals on the stalks.*

**8.** *The day after canning, it was discovered that two jars of peaches had not sealed.*

**9.** *There are black spots underneath the lids on canned tomatoes.*

**10.** *It was discovered that the liquid in the jars of canned peas was cloudy.*

**11.** *I found cloudy liquid and a white sediment in jars of carrots 2 months after canning.*

**12.** *The liquid siphoned out of jars of canned beets during processing so each jar now has little liquid.*

**13.** *Meg cans strained pumpkin to use for pies. She processed it at 12 pounds pressure for 80 minutes for quart jars (altitude, sea level).*

**14.** *I used a steamer to process my hot-packed quarts of canned pears. I processed them for 33 minutes.*

**15.** *Grandmother's pressure canner leaks steam, so she adds extra water at the beginning of processing. Is the heat treatment still adequate?*

## Survey of Canning Literature

**1.** Skim-read the USDA 1988 publication *Complete Guide to Home Canning.* Note any questions you may have and bring them to class.

**2.** Read a scientific journal article of your choice on home canning. Prepare a short (1–2 pages) written report and a 5-minute oral presentation for class. Post the written report in class for all to share. If the authors made recommendations for householders to follow, do you feel they are suggested by research?

**3.** Photocopy a popular press recipe for a home canned product. Analyze the various procedures in it according to scientific theory. What are the author's credentials? How does this recipe differ from processing procedures recommended by the USDA for this same product? Prepare to discuss the recipe in class and post it with your comments for all to share.

**4.** Read a scientific journal article of your choice on home canning of tomatoes. Be prepared to share your knowledge of tomatoes during class and bring the citation to share with others.

**5.** Bring any canning questions you may have encountered while working with consumers, or inaccurate procedures that circulate in your neighborhood. What resources do you have to answer such questions after graduating from this class? Is your sound scientific advice likely to be followed? Why?

# PRACTICAL EXPERIENCES

When several lots of one product are being processed, the jars must be labeled for identification after heating. Tying a string around the neck of the jar below the threads and then adding knots in the end of the string (1 knot for treatment one, 2 knots for treatment two, etc.) works well.

A maximum temperature thermometer, such as the Taylor 21811, is a good teaching tool for those working with householders. Ask your laboratory guide for a local source of these thermometers or see Appendix A.

## Rate of Heat Penetration

Factors included in this experiment are type of food, type of pack, size of container, type of container, heating medium, and the temperature of the heating medium. Please note the effect of each of these variables. (Note: These are not recommended processes.)

### Packing the Jars

#### Spinach

Wash about 4 pounds of spinach and steam blanch 3 minutes. Cool to room temperature in cold water. Drain. Weigh spinach and divide into 4 lots, 1 weighing 1/7 of the total sample weight and the other 3 weighing 2/7 each. Pack into four half-pint jars. Add tap water leaving 1/2-inch headspace. Place a maximum temperature thermometer inside one of the firmly packed jars, then close with the flattop lid and ring. The other three jars should be closed with lids that have a hole in the center for a laboratory-type thermometer to protrude through. Adjust the thermometer so the bulb is in the center portion.

#### Beans

Wash and cut about 1/2 pound of snap beans. Blanch 2 minutes, cool in tap water, and drain. Pack in a half-pint jar and add tap water to within 1/2 inch of top. Put a lid prepared with a thermometer protruding, on the jar.

#### Pumpkin

Fill each of two half-pint jars, one pint jar, and one metal can to within 1/2 inch of the top with canned pumpkin puree. Put lids with thermometers on the jars and the metal can.

### Heating the Jars

#### Boiling-Water Canner

Fill kettle with water 2 inches above the rack and heat to boiling. Place the metal can and all of the jars except the one spinach with the maximum thermometer, one of the half-pints of pumpkin, and one of the pints of pumpkin into the bath at the same time. Add boiling water to reach the neck of the jars. Cover and boil 10 minutes. Record temperature reached in each jar and in the tin. Prepare a similar boiling-water canner for the pint jar of pumpkin and proceed as for the half-pint jars above. Record temperatures for each container after 10 minutes of boiling-water processing. Repeat for three readings.

#### Oven

Heat to 212F (100C). Set the half-pint jar of pumpkin in oven and heat 10 minutes. Record temperature. Repeat for three readings.

#### Pressure Canner

Put 1 inch of water in a pressure canner and heat to boiling. Place the half-pint jar of firmly packed spinach with the maximum temperature thermometer into the canner and clamp on the lid. Follow canner instructions for venting and bringing up to pressure. Bring up to 10 pounds and hold there for 10 minutes. Remove from heat, cool until pressure reads zero, open cooker, and record temperature in the jar.

Make a chart of the internal jar temperatures after 10 minutes for each treatment. What differences are expected between dry-heat and moist-heat transfer? What other factors affect the rate of heat penetration? How do these principles apply to processing recommendations?

## ■ Factors Affecting Quality of Tomatoes

For this experiment, tomatoes with added liquid, either tomato juice or hot water, will be used according to directions on pages 9 and 10 in *Complete Guide to Home Canning* by USDA, 1988.

**1.** Wash, rinse and drain pint jars and prepare flattop lids as directed for the brand being used.

**2.** Fill boiling-water canner with hot water and bring to a boil.

**3.** Fill kettle with enough hot water to cover the amount of tomatoes to be scalded at 1 time. Bring to boil. Partially fill sink with cold water.

**4.** Wash and remove stem ends from 5 pounds of tomatoes. Scald as directed.

**5.** Vary the pack as follows:

A. Raw pack: One pint without acidification, 1 pint with an added 1 tablespoon lemon juice, and 1 pint with added 1/4 teaspoon citric acid.

B. Raw pack with calcium: Same as in step A only add 1 teaspoon $CaCl_2$ solution (2 1/4 oz anhydrous calcium chloride per pt distilled water).

C. Hot pack: One pint without acidification, 1 pint with an added 1 tablespoon lemon juice, and 1 pint with added 1/4 teaspoon citric acid.

D. Pressure canner hot pack: One pint without acidification, 1 pint with an added 1 tablespoon lemon juice, and 1 pint with added 1/4 teaspoon citric acid.

**6.** Place prepared lids on top and process as directed. Then record your evaluations.

## ■ Type of Tomato Pack and Processing Times

**1.** Wash, rinse, and drain three pint jars and prepare flattop lids.

| Factor | Food | Heating Medium | How and Why Rate Affected |
|---|---|---|---|
| Type food | Beans | Boiling water | |
| | Pumpkin | Boiling water | |
| Pack | | | |
| loose | Spinach | Boiling water | |
| firm | Spinach | Boiling water | |
| Jar size | | | |
| 1/2 pt | Pumpkin | Boiling water | |
| pt | Pumpkin | Boiling water | |
| Container | | | |
| tin | Pumpkin | Boiling water | |
| glass | Pumpkin | Boiling water | |
| Heating | | | |
| medium | Pumpkin | Boiling water | |
| | Pumpkin | Oven | |
| Temperature | | | |
| 212F | Spinach | Boiling water | |
| 240F | Spinach | Pressure cooker | |

| Type Pack | Variable | Shape and Texture | Flavor | pH |
|---|---|---|---|---|
| | W/o acid | | | |
| A | Lemon juice | | | |
| | Citric acid | | | |
| | W/o acid | | | |
| B | Lemon juice | | | |
| | Citric acid | | | |
| | W/o acid | | | |
| C | Lemon juice | | | |
| | Citric acid | | | |
| | W/o acid | | | |
| D | Lemon juice | | | |
| | Citric acid | | | |

**2.** Fill boiling-water canner with hot water and bring to a boil.

**3.** Prepare and scald 4 pounds of tomatoes according to directions on p. 9, Guide 3, USDA, 1988.

**4.** Vary the pack as follows: Use either hot or raw for all treatments.

a. Whole, water packed (Guide 3, USDA, 1988, pp. 9–10).

b. Whole, tomato juice packed (Guide 3, USDA, 1988, pp. 10–11).

c. Whole tomatoes without added liquid (Guide 3, USDA, 1988, pp. 11–12).

**5.** Place prepared lids on jars and process for the times recommended.

**1.** How and why does the pack influence the processing time?

**2.** With which medium did the tomatoes retain their shape best?

**3.** What is the range in pH of tomatoes?

**4.** How do the processing times for tomatoes and pears compare? Why?

**5.** Is it safe to process tomatoes in a boiling-water canner? Why?

**6.** Why do the skins slip on scalded tomatoes?

## Factors Affecting Quality of Tomato Juice

**1.** Wash, rinse and drain four pint jars and prepare flattop lids.

**2.** Fill boiling-water canner with hot water and keep hot.

**3.** Wash and remove stem ends from 5 pounds of tomatoes.

**4.** Cut each tomato into thirds. Place one-fourth of each tomato into 1 of 4 lots.

**5.** Treat the lots of tomatoes as follows:

*LOT 1.* Quickly, as soon as quarters are cut, put into saucepan, heat to boiling while crushing. Simmer 5 minutes.

*LOT 2.* Prepare as for Lot 1 but bring to a boil *slowly* (20 min) and then simmer 5 minutes. Stir.

*LOT 3.* Put into jar of blender and blend for 3 minutes. Then place into pan, bring to boil and simmer 5 minutes.

*LOT 4.* Leave this lot unheated. Simply put fresh tomatoes through a food mill.

**6.** Strain the juice, add 1/2 teaspoon of salt and 1 tablespoon lemon juice per pint, bring the 3 heated lots back to a boil and at once fill jars to 1/2 inch of the top.

**7.** Process as directed in boiling-water canner for 35 minutes (Guide 3, USDA, 1988, p. 6). Evaluate and fill out table.

**1.** How and why do the processing times for tomatoes and tomato juice differ?

**2.** Account for the differences in the appearance of the juices.

**3.** Account for any differences in flavor.

**4.** What "hidden" differences might you

| Pack | Time | Shape | Texture | Flavor |
|---|---|---|---|---|
| Water | | | | |
| Tomato juice | | | | |
| None | | | | |

| Treatment | Processing Time | Appearance | Flavor |
|---|---|---|---|
| Rapid boil | 35 min | | |
| Slow boil | 35 min | | |
| Cold crush | 35 min | | |
| Uncooked | 35 min | | |

expect? Why? Do these principles apply to other foods also?

## Factors Affecting Quality of Home Canned Pears

**1.** Arrange equipment with work simplification in mind. Wash, rinse, and drain seven wide-mouth pint jars. Prepare flattop lids.

**2.** Make sufficient medium syrup (Guide 2, USDA, 1988, p. 5) to cover fruit in five pint jars, and sufficient very light syrup to cover fruit in two pint jars.

**3.** Fill boiling-water canner with hot water and bring to a boil.

**4.** Pare, halve, and core (use corer or measuring spoon) 14 medium-sized pears. Place in bowl of water containing 2 teaspoons of salt per gallon.

**5.** Vary the pack as follows:

a. *Hot pack.* Place four halves in a saucepan with sufficient medium syrup for 1 pint. Repeat with sufficient very light syrup. Follow directions on p. 17, Guide 2, USDA, 1988.

b. *Raw pack.* Pack four halved pears into each of 5 pints.
   1) Fill one jar with boiling very light syrup.
   2) Fill one jar with lukewarm medium syrup.
   3) Fill two jars with boiling medium syrup.
   4) Fill one jar with boiling medium syrup to which has been added 300 mg of ascorbic acid.

It is recommended that the jars be coded by tying a string around the neck. Then tie one knot in the end of the string for treatment one, two knots for treatment 2, etc.

**6.** Put on lids and process for the recommended time except double the processing time for one of the jars packed with medium boiling syrup. Record your observations.

**1.** How much syrup was needed to fill a pint jar? What are the proportions of sugar to water for a very thin syrup? Medium syrup? Heavy syrup?

**2.** Why were the jars completely tightened before processing—as opposed to the glass closures?

**3.** When was the seal tested? Why?

**4.** How much headspace was allowed in the jars? Why?

**5.** What is the average pH of pears? Why can they be processed in a water bath?

**6.** How did the concentration of the syrup affect the fruit? Why?

**7.** How did the temperature of the syrup affect the processing time and the fruit? Why?

**8.** How did the type of pack affect the fruit? Why? Advantages of each type?

**9.** Account for the effects on the pears of

| Pack | Syrup | Temp. | Time | Color | Flavor | Texture |
|---|---|---|---|---|---|---|
| Hot | Med. | Boiling | 20 | | | |
| Hot | V. light | Boiling | 20 | | | |
| Raw | V. light | Boiling | 20 | | | |
| Raw | Med. | Lukewarm | 20 | | | |
| Raw | Med. | Boiling | 20 | | | |
| Raw | Med. | Boiling | 40 | | | |
| Raw | Med. w/acid | Boiling | 20 | | | |

doubling the processing time.

**10.** How did the addition of ascorbic acid affect the fruit? Why?

**11.** Which fruits are handled similarly to pears for processing?

**12.** Did any pink pears result from this experiment? What are possible causes of this color change?

## Factors Affecting Quality of Home Canned Snap Beans

**1.** Wash, rinse and drain three pint jars and prepare flattop lids.

**2.** Prepare the pressure canner for processing.

**3.** Wash and drain 3 pounds of beans. Trim ends and cut into 1-inch lengths.

**4.** Treat beans as follows:

A. Pack 1 pint (1 lb) raw.

B. Blanch 1 pound by placing in 1 gallon of rapidly boiling water on a *hot* unit—preheating the burner is important for electric ranges. Boil 5 minutes.

C. Blanch 1 pound by placing in 1 gallon of water at 180F (82C) and holding for 5 minutes.

**5.** For each treatment, fill the jars with beans and then add hot liquid, leaving 1-inch headspace. Process as directed on pages 8 and 9, Guide 4, USDA, 1988. Code the jars before they are placed in the canner. Process all at the same temperature for 20 minutes. Record your observations.

**1.** How do the processing times recommended for raw and hot pack differ?

**2.** What factors influence the amount of liquid lost from the jar?

**3.** Which pretreatment gave the softest beans? The firmest? Why?

**4.** What is the pH of snap beans?

**5.** What is the purpose of the 2 to 3 inches of water in the pressure canner?

| Pretreatment | Loss of Liquid | Texture |
|---|---|---|
| Raw pack | | |
| Boiling water blanch | | |
| Simmering water blanch | | |

**6.** Why is the pressure canner exhausted (Guide 1, USDA, 1988, p. 21)?

**7.** How is the pressure in the canner related to the effectiveness of the process?

**8.** Why should the pressure be held steady?

**9.** Pressure saucepans are no longer recommended for processing canned foods. In the past, when a pressure saucepan was used for canning instead of a pressure canner, the processing time was increased even though both pressures were the same. Explain both the previous and the current recommendations for use of pressure saucepans in canning. If the class has extra product, pack two jars of low-acid food with maximum temperature thermometers, then process one in a pressure canner and one in a pressure saucepan for equal times.

**10.** What adjustment for altitude should be made when food is processed in a pressure canner? Why?

**11.** Is canned food sterile?

## Effect of pH on Processing Time for Beets

**1.** Wash, rinse, and drain two pint jars.

NOTE: Some may wish to use glass closure jars on the pickled beets to gain experience with this type. Many young householders tighten flattop lids after processing, as their mothers did when using the old glass and ring type. Seeing the differences in the lids helps in understanding the differences in procedures. Tightening the new lids after processing may break the seal. The old glass closure jars have become recently available for craft projects. Householders are using them to can nonfood gift items and there are questions about their use. They are not recommended for use in canning foods. See Appendix A for a mail order source of these supplies.

**2.** Prepare a boiling-water canner and a pressure canner for processing.

**3.** Prepare pickling syrup.

**4.** Prepare 2 pounds of beets for canning as directed on p. 9, Guide 4, USDA, 1988. Beets may be precooked in pressure saucepan for 5 to 15 minutes or follow the boiling instructions.

**5.** Slice the beets and pack 2 pints as follows:

A. Can as directed on p. 10, Guide 4, USDA, 1988.

B. Pack with pickling syrup and process 30 minutes (0-1,000 ft altitude) in boiling-water canner.

**6.** Record your observations.

**1.** Why are stems left on beets when they are cooked?

**2.** Why are the processing times the same for the two packing media?

**3.** What is the pH of beets? Of vinegar?

**4.** Account for the differences in texture and color you observe in the beets.

**PICKLED BEETS**

Pickling brine (for 1 pt, adapted from USDA Home and Garden Bull. 92, 1978)

½ tsp allspice
⅓ stick cinnamon
⅓ c sugar

| Style | Medium | Temp. | Time | Color | Texture |
|---|---|---|---|---|---|
| Sliced | Water | | | | |
| Sliced | Acidified | | | | |

¼ tsp salt
½ c plus 1½ Tb vinegar
¼ c water

Loosely tie allspice and cinnamon sticks in clean cheesecloth or spice bag. Combine sugar, salt, vinegar, and water; add spice bag. Bring to a boil; then simmer for 15 minutes.

Remove spice bag. Pack beets into hot pint jar. Cover with hot liquid, leaving ½-inch headspace. Adjust jar lid.

Process in boiling water for 30 minutes. Remove jars and complete seal if glass closure was used. Cool and label.

## Effect of Container on Canning Asparagus

There are currently no recommended processing times for cans, however householders with can sealers continue to use them and the USDA may revise the old processing times. This experiment is designed to provide exposure to this method of home canning. See Appendix A for a source of these materials.

1. Prepare 2 pressure canners for processing.
2. Wash, rinse, and drain one pint jar and rinse 1 metal can and lid.
3. Set up the metal-can sealer.
4. Prepare 3 pounds of asparagus as directed on p. 5, Guide 4, USDA, 1988.
5. Pack hot, as on p. 5.
6. Process as directed.

1. Account for the difference in processing time.
2. After a storage period, open the can. Account for the difference in appearance between asparagus canned in glass and asparagus in metal.
3. How and why does the processing time of corn differ from that for asparagus?
4. To what temperature should the contents of metal cans be heated before they are sealed? Why?
5. How is the same problem addressed in a glass canning jar?

| Container | Processing Time | Appearance |
|---|---|---|
| Glass jar | | |
| Metal can | | |

## Canning Meat, Fish, and Poultry

### Ground Meat

**1.** Shape 1 pound of ground meat into flat cakes approximately 1/2 inch larger in diameter than that of the canning jar. Cook in skillet. Pack hot into the jar. See p. 6, Guide 5, USDA, 1988.

**2.** Prepare pressure canner for processing.

**3.** Put lid on the jars, label, process as directed, and cool. Check seal carefully before storing since fat may have come between the jar rim and the sealing compound of the lid and cause the lid to not seal.

### Poultry

**1.** Can hot pack, with bone, one-half of a chicken in a pint jar.

**2.** Follow directions on p. 5, Guide 5, USDA, 1988.

**3.** If desired, also hot pack, without bone, one-half of a chicken in a no. 2 can. The contents of the can should be 170F (77C) when it is sealed.

**4.** Pressure process the can for 90 minutes at 10 or 11 pounds and the pints for 65 minutes at 10 or 11 pounds. Clean outside of can if needed, label, and cool.

### Salmon

**1.** Scale, clean, rinse, and cut the fish as directed on p. 10, Guide 5, USDA, 1988.

**2.** Prepare pressure canner for processing.

**3.** Pack fish in jar solidly.

**4.** Process in pressure canner at 10 (weighted gauge) or 11 (dial gauge) pounds for 100 minutes. Wash jars if needed, label, and cool.

Notice the difference in pressure recommendations (10 or 11 lb) but for the same time of 100 minutes. How would you explain this difference to a concerned householder who owns only a weighted-gauge canner? In what other foods does this situation occur? (Note: The USDA reasoning is reviewed in USDA, 1988.)

## Thickened Pie Fillings

Canning these products has just recently been made possible for householders by the availability of a specialty starch which has been used commercially. Clear Jel is a modified starch which does not break down when heated, as cornstarch or the starch in flour would. The thickening power of supplies on the market varies, so an experimental jar is recommended first for any filling. The amount of Clear Jel can be adjusted without compromising the safety of this recipe. Consult your Extension Service for a local source, or see Appendix A for a mail order source.

**Blueberry Pie Filling**

1¾ c blueberries, thawed or fresh
7 Tb sugar
3 Tb + 1½ tsp Clear Jel
½ c cold water
1¾ tsp bottled lemon juice

Fill boiling-water canner and heat. Prepare lids in boiling water as package directs. Wash and drain berries. In a dry, heavy saucepan thoroughly combine sugar and Clear Jel. Add cold water. Thoroughly combine water and dry ingredients. A blender may be used briefly if necessary. Cook on medium high heat with stirring until mixture boils. Add bottled lemon juice and cook 1 minute, stirring constantly. Remove from heat and fold in berries. Quickly fill one pint jar, leaving ½-inch headspace. Process 30 minutes at sea level in boiling-water canner. Recipe adapted from USDA, 1988.

**1.** After processing, open jar and serve it beside commercially canned blueberry pie filling. What differences do you notice?

**2.** Did you experience difficulty in evenly dispersing the Clear Jel with water? What recommendations do you have for dispersing it?

**3.** Consider the thickness of your pie filling. Would you adjust the amount of Clear Jel?

**4.** Is it more expensive to prethicken canned pie fillings with Clear Jel than to can fruit in syrup and thicken with cornstarch just before using?

**5.** What is the purpose of bottled lemon juice? This addition makes the filling taste too tart, what do you recommend?

# 10 Root Cellar and Dry Storage

## Typical Consumer Questions

Give both a consumer answer and a technical answer to the following questions.

**1.** *Why do apples so often have brown flesh when purchased after January?*

**2.** *Can I determine if an apple has been stored too long in a commercial warehouse before purchasing it?*

**3.** *I prefer natural-type foods. How do I know if produce in stores has been held in controlled atmosphere storage?*

**4.** *The apples on my neighbor's tree do not drop in the fall. They are still in the branches this November. Are they safe to consume?*

**5.** *There are small moths and little worm-like creatures in my flour, cornmeal, jello, and just about everything else in my kitchen! What do I do?*

**6.** *We operate a small food bank for our rural community. We received several 50-pound sacks of pinto beans. The bottoms of some of the sacks appear to have watermarks on them. Are the beans safe to distribute?*

**7.** *I found a large box of nonfat dry milk in the back of a cupboard. I don't know when it was purchased. It is slightly tan and smells a little funny. Do I have to discard it?*

**8.** *A lady in her late 80s brought a bushel of carrots in to the senior center as a donation for our annual fund-raiser dinner. They are pretty dirty and I even saw pieces of wet sawdust on them. This is February so I know they are old carrots. She said they were in her cellar. What do we do?*

**9.** *I am concerned about purchasing dry pasta such as macaroni from the bulk barrels in our supermarket. Is it going to be safe to serve my family with so many hands touching it first?*

## PRACTICAL EXPERIENCES

### Survey of Local Cellaring and Dry Storage Practices

**1.** Contact your local Extension Service for a referral of a householder who uses a root cellar. Would a class tour be possible? If none is available, try to arrange a visit to a commercial produce warehouse or the postharvest physiology laboratory on a university campus.

**2.** Interview a householder about dry storage. To best achieve an unbiased interview, do not be judgmental. Be a good listener.

Ask the householder about insects in food. What are the major infestations in your area? What does the major food preparer do concerning insects in flour and rice? Why? Do you agree with the way this person handles infested food? Why?

Has the householder ever had nonfat dry milk or other food become yellow and acquire an off-flavor? Why does the householder think this occurred? What became of the remaining milk powder? Why? How would you handle a similar Maillard browning situation? Why?

Has rancid food ever been a problem in this kitchen? How was it detected? What was done with it?

How long does the householder feel dry macaroni can be kept for safety? for quality? What about cake mixes, chili powder, vanilla extract, coconut, cornmeal, sugar, oil, and potato chips? How does the householder determine appropriate storage times? Is this a scientifically sound method?

Which dry-stored foods are discarded fairly often in this household? Why? Is there a practical solution to this?

## Storage of Carrots

Trim tops of carrots leaving 1 inch of green stems. Do not trim the root end. Do not scrub or peel. Place carrots in a small crate with 3 inches of wet sawdust, sand, or dirt in the bottom and 1 inch of this wet material in between layers. Top the crate with another 3 inches of wet material. If space permits, store the crate at temperatures just above freezing until next quarter's class for evaluation after storage.

**1.** Are the bottom carrots likely to be bruised by the weight above them?

**2.** Would you feel comfortable eating carrots stored in dirt, sand, or sawdust? Explain.

**3.** Many householders successfully store carrots packed loosely in crates without layers of wet material in between, and a piece of damp burlap covering the top. Which storage method do you feel is more likely to result in longer shelf life?

**4.** Given the price of carrots available year-round in retail markets, do you feel root cellaring this vegetable is practical from an economic perspective? How about potatoes? Explain your point of view.

## Sugar Accumulation in Potatoes

Start the storage procedure 4–6 weeks before the root cellaring laboratory.

Select medium-sized russeted baking potatoes. Divide into 5 equal lots. Put all in perforated plastic bags, or store a more suitable way if possible, for moisture control. Store 1 lot at 32F (0C) for 4 weeks, store 1 lot at 32F (0C) for 4 weeks then at room temperature for 1–2 weeks, 1 lot at 38–40F (3–4C) for 4 weeks, 1 lot at 50F (10C) for 4 weeks, and 1 lot at 68–70F (20–21C) for 4 weeks.

To test for sugar content, pour 1/4 inch fresh oil into electric skillet. Heat to 375F (190C). Cut potatoes into even slices and fry all lots for equal time (3 min per side for 1/4-in. thick slices). Drain on paper towels and serve. If lots must be done in batches, do not re-use the oil as it becomes darker with continued heating. Evaluate color and sweet taste of the lots. Sugar accelerates browning (Maillard reaction). Which storage temperature resulted in greatest sugar accumulation? Why?

What is the usual storage method of potatoes in your locale? Is it a sound practice? If not, recommend a good method that is likely to be followed by most householders.

## Relative Humidity and Condensation

Controlling plant respiration and storage room humidity are the two most important factors in increasing storage life. Household root cellars usually achieve 90% humidity in the air directly surrounding produce through use of damp sawdust or burlap, but when condensation problems occur, and in the more elaborate cellars, there must be adjustment of humidity throughout the storage room. Ninety to 95% relative humidity (rh.) is recommended for most vegetables to retard moisture loss and not encourage rots—microorganism growth. Regulation of rh. starts with its measurement.

A small sling psychrometer works well for households. One can be purchased inexpensively at refrigeration supply outlets, refrigeration repair shops, and some hardware stores. The psychrometer has two thermometers, one bulb is left dry and the other is covered with a wick wetted with distilled water. As the psychrometer is spun in the air, water evaporates from the wick—into the air—and the wet bulb cools. The bulb cools to a specific temperature related to the amount of water that evaporates. The amount of water that evaporates is directly related to the amount of water already in the air—the drying power of the air. Thus the sling psychrometer measures relative humidity.

When the wet bulb and dry bulb temperatures are known, rh. can be read from a psychrometric chart. Figure 10.1 is a simplified psychrometric chart. Practice reading it. The wet and dry bulb temperatures are listed along the edges. Lines drawn from these coordinates until they meet in the center area will intersect at the relative humidity value. Dry bulb temperature is the same most thermometers read. The wet bulb temperature curve is also 100% humidity and saturation temperature.

**1.** If the humidity is near saturation and the temperature drops, what happens?

**2.** Practice using a sling psychrometer and other types of psychrometers if available. What are the advantages of each? Do you feel householders can accurately determine rh. using a sling psychrometer? Do discard savings by using this piece of equipment justify its cost?

## Condensation

Relative humidity and temperature determine condensation. Together as a group, read the psychrometric chart.

**1.** At 90% humidity and 32F (0C), is condensation likely?

**2.** When the storage temperature is 85F

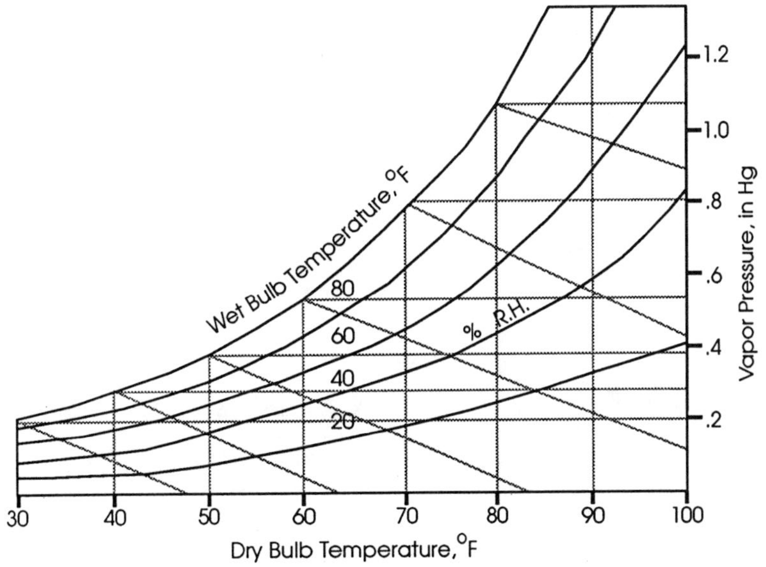


**10.1.** Psychometric chart for determining percent relative humidity on the basis of wet bulb temperatures.

(29C), and humidity is 90%, what happens if the temperature drops 5F (1C)?

**3.** Consider a situation in which a box of individually wrapped apples has been stored in a garage for 3 months. Rots are discovered on some and the entire lot needs to be checked. The apple flesh is 40F (4C). The box is taken inside to 72F (22C) for comfortable sorting. Is condensation likely either during sorting or when the apples, warmed to 45F (7C), are returned to a 40F (4C) garage?

**4.** Root cellar humidities over 95% usually result in condensation, so are not recommended. Do you think this caution is warranted?

**5.** Why are temperature changes in autumn a concern in root cellaring?

# 11 Preservation with Sugar: Jam and Jelly

## Typical Consumer Questions

Divide the following questions among the class members then share the answers. Give a consumer answer plus a technical answer. If possible, provide a reference(s) for the latter.

**1.** *Freezer-type (uncooked) jams have so much sugar. Just look at the amount you add to red raspberries. Why do they have to be so sweet? Why do they contain more sugar than boiling-water processed jams?*

**2.** *My homemade grape jelly has gritty lumps in it. How could I prevent this next time?*

**3.** *We have bees and produce our own honey. To save money, I want to substitute honey for sugar in making jelly. What proportions do I use?*

**4.** *Should moldy jam or jelly be discarded?*

**5.** *My strawberry jelly is usually successful. However, this year it is too soft. What causes this?*

**6.** *An entire batch of jam has fermented. What caused this?*

**7.** *My apple jelly has crystals in it. I think they are sugar crystals. What happened?*

**8.** *My great-aunt makes the best-tasting crabapple jelly. She always seals it with paraffin. Should I discard the jars she shares with us, or is it safe to consume? She would be very offended if I suggested she was not processing it properly and she does not have the equipment to water bath anyway.*

**9.** *I found a recipe for corn silk jelly in a cookbook my grandmother used during the depression. I would like to recreate some of these memories for my family. The ingredients are corn silks and water cooked to make 3 cups juice, one package pectin, 2 tablespoons lemon juice, and 4 cups sugar; all boiled for 1 minute. Is this going to be safe to eat since it is not made from a fruit? Will the recipe turn out?*

## Gelling Literature and Household Practices

Read one of the following journal articles. Write a 1 page abstract and post the abstract and citation in lab for all to share.

Gilpin, Gladys, Lamb, Jessie C., Staley, M. and Dawson, Elsie, H. 1957. Development of jelly formulas for use with fully ripe fruit and added pectin. Food Tech. 11:323-328.
Goldthwaite, N.E. 1910. Contribution on jelly making. Ind. and Eng. Chem. 2:457-462.
Kramer, Mary A. and Sunderlin, Gertrude. 1953. The gelation of pectin in uncooked jam from frozen red raspberries and strawberries. J. Home Econ. 45:243-247.
Olsen, A.G. 1934. Pectin Studies III. General theory of pectin jelly formation. J. Phys. Chem. 38:919-930.

### Homemade Jelly Analysis

Analyze the following recipe and briefly write the reason for each of the steps. You do not need to use complete sentences; phrases are fine—Examples: Firm apple—overripe fruit contains less pectin. Crush grapes—so fruit interior is exposed.

**CONCORD GRAPE JELLY**

4 lb grapes
1 firm, unpeeled, uncored apple
½ c water
3 c sugar

Rinse grapes, removing any stems or leaves, Put them in a large pan and crush. Cut the apple into eight pieces and add along with water.

Stir while bringing to a boil, then simmer approximately 10 minutes until apple is soft. Pour into a damp jelly bag and let drain without squeezing. Refrigerate this juice overnight and then strain again.

Measure 1 quart of this juice into a large pan and add sugar. Water may be added if needed to provide 1 quart of liquid. Bring to a boil quickly and boil until it reaches 220F (104C). It will sheet from a cold spoon at this point. Remove pan from heat, do sheeting test, and quickly skim off the foam. Pour into clean half-pint jars leaving ½-inch headspace. Process in a boiling-water canner for 5 minutes, at altitudes of 1,001 to 6,000 feet, process for 10 minutes, above 6,000 feet for 15 minutes.

## PRACTICAL EXPERIENCES

### Gelling Experiment

**BASIC JELLY**

2 c water
1¾ oz powdered pectin
2 Tb lemon juice

Heat above three ingredients to boiling, then stir in

3¼ c sugar
color and flavor

Boil all for 1 minute stirring constantly. Fill jars that are tall and narrow and seal; shot glasses or juice glasses will also work. When completely cooled to room temperature (another lab period), turn molded jelly out onto a flat, hard surface. Estimate the percent sag and gel strength for all variations made.

VARIATIONS

1. Substitute 3 c sugar for the original 3¼ c
2. Substitute 2½ c sugar for the original 3¼ c
3. Substitute 1¾ c sugar for the original 3¼ c
4. Substitute 4 c sugar for the original 3¼ c
5. Omit lemon juice
6. Add ¼ tsp sodium bicarbonate

For gel strength: Measure using a penetrometer, if available.

To measure percent sag: Measure the height of the jelly *in* the jar; turn out carefully and measure height after 1 minute. The percent sag is the one minute height in cm divided by the original height in cm times 100.

Explain the gelling differences in the variations. Do you have a personal preference? Which do you think is the closest to the texture of commercial jelly?

### Conventional Apple Jelly Experiment

**CONVENTIONAL APPLE JELLY**

3 lb tart apples (¼ under-ripe)
3 c water
3 c sugar
1 lemon

Wash the apples and remove stems and blossom ends. Do not pare or remove the core. Cut each apple into thin slices. The yield should be about 6 cups. Remove the yellow rind from 1 lemon, cut lemon in thin slices and add to the apple slices. Add 3 cups water, bring to a boil, cover, and simmer until fruit is tender (20 to 25 min). Extract juice by putting apples in a moist jelly bag and allow juice to drip freely into a glass bowl or stainless steel saucepan. Do not squeeze the bag. The yield should be approximately 4 cups.

Use a Jelmeter to measure the concentration of pectin in the juice. Two tablespoons of juice will be required for the test.

Combine the fruit juice and sugar in the ratios given below to make 3 different lots of jelly. Use sufficient juice and sugar to yield 3/4 to 1 cup of jelly. The volume of jelly will be slightly greater than the volume of sugar. Use a 1-quart saucepan for each lot and put a masking tape label on the handle of each pan to keep the lots straight.

*LOT 1.* Combine sugar and juice in the proportions indicated by the Jelmeter test.

*LOT 2.* Use the same volume of juice as in lot I and reduce the volume of sugar by 10%.

*LOT 3.* Use the same volume of juice as in Lot I and increase the volume of sugar by 10%.

Wash, rinse, cover with boiling water, and boil for 10 minutes three 8-ounce jelly glasses and three test glasses. Remove from water and drain. NOTE: Glasses are sterilized because boiling-water processing will not be used. Bring the jelly to a boil and keep at a full rolling boil. Stir down foam if necessary. When the temperature of the boiling jelly reaches 217F (103C), use the sheet test to decide when the jelly is done. If uncertain about the doneness, remove jelly from the unit. This is usually at 220F (104–105C). As soon as the jelly is done, fill one test glass and pour the remainder into a jelly glass. Avoid splashing jelly on the sides of the glass. Pour a *thin* layer of melted paraffin on the top of two of the 8-ounce glasses. Pour a *thick* (1/4-in.) layer of paraffin on the top of

| Variation | Sag | Gel Strength | Mouthfeel | Comments |
|---|---|---|---|---|
| Original | | | | |
| 1. | | | | |
| 2. | | | | |
| 3. | | | | |
| 4. | | | | |
| 5. | | | | |
| 6. | | | | |

the remaining 8-ounce glass. NOTE: To melt paraffin, put metal container—a clean can will work—in the bottom of a saucepan containing boiling water.

When the jelly has thoroughly cooled, probably next lab period, evaluate each product and answer the following questions.

**1.** Remove the paraffin from each 8-ounce glass. Describe the seal on each. Why is paraffin no longer recommended for sealing jelly?

**2.** Turn out each of the sample glasses and measure the percent sag for each lot. Turn out each of the 8-ounce glasses and display with a label and bread or crackers for sampling. The standard for good jelly is that it will hold its shape when turned out. Explain any deviations from the standard.

**3.** What is the cost of the conventional jelly you made? How does this cost compare to commercial apple jelly? How does the quality compare? How long did it take to make this jelly; count separately the time you were involved in the actual preparation and the time spent waiting.

**4.** Estimate the amount of sugar in a serving of conventional jelly or jam and that in a pectin-added or a frozen product. Is the difference nutritionally significant or economically significant to an average household?

**5.** Ask the students making conventional jelly for the first time if they felt experience is important in this recipe. Were they able to make a satisfactory jelly from the written instructions? Would they try the conventional method again?

## Use of Jelmeter

Jelmeters are difficult to find in retail markets now. If a Jelmeter is available to you, this experiment shows its value for certain types of products. The jelly recipes being published now often only specify length of boiling time and use the sheet test to determine doneness. The Jelmeter is a tube with a larger diameter at the top and a fine capillary tube diameter at the bottom. Pure water moves rapidly through the capillary tube as columns of round water molecules slide past each other in a telescopic manner (laminar flow). Large pectin molecules with their kinks and curves impede this flow and result in longer times for the liquid to exit. Short segments of pectin do not disrupt this flow markedly, nor do they contribute significantly to gel formation. Therefore the Jelmeter does not measure the total amount of pectin; it measures the amount of good-quality pectin or the gelling power of a juice. Both the Jelmeter and the juice should be at room temperature as viscosity is temperature dependent. Fill the Jelmeter with the juice—a pectin sol. Allow a drop or two to fall to remove air bubbles, then fill completely with your finger over the narrow bottom. Remove your finger from the bottom of the capillary tube and allow the juice to flow out for 1 minute then stop the flow and take a reading.

NOTE: If the juice did not flow below the top mark (1 1/4 c), then dilute the entire batch with water or a juice without pectin and take another reading. Do this gradually. If the juice flowed so rapidly that the level was below the lowest mark (1/2 c), then there is not enough pectin to make a good gel. However this juice may produce an excellent flavored pectin-added jelly. Another alternative may be to concentrate the pectin present by boiling the juice. This will also concentrate the other flavor—sugar, acids—compounds. Before boiling, consider whether juice from this type of fruit is likely to contain adequate pectin even if boiled to concentrate. (See Table 11.2 in Book I.)

## Lower-Sugar Jam and Jelly

NOTE: Traditional jams and jellies are preserved by sugar; these products contain large amounts (55 to 85%) and they are thickened by a pectin-sugar-acid mechanism. The lower-sugar products contain significantly less sugar than the traditional so are not preserved by low $a_w$; however, any concentration of fruit will be a food with at least moderate amounts of sugar.

For the following experiments, use either no added sweetener or add similar amounts to each product. Use the same fruit for each type thickener if possible. Display the products for sampling with their recipe instructions. After all have sampled, the students who prepared each recipe should explain any unusual procedures or difficulties they had.

#### Low-Methoxyl Pectin (LMP)

Choose two brands of LMP on the market and prepare the manufacturer's recipe for both a jam and a jelly. Follow their directions for boiling-water processing. (See Appendix A for mail order source of LMP pectin in which the calcium is separately packaged for addition by consumer. This type can provide experiences in the effect of calcium in gelling.)

#### LMP and Artificial Sweetener

Prepare jam using an artificial sweetener and following the package directions.

#### Reduced Sugar

Use a reduced sugar pectin such as Sure-Jell Light, following package directions for a jam.

#### Gum

Choose a gum-based preparation for making lower-sugar gelled foods at home. Follow package instructions for both a jam and a jelly, and for boiling-water processing.

#### Gelatin

Prepare the following recipe which is based on one from the Oregon State Extension Service.

**STRAWBERRY JAM WITH GELATIN**

1½ tsp unflavored gelatin
1½ Tb cold water
3 c strawberries, crushed
1 Tb lemon juice

Soften gelatin in cold water. Combine fruit and sweetener if desired (to taste) in a saucepan. Place over high heat and stir constantly until mixture boils. Remove from heat; add softened gelatin; cook 1 minute more. Remove from heat and add lemon juice. Ladle into clean jars leaving ½-inch head space, cover, and store in refrigerator.

**1.** Why was the gelatin jam not heat processed?

| Brand | Texture | Comments |
|---|---|---|
| 1. LMP jelly | | |
| 2. LMP jelly | | |
| 3. LMP jam | | |
| 4. LMP jam | | |
| 5. LMP jam w/ sweetener | | |
| 6. Reduced sugar | | |
| 7. Gum jelly | | |
| 8. Gum jam | | |
| 9. Gelatin jam | | |

**2.** Did the processing times and instructions vary significantly among the products? Do they differ from the USDA recommendations?

**3.** Was the firmness appropriate for the product standard?

**4.** Ask those who often consume artificially sweetened foods to comment on the flavor of this jam. Was the sweetener heated?

**5.** For each gelling agent used, copy the ingredients from the label and indicate the purpose of each.

# APPENDIX A
# Sources of Additional Information and Hard-to-Find Preservation Supplies

*Resources are grouped by chapter*

## FOODBORNE ILLNESS

***Sources of Food Additive Information***

Pesticide registration and pesticide residue limits: Environmental Protection Agency, Public Information Center, 401 M Street SW, Washington, DC 20460. Phone: (202) 475-7755. Ask for *A Citizen's Guide to Pesticides.*

Animal drug registration, residue limits for animal drugs, animal feed, marketbasket surveys: Call the Food and Drug Administration (FDA) nearest you and ask for the FDA Consumer Affairs Officer. For FDA publications, write: Food and Drug Administration, HFE-88, Rockville, MD 20857.

***Practical Food Handling Information for Meat and Poultry***

USDA Meat and Poultry Hotline, 1-800-535-4555 in the 50 states, (202) 447-3333 in the Washington, DC, area. FDA Seafood Hotline, 1-800-332-4012.

### Microbiological Examinations of Food

Petri dishes, streaking loops, alcohol burners, and media are available from many sources. A company that routinely sells by mail order is VWR Scientific, 1-800-932-5000 to order from your closest office.

## FERMENTATION

Salometers are commonly available from laboratory supply outlets located in metropolitan areas. A source near you may carry a less expensive, imported salometer as these are widely distributed. However, the following is an outlet that ships nationwide. They offer an accurate, easy-to-read salometer, stock number 4260, for $17.00 plus $3.00 shipping and handling. This salometer's 0 reading is distilled water at 60F, and its 100 reading is 100% salt concentration. Nurnberg Scientific, 6310 SW Virginia, Portland, OR 97201. 1-800-527-8594 in Oregon, 1-800-826-3470 outside of Oregon.

## CURING

Hygrometers can often be purchased locally, but a mail order source of one that works well for brines is available from: Nichols Garden Nursery, 1190 North Pacific Hwy., Albany, OR 97321-4598. Phone: (503) 928-9280. FAX: (503) 967-8406.

The Morton pickle pump and its repair parts, smoking nets, and curing mixtures are available by mail order from Cumberland General Store if your local supermarket does not carry them. Cumberland General Store, Route 3, Crossville, TN 38555. Phone: 1-800-334-4640. FAX: (615) 456-1211.

An assortment of sausage stuffers is widely available in department stores and mail order catalogues. They range from electric ones with plastic dishwasher-safe parts, to those with hand cranks and metal which must be oiled to prevent rusting between uses.

A variety of curing mixes, some with smoke flavor, and pickle pumps are available from: Embarcadero Home Cannery, 2026 Livingston St., Oakland, CA 94606. Phone: (510) 535-2311.

## REFRIGERATION

There are many good sources for the plate count media used in the practical experience section. BBL Microbiology Systems, P.O. Box 243, Cockeysville, MD 21030. Phone: (301) 771-0100. Difco media are readily available nationwide. The *Difco Manual* is updated periodically and contains accurate procedures for using media. Many libraries carry the *Difco Manual.* Difco Laboratories, Inc., P.O. Box 1058, Detroit, MI 48232.

## CANNING

Maximum temperature thermometers are excellent tools for teaching heat penetration. These thermometers are approximately 3 inches long and fit easily inside of canning jars. They also accurately record the maximum temperature reached inside a dishwasher; a good experiment to explain why vacuum sealing of jars inside dishwashers is not a lethal heat treatment for microorganisms. The cost is approximately $50 each. Call the company for address of a local distributor: Taylor Company, Phone: 1-800-438-6045 or (704) 684-5178.

A distributor for the western United States and Canada is: Branom Instrument Co., 5500 4th Ave. South, Seattle, WA 98108. Phone: 1-800-767-6051.

A distributor for the eastern United States and Canada is: Green Equipment and Supply, 906 Brook Rd., Conshohocken, PA 19428. Phone: 1-800-227-0286.

The following is a mail order source for glass jars with glass dome lids and rubber jar rings. A metal can sealer, metal cans, replacement parts for pressure canners, and many other home preservation tools are also offered. Some of the items for sale in this source and some of the mentioned processes—such as canning 1/2-gallon jars, and using metal cans—are currently not recommended for safety reasons. This catalogue is listed here as an example of the availability of such items, and because it is a source of some hard-to-find recommended utensils: Cumberland General Store, Crossville, TN 38555. Phone: 1-800-334-4640.

The Nichols Nursery catalogue is a source for Clear Jel, a modified cornstarch, for safely canning fruits in thickened juice. Clear Jel is approved by the USDA and its use and recipes are discussed in USDA, 1988. The instant form of Clear Jel is not used in canning. Nichols Garden Nursery, 1190 North Pacific Hwy., Albany, OR 97321-4598. Phone: (503) 928-9280. FAX: (503) 967-8406.

The Embarcadero Home Cannery offers dial gauge testing by mail and is a source of canner parts and canning supplies. They carry a large assortment of pressure canners, including an electric model and a model to be used with propane gas. They are also a source for food-grade lime and 4-ounce jars which they market for jam and jelly: Embarcadero Home Cannery, 2026 Livingston St., Oakland, CA 94606. Phone: (510) 535-2311.

## JAM AND JELLY

The following mail order catalogue is a source of LMP pectin with calcium in a separate package. This is useful for experiments on the effect of calcium in gelling, and also provides experience with this type of LMP which is available in health food stores in metropolitan areas: Nichols Garden Nursery, 1190 North Pacific Hwy., Albany, OR 97321-4598. Phone: (503) 928-9280. FAX (503) 967-8406.

# APPENDIX B
# Suggestions for Answering Typical Consumer Questions from Book II

Making food safety decisions and food quality predictions when standard recommended procedures were not followed, is the mark of an accomplished food preserver. To solve the typical consumer questions at the beginning of each practical experience section you must put several food principles together and apply them to a household situation. When you are able to do this, you truly understand the principles of food preservation. Such decision making is important for householders when they make errors but do not wish to discard the food if it can be safely consumed, and these judgments are a critical part of any food professional's job.

The following suggestions for approaching food preservation questions follow scientifically sound principles. They are a guide to steer your thinking in the right direction, but complete answers require elaboration. Some of the chapter questions do not provide you with enough information to make a decision. For these questions, state facts you assume to be present and base your decision on them. Some chapter questions present facts that are irrelevant to making a judgement of the food's safety. A thorough answer will also note which information is extraneous.

Sometimes safety changes do not accompany quality changes. You *must* recommend discard when the food's safety is in question, but we encourage you to let the individual consumer determine when quality changes are acceptable. For example, when advising a diabetic home canner on preserving peach halves in plain water it is best to state that the fruit texture will be soft and perhaps even mushy with many fruit particles in the liquid, instead of stating, "If peaches aren't canned in heavy syrup they fall apart and aren't worth eating." Personal preference decisions can only be made by consumers themselves. For more complete information, refer to each chapter.

## 1. FOOD QUALITY

**1.** Pasteurized dairy products are perishable, but at 38–40F, growth of organisms that cause illness is not a concern. The liquid is due to a simple separation—physical change. Molds and bacteria that are able to grow at refrigerator temperature may have produced changes during storage. Off-odors and -flavors would then be present and reason for discard because of poor quality.

**2.** Quality change in texture—broken emulsion—during freezing is common in cheeses of all types.

**3.** Quality change: relocation of water.

**4.** Quality change: oxidative browning. Off-flavors may be present also. This is a chemical reaction and not harmful. Because it is unattractive, this layer is generally discarded.

**5.** Quality change: Maillard browning continues during storage.

**6.** High-fat foods such as snack crackers are prone to oxidative rancidity: discard for off-flavors. Crispness may also be lacking due to moisture uptake in opened packages.

**7.** Quality change: cake stales (changes in starch due to retrogradation) quickly at refrigerator temperatures. It is necessary to refrigerate custard fillings for safety. Combine the two products just before serving.

**8.** Quality change: physical separation. Try stirring gently.

**9.** Quality change: oxidative rancidity. This happens less quickly if meats are not sliced or are covered with gravy.

**10.** Quality change: water from filling softened crust. Generally, pies are baked the same day as served for top quality.

**11.** Quality change: add dressing immediately before serving.

**12.** Quality change: changes in starch (retrogradation) at refrigerator temperatures. Reheat with stirring.

**13.** Safety concerns: discard. Before adding water, rice is preserved by its dryness. After cooking (in water), it becomes perishable. Householders disliking the quality of refrigerated rice leftovers should prepare an amount which will all be eaten.

**14.** Quality change: pectin gels become thicker with time.

**15.** Quality change: Maillard browning. Taste to decide if discard is necessary.

## 2. FOODBORNE ILLNESS

**1.** Warm temperatures cause more rapid multiplication of bacteria in foods. Outdoor and water activities increase. Fish and animal feces are excellent sources of salmonellae, and human exposure to this bacteria is more common in summer. Barbeçued foods are often undercooked. Cross-contamination at picnic sites without good dish- and hand-washing facilities is likely.

**2.** Leftover pot pie improperly handled—not refrigerated, not thoroughly reheated. Interior of pie is anaerobic. Spores survive baking.

**3.** *S. aureus* has fewer competitors in salty foods and when it is introduced to a food after cooking. Widespread use of convenience foods.

**4.** Put a thermometer in the refrigerator section; the reading should be 40F (4C) or below. As long as the food is frozen at all, it is safe but quality loss will be rapid.

**5.** *S. aureus* thrives on unrefrigerated cooked meats. Ham is salty which inhibits growth of many other bacteria. *Salmonella* is commonly found on raw turkey, but would not survive cooking. The short time between eating the food and the illness is typical of staphylococcal food poisoning. Use an insulated lunch box with a frozen solution container.

**6.** Avoid parasitic infections by eating thoroughly cooked, hot foods; canned fruits or those that you peel yourself; and bottled beverages or boiled water.

**7.** Choose nonperishable foods. Tuna fish in an unopened can with a jar of mayonnaise and package of diced celery could be made into tuna salad sandwich fillings. Look for other ideas in the canned food section. Cheeses and assorted breads add variety. Fresh fruits and cookies are a wiser choice than cream pies.

**8.** Do not reduce vinegar. Add a small amount of sugar.

**9.** Question - Has it been on the kitchen counter? If so, not safe because of risk of staph food poisoning.

**10.** Report the incident to the local health department. They will then interview the children to see if a common food item was eaten by those ill and also will review food preparation practices where the lunch was prepared. Samples may be examined. When excited children overeat then ride in a vehicle that sways, stomach upset may also result.

**11.** Prevent cross-contamination.

**12.** Is the guest an acknowledged expert in identifying wild mushrooms? If not, then there would be a risk. A leading mycologist, who is faculty at a university in Washington State, is adamant that even he does not eat most types of mushrooms that he collects himself without first making a spore print for positive identification.

**13.** Paralytic shellfish poisoning (PSP) occurs in some seasons on some beaches. The shellfish are monitored and if a problem is identified, the harvesting is closed.

**14.** Open several packages that were on the outside of the stack. If these are acceptable in odor and appearance, they may be distributed. However, the turkeys will spoil quickly even when re-refrigerated. Since the birds were raw, bacteria that cause foodborne illness were competing with other types; those that could grow will be killed during cooking of the turkey.

**15.** No. Certified establishes the maximum number of microorganisms, but some of these may be those that cause foodborne illness. Risk of salmonellosis, campylobacteriosis, and listeriosis, especially.

**16.** Raw or rare hamburger may be contaminated with pathogenic bacteria; risk of the presence of salmonellae, *Campylobacter*, and hemorrhagic *E. coli*.

**17.** *Listeria*'s most common food vehicles are perishable foods that are purchased precooked, ready-to-eat, and then stored in refrigerators before eating. *Listeria* can multiply at refrigerator temperatures; the numbers may increase within a few days and be a risk. Sliced deli meats have been one of the most common problem foods, but other ready-to-eat items have also been implicated. Although the food industry, in general, has effectively decreased contamination of these foods, restaurants and delis have fewer resources for doing this.

As an extra margin of safety during pregnancy, because the consequences are so serious (possible miscarriage), it is prudent for pregnant women to reheat such foods before consumption. Thoroughly heated foods do not contain live *Listeria* organisms. Ordering steaming-hot foods (soup, casseroles, well-done meats) in restaurants instead of cold perishable foods (submarine sandwich, macaroni-meat salad) is a safer alternative when dining out.

## 3. DRYING

**1.** Yes, wood is a safe material. Visible dirt should be removed from the walls, the screens thoroughly cleaned and, if metal, examined for signs of corrosion.

**2.** Safe. Quality change: Maillard browning.

**3.** Commercial jerky is properly dried. Unless it has absorbed moisture, it is safe. It is unlikely, but possible, for some microorganisms to contaminate the surface of pieces in the jerky container from other human hands, just as this contamination is possible in restaurant and deli service if precautions are not taken. Health departments do not believe this is likely enough to forbid these displays.

**4.** Both home and commercially dried infant foods are convenient for preparing small portions of a variety of fruits and vegetables. The home dried may have greater nutrient loss, though they may be more economical. Home dried foods have varying nutrient retention, but usually the vitamin loss is greater than with other methods of preservation. The flavor of some dried infant foods may be inferior to canned.

**5.** For specific recipes on foods, consult Cooperative Extension Service or a drying "how-to" book.

**6.** Root cellaring until December would preserve more of the original qualities than drying.

**7.** Drying herbs with heat may result in poorer flavor, but some householders find them acceptable.

**8.** Ask for more detailed description if this isn't your own question. Is it spoiled, or typical fish jerky? It is consumed as a snack, similarly to beef jerky, and should be dry. The fat in fish is also readily oxidized so there may be a rancid odor.

**9.** Rehydrate in cooked dishes such as soups. Zucchini chips may be a more appetizing snack with a dip or spread.

**10.** Poor quality from enzyme activity.

**11.** Typical onion flavor is present only in freshly cut samples. However some householders find storing dried or frozen chopped onions acceptable; more is used than of the fresh.

**12.** Expect quality changes, but food will be safe.

**13.** Stainless steel are the only metal screens recommended for dryers. A brush may help in cleaning them.

**14.** Not food poisoning. Dried foods can rehydrate in the stomach after consumption. If equivalent of several apples was consumed, problems are not unexpected.

**15.** Fruits contain significant amounts of sugar in their natural state and leathers are concentrated fruits. Leathers may also stick to teeth more readily than fresh fruits.

**16.** With steady sunshine. There are several designs available.

**17.** Personal preference. Even nectarine peels may be too gritty in leather.

## 4. QUICK PICKLES

**1.** If not properly acidified, cucumbers could support growth of pathogenic bacteria. Though highly unlikely, mishandling by recipients (a sequence of mold growth, decreased acidity, pathogen growth) could result in an unsafe product. Commercially canned foods have documented records of processing and periodic inspections to shift liability away from manufacturer and distributor; home canned foods lack this. Therefore, in most states regulations specify that home canned foods cannot be served in group feeding sites or restaurants.

**2.** Check beginning cucumber quality and brine strength. More-mature cucumbers and those stored after picking may be a problem.

**3.** Yes. The peppers are only for flavor.

**4.** Pasteurization in a boiling-water process inactivates enzymes and results in a strong vacuum seal so that the pickles won't mold.

**5.** If vinegar:water ratio is at least 1:1 or the recipe is from a reliable source. Experiment with small batches to test flavor and taste after at least 3 weeks.

**6.** Decrease no more than 1:1 for household experimenting. Any less vinegar may result in a pickle that spoiled. Botulism would be possible.

**7.** Sweet pickles are as acid as dills, but contain more sugar which masks sourness.

**8.** Freshly harvested cucumbers give the best product but the quality will be acceptable if the pickles in the store appear to be free from defects.

**9.** Use top quality cucumbers, work quickly, follow recipe precisely, do not overheat. Experiment with recipes which have calcium (lime) as an ingredient.

**10.** As long as vinegar:water ratio is 1 part vinegar to 1 part water and vegetables are very small or in pieces.

**11.** Use litmus paper purchased from a drug store to test acidity; or a high school or college chemistry teacher may be willing to test acidity with a pH meter. The pH should be 4.6 or below. Pickles are usually about pH 4.0.

**12.** Yes, both vegetables will be preserved by the vinegar.

**13.** They cannot be used if a standard jar lid doesn't fit. Even so, heat penetration during the pasteurization can be a problem with unusually shaped jars. For pickles (and also jams and jellies), pasteurization in a boiling-water process can be estimated from the next largest jar size.

**14.** Discoloration probably from iron in water or a change in the white pigment to pink. They are safe.

**15.** Apples are acidic, even without the addition of vinegar. If there are no signs of spoilage, they are safe.

**16.** The label should state name and address of manufacturer. Commercially sold foods, even those from home kitchens, are required to pass an approval process for the production site in most states. Ask supermarket manager and/or state licensing agency.

## 5. FERMENTATION

**1.** Genuine dills are fermented, but the new formulas used produce different pickles from those made at home a generation ago. Try different brands.

**2.** It is very rare for fermentations to be unsafe. However products appearing spoiled should not be eaten.

**3.** Low-salt and no-salt sauerkrauts often spoil early in the fermentation process, but some batches may result in acceptable kraut. It is especially important to keep the kraut at 70–75F (21–24C) during the fermentation.

**4.** Kimchi usually contains garlic. The fermentation is also slightly different.

**5.** Poor texture is likely; spoilage may occur.

**6.** Add brine (1 1/2 Tb salt per qt of water) so there is enough liquid to cover the kraut. Stir during heating. Dry cabbage should not be canned (too poor heat penetration and oxidation).

**7.** Oxidation may occur if the brine doesn't cover the kraut.

**8.** Start with top quality, fresh cucumbers, prepare brine strength accurately, pasteurize only with the amount of heat required, work quickly.

**9.** Yes, but texture will be softer with these more-mature gherkin cucumbers. Next season, harvest when they are little.

**10.** Food-grade plastics do not contain such contaminants. Did the pail previously hold food?

**11.** If there are no chips or scratches in the enamel it is safe.

**12.** Fermentation brine often is cloudy. Filter it through a coffee filter if you want a clearer brine.

**13.** Compare your recipe to others. Check vinegar, salt, and spices. Check your ingredients; are spices old?

**14.** It is safe. The pickle may be somewhat different in quality.

**15.** Recipes for home production of black olives may be available from Extension offices in the states where olives are grown.

**16.** Yes.

**17.** You may reprocess, adding more liquid if necessary in reheating the kraut and using a new jar lid. Alternatively, store in refrigerator.

**18.** The texture of pickles will vary.

**19.** Be sure the plants have not been sprayed, and are not toxic. A Poison Control office will have a list of poisonous plants for local areas. Cooperative Extension offices usually have lists of edible plants for local areas. Day lilies pickle well. Herbs will lend some flavors. Edibles that work well in salads and punch bowls such as nasturtiums and pansies may be too fragile to pickle.

**20.** Yes, if glaze is intact. Wash well with soap and water to remove all visible dirt, then rinse with a 10% bleach solution.

## 6. CURING

**1.** Refer to nitrite section in text. Very high consumption of cured meats, such as bacon, by those in some countries have been associated with higher incidence of stomach cancer.

**2.** Generations ago, the hams had a much higher salt content. Today's lightly cured hams should be refrigerated for safety.

**3.** During manufacture—salting, heating—of ham, bacteria that would compete with *S. aureus* are eliminated, then staph are introduced from a human source and can grow in the salty environment. Ham is also more likely to be unrefrigerated (sandwiches, spreads) which encourages bacteria growth.

**4.** Refrigerate or freeze. The amount of drying different smoked salmon batches receive can be variable, so a second preservation method is needed. There are USDA directions for canned smoked salmon that are safe. A slightly longer processing time is required because heat transfer is slower than for fresh salmon.

**5.** These recipes are not widely in demand, however some local areas do address preparation of these foods. Try writing to the Extension Food Specialist at the state university, in a state where such foods are popular.

**6.** The use of nitrite and salt are essential to provide the typical cured ham. The process for a corned beef product prepared with refrigeration could be tried.

**7.** Salmonellae do not survive in frankfurter production. An extra margin of safety is provided when they are heated before serving.

**8.** The curing properties of saltpeter do not deteriorate with age; however if it was part of a curing mix, it shouldn't be used now. Nitrosamines may be formed in such combinations during long storage.

**9.** Yes, it would cool to refrigerator temperature and be safely stored.

## 7. FREEZING

**1.** Freezer burn and rancidity cause quality loss but the food is safe.

**2.** Green peppers are a vegetable that freezes well without blanching because the enzymes that cause quality deterioration are not present. True onion flavor is due to an enzyme-catalyzed reaction that occurs only when the tissue is freshly cut.

**3.** Wrap well as with any food. The quality and texture will be best if it is stored for a short time (3 mo).

**4.** Cheeses do not freeze well. However it may be used as an ingredient in recipes where texture will not be important.

**5.** The quality becomes poorer but the safety doesn't change.

**6.** Power outage for 4–6 hrs will not markedly affect a fully loaded freezer. The beef will be solidly frozen for several days with the quantity of frozen product in it.

**7.** Yes, there may be some darkening.

**8.** Unlikely if both are well wrapped and at 0F. Some cakes can develop off-odors from rancidity.

**9.** Yes. Texture does not change as much for bacon as for ham.

**10.** It will remain safe, but quality may deteriorate. The colder the freezer, the slower the changes in quality.

**11.** Some pies freeze better than others, so experiment. Cream and custard pies do not freeze well. For these, freeze unbaked pie shells.

**12.** The original, unopened container does not have enough room for expansion of the ice so some milk must first be removed. Thawed quality will be unacceptable as a beverage to most people, but it can be used in most baked products. However, milk keeps well in the refrigerator; you might plan to use it in sauces and desserts.

**13.** Boxed frozen foods are often impossible to evaluate before thawing. Purchasing reputable brands may help.

**14.** Refer to the latest recommendations from the state Extension Service or other reliable sources. Either steam or boiling water may be used for blanching.

**15.** See the question above. It may not be possible to heat the cob thoroughly enough to inactivate enzymes without over cooking the kernels. Frozen cobs also take more space in freezer than cut-off corn does.

## 8. REFRIGERATION

**1.** Assume refrigerator temperatures in auto during work shift, but the surface warmed slightly during 30 minute drive home. Assume safe handling before it reached the employees. Ham will keep refrigerated.

**2.** Physical change.

## 9. CANNING

**1.** No processing time has been found that will permit safe processing of low-acid vegetables at the temperature of boiling water. These have a high risk of being unsafe.

**2.** Safe. Lima beans require longer processing than whole kernel corn.

**3.** Salmon processing times are designed to be continuous. If pathogens were not killed in the first 90 min, 10 min of heat would certainly have no effect. New lids should have been used and the jars then processed for the total correct time.

**4.** Maybe. Those jars that are similar to typical pint or quart jars but with designs may be used. Excessive breakage is a problem with older jars so use in the pressure canner is not recommended. Those that new lids do not fit or that are odd shapes should not be used.

**5.** To adjust for a lower boiling temperature at that altitude, she must use the 15-lb weight.

**6.** Discard. Molds may have resulted in loss of safety.

**7.** This is a quality change due to precipitation of rutin, a pigment in asparagus. The crystals are yellow.

**8.** Put on new lids and process for total time again.

**9.** A chemical reaction can occur between components used on some jar lids and the volatile acid.

**10.** This could be starch from mature peas or the cloudiness could be from microorganisms. Need more-detailed description of appearance and of canning procedure.

**11.** Probably due to microbial change, discard.

**12.** Siphoning occurs if the pressure fluctuates widely. Was the processing adequate or did the pressure fall? If there is a question, either reprocess or refrigerate the jars until they are used. Often these jars do not seal because of pieces of food on the rim.

**13.** Pumpkin in discreet pieces is recommended for canning. Strained pumpkin does not heat evenly and should not be canned.

**14.** A steamer is not recommended as heat transfer is less efficient—heat may be uneven inside the steamer—than in the boiling-water process.

**15.** If gauge is accurate and maintains pressure, then food is adequately heat processed. However, the gasket should be replaced.

## 10. ROOT CELLAR AND DRY STORAGE

**1.** Storage disorders occur.

**2.** Some retail outlets allow cutting of a sample.

The first signs of storage disorders are internal.

**3.** Controlled atmosphere storage does not include the use of additives. If the produce in question is out of season or not grown locally, it is likely that controlled atmosphere storage has been used. These warehouses have lower oxygen and higher carbon dioxide levels than air to slow ripening, but do not impart residues to the fruit. If concern is about fungicide dips and waxes, these may be on produce. Some must be listed on the label or on the shipping box; ask to see that.

**4.** The fruit is safe because of its acidity. Quality will vary.

**5.** They are not harmful to consume; but, if left with the food supply, there will be economic loss. To eradicate, all infested food must be removed, the kitchen thoroughly cleaned to eliminate insect eggs, and cracks in the floor or walls sealed to prevent reentry.

**6.** Check for moistness and molding in the beans in that part. If mold is found, discard.

**7.** Maillard browning has caused quality loss. The dried milk will not rehydrate well. Although it is safe, it will not give a quality product.

**8.** Wash the dirt off in several washings as a part of preparation. Carrots store well.

**9.** Contaminants from hands would not grow on dry pasta. Bacteria and viruses would be killed during the boiling in preparation.

## 11. PRESERVATION WITH SUGAR

**1.** The preparation steps are different from the traditional jams which are boiled to reduce the water content and concentrate the pectin and sugar. Final proportion of sugar is the same.

**2.** Tartaric acid crystallizes from the juice. It can be removed first by chilling and straining the juice. If grape jelly is stored for long periods, some crystals may form.

**3.** Honey gives a different flavor and consistency to jellies and jams. Try substituting no more than one-half honey for the sugar (a maximum honey:sugar ratio of 1:1). Even mild-flavored honeys tend to mask the fruit flavor.

**4.** There is little evidence as to the safety or unsafety of moldy jellies and jams. Therefore, to reduce the risk, the product should be discarded since this is a soft product that would allow compounds to diffuse throughout.

**5.** Strawberries which are overripe contain less pectin and acid. In some seasons, all of the fruit has these characteristics. If the fruit has molded, there is also a loss of pectin.

**6.** This may be a problem with uncooked types such as freezer jam. If the berries are soft and of poor quality, hold in refrigeration instead of room temperature for the recommended time before freezing.

**7.** Check proportion of sugar to fruit. Cooking should be done at a medium rate. If apples are lower in acid, sugar crystallization is more apt to occur.

**8.** It is safe if there are no signs of spoilage. Paraffin seals are not as strong as vacuum-sealed jars, so the seal is more likely to be broken and molds could then grow.

**9.** It is safe to make typical jellies from vegetables. Pepper jelly is an example. This recipe is acidified and contains sugar for preservation. To determine gel strength, check proportion of ingredients.

# INDEX

TO CONSUMER QUESTIONS

# INDEX

# ABOUT THE AUTHORS

**Shirley J. VanGarde** received her BS and MS in food, nutrition, and institutional management from Washington State University at Pullman and her PhD from Oregon State in consumer food science. Besides food safety, her research interests include beef additives, shelf life, and quality evaluation of foods. Her articles appear in *J. Am. Dietetic Assoc.*, *Home Economics Research J.*, *and J. Food. Prot.* She has taught at university and community college level; been an employee of the Continental Can Co., and spent time in restaurant management.

Born and raised in rural Kansas, where she is still part owner of the family farm, Dr. VanGarde is currently an Oregon State University Cooperative Extension agent with the Metro Food Preservation Hotline. She also works as a volunteer with 4-H, with groups of householders at community canning kitchens, and assists food banks when they need her advice. With her husband and two daughters on their small acreage in Oregon, VanGarde raises fruits and vegetables and animals to preserve for later consumption.

**Margy Woodburn** is professor and head of Oregon State University's Nutrition and Food Management Department. During her 25 years there, Dr. Woodburn has been a leader in food safety issues. She teaches undergraduate and graduate courses in food science and has an active role in Extension, including the training of Master Food Preservers. Woodburn's research has included common causes of foodborne illness, including botulism and salmonellae. She has published numerous articles in such journals as *J. Food Prot.*, *J. Food Sci.*, *Applied and Environmental Microbiology*, and *Home Economics Research J.*

Dr. Woodburn received her BS in Home Economics Education from the University of Illinois, her MS in Foods and Nutrition, and her PhD in Food Science and Microbiology from the University of Wisconsin. She has been a member of faculty at the University of Wisconsin, Madison, and Purdue University and an active member of many professional organizations, including the American Dietetic Association, American Home Economics Association, American Institute of Nutrition, and the International Association of Milk, Food and Environmental Sanitarians.